Invitation to
Real Analysis

Pure and Applied
UNDERGRADUATE TEXTS · 36

Invitation to Real Analysis

Cesar E. Silva

AMERICAN MATHEMATICAL SOCIETY
Providence, Rhode Island USA

EDITORIAL COMMITTEE

Gerald B. Folland (Chair)   Steven J. Miller
Jamie Pommersheim   Serge Tabachnikov

2010 *Mathematics Subject Classification.* Primary 26-01, 26A03, 26A06;
Secondary 26A24, 26A42.

For additional information and updates on this book, visit
www.ams.org/bookpages/amstext-36

Library of Congress Cataloging-in-Publication Data
Names: Silva, Câesar Ernesto, 1955– author.
Title: Invitation to real analysis / Cesar E. Silva.
Other titles: Real analysis
Description: Providence, Rhode Island : American Mathematical Society, [2019] | Series: Pure and applied undergraduate texts ; volume 36 | Includes bibliographical references and index.
Identifiers: LCCN 2018058742 | ISBN 9781470449285 (alk. paper)
Subjects: LCSH: Mathematical analysis. | Numbers, Real. | AMS: Real functions – Instructional exposition (textbooks, tutorial papers, etc.). msc | Real functions – Functions of one variable – Foundations: limits and generalizations, elementary topology of the line. msc | Real functions – Functions of one variable – One-variable calculus. msc | Real functions – Functions of one variable – Differentiation (functions of one variable): general theory, generalized derivatives, mean-value theorems. msc | Real functions – Functions of one variable – Integrals of Riemann, Stieltjes and Lebesgue type. msc
Classification: LCC QA300 .S5257 2019 | DDC 515/.83–dc23
LC record available at https://lccn.loc.gov/2018058742

Copying and reprinting. Individual readers of this publication, and nonprofit libraries acting for them, are permitted to make fair use of the material, such as to copy select pages for use in teaching or research. Permission is granted to quote brief passages from this publication in reviews, provided the customary acknowledgment of the source is given.

Republication, systematic copying, or multiple reproduction of any material in this publication is permitted only under license from the American Mathematical Society. Requests for permission to reuse portions of AMS publication content are handled by the Copyright Clearance Center. For more information, please visit www.ams.org/publications/pubpermissions.

Send requests for translation rights and licensed reprints to reprint-permission@ams.org.

© 2019 by the American Mathematical Society. All rights reserved.
The American Mathematical Society retains all rights
except those granted to the United States Government.
Printed in the United States of America.

∞ The paper used in this book is acid-free and falls within the guidelines
established to ensure permanence and durability.
Visit the AMS home page at https://www.ams.org/

10 9 8 7 6 5 4 3 2 1   24 23 22 21 20 19

Contents

Preface	vii
Chapter 0. Preliminaries: Sets, Functions, and Induction	1
0.1. Notation on Sets and Functions	1
0.2. Basic Logic: Statements and Logical Connectives	4
0.3. Sets	10
0.4. Functions	19
0.5. Mathematical Induction	28
0.6. More on Sets: Axioms and Constructions	39
Chapter 1. The Real Numbers and the Completeness Property	47
1.1. Field and Order Properties of $\mathbb{R}$	48
1.2. Completeness Property of $\mathbb{R}$	53
1.3. Countable and Uncountable Sets	62
1.4. Construction of the Real Numbers	71
1.5. The Complex Numbers	74
Chapter 2. Sequences	79
2.1. Limits of Sequences	79
2.2. Three Consequences of Order Completeness	92
2.3. The Cauchy Property for Sequences	104
Chapter 3. Topology of the Real Numbers and Metric Spaces	109
3.1. Metrics	109
3.2. Open and Closed Sets in $\mathbb{R}$	115
3.3. Open and Closed Sets in Metric Spaces	120
3.4. Compactness in $\mathbb{R}$	125

3.5.	The Cantor Set	130
3.6.	Connected Sets in $\mathbb{R}$	136
3.7.	Compactness, Connectedness, and Completeness in Metric Spaces	137

Chapter 4. Continuous Functions — 145
- 4.1. Continuous Functions on $\mathbb{R}$ — 145
- 4.2. Intermediate Value and Extreme Value Theorems — 152
- 4.3. Limits — 157
- 4.4. Uniform Continuity — 164
- 4.5. Continuous Functions on Metric Spaces — 167

Chapter 5. Differentiable Functions — 175
- 5.1. Differentiable Functions on $\mathbb{R}$ — 175
- 5.2. Mean Value Theorem — 183
- 5.3. Taylor's Theorem — 190

Chapter 6. Integration — 197
- 6.1. The Riemann Integral — 197
- 6.2. The Fundamental Theorem of Calculus — 213
- 6.3. Improper Riemann Integrals — 219

Chapter 7. Series — 223
- 7.1. Series of Real Numbers — 223
- 7.2. Alternating Series and Absolute Convergence — 232

Chapter 8. Sequences and Series of Functions — 241
- 8.1. Pointwise Convergence — 241
- 8.2. Uniform Convergence — 243
- 8.3. Series of Functions — 254
- 8.4. Power Series — 259
- 8.5. Taylor Series — 265
- 8.6. Weierstrass Approximation Theorem — 276
- 8.7. The Complex Exponential — 280

Appendix A. Solutions to Questions — 283

Bibliographical Notes — 293

Bibliography — 295

Index — 299

Preface

Real analysis is a beautiful subject that brings together several areas of mathematics. Its origins go back to calculus, a subject that is certainly at the center of our scientific understanding of the world. When trying to justify and give a firm foundation to the fundamental concepts and theorems of calculus, one is faced with having to understand the real numbers, functions, and their properties. The properties of functions include continuity, differentiation, and integration, which lead us to study the convergence of sequences and of series (of numbers and of functions). To understand the real numbers, one has to study properties such as compactness. Compactness also plays a crucial role when studying functions and sequences of functions. This in turn is expressed in the language of set theory, thus one needs a basic acquaintance with properties of sets.

One of the challenges in teaching and learning real analysis is that the subject was developed about 150 years after calculus. We know that calculus is largely the result of the work of Newton and Leibniz in the second half of the seventeenth century, though some of its ideas, such as the approximation methods of Archimedes, go back much earlier. Analysis as we know it today is the result of the work of mathematicians at the start of the nineteenth century. Students are confronted with concepts such as continuity, differentiation, and integration, and wonder why they have to relearn them when they have already seen them in calculus. A challenge for the instructor is to make these ideas come alive in new ways that shed new and important light on an older subject.

Audience

This book is addressed to students who have completed calculus and linear algebra courses as they are normally taught at colleges and universities in the United States. At the same time, all the notions that are used are defined, and linear algebra is not used explicitly. The book covers the basic theorems of calculus and also introduces students to more modern concepts such as open and closed sets, compactness, and uniform convergence. Metric spaces are introduced early in this book, but the instructor who wishes to avoid them or cover very little of them may do so.

Chapter 0 covers basic logic, sets, an introduction to proof techniques, functions, and mathematical induction. This chapter can be used as an introduction to proof-writing techniques, for self-study, or it can be omitted by students who have had a course covering sets, proof-writing, and induction. Readers who plan to omit this chapter should review Section 0.1, where all the notation used in the book regarding sets and functions is covered. On the other hand, students who are not familiar with these concepts or who have no or little experience with proof-writing and logic should cover Chapter 0 starting with Section 0.2. Depending on the class, the instructor may decide to spend one or more weeks with Chapter 0. The chapter ends with an optional section, which could be assigned for independent reading or covered later in the course, on the axioms of set theory and the construction of the natural numbers, the integers, and the rational numbers from these axioms.

Chapter 1 introduces the real numbers, including the notion of (order) completeness. It covers the field and order properties of the real numbers in detail, with an emphasis on understanding the various implications among these properties. The notion of supremum is introduced and used to define the completeness property. There is also an optional section on the construction of the real numbers that may be omitted as it is not needed later in the book.

Chapter 2 covers sequences of numbers, including the Cauchy property. This chapter introduces the formal definition of the limit of a sequence and emphasizes proof techniques dealing with limits.

Chapter 3 is probably the chapter with the largest number of concepts that may be new to the reader as they are not typically covered in a calculus course. These include open and closed sets, accumulation points, compactness, and the Cantor set. This chapter also covers those concepts in the context of metric spaces, but those sections may be omitted. The n-dimensional Euclidean space is treated as an example of a metric space, and the metric space sections are where the notions of open and closed sets are discussed in $\mathbb{R}^n$. However, the chapter is written so that the reader may omit the sections dealing with metric spaces and study these notions only in the context of the set of real numbers $\mathbb{R}$, or come back to them later. For example, it is possible to start with Section 3.2 instead of starting with Section 3.1.

Chapter 4 treats continuity, including uniform continuity, and it has a discussion of proof techniques for continuity. This chapter also introduces limits of functions and ends with continuity in the context of metric spaces.

Differentiation and integration and treated in Chapters 5 and 6, respectively. While they cover concepts from calculus, the approach is different and probably new to the reader; these should be covered in detail.

Series are the subject of Chapter 7. Chapter 8 introduces sequences and series of functions, and it brings together all the notions that have been studied earlier.

The book's emphasis is on understanding these ideas and making connections between the different concepts. Sometimes we give more than one proof of a result when the second proof brings out new ideas. Students are expected to write complete and careful proofs for the exercises.

Preface

This book contains two types of exercises. Regular exercises at different levels of difficulty are at the end of each section: those that may be particularly hard are marked with a star ($\star$); all exercises are numbered within their chapters. The second type, called "Questions", are interspersed throughout the exposition so that readers can verify their understanding of the material; solutions to all of the Questions are supplied in Appendix A.

The end of a definition is marked by the symbol "$\Diamond$".

Convention for Numbering Theorems and Exercises

Chapters are divided into sections. Corollaries, definitions, lemmas, propositions, questions, remarks, and theorems are numbered consecutively within each section. Figures are numbered consecutively within each chapter. Some sections have subsections, but they don't change the numbering. Exercises are at the end of each section and are numbered within that section.

Building a Course

The book has been designed and used for a one-semester course in real analysis, but there is more material than would typically be covered in a semester. At Williams College, in our regular course we start with Chapter 0 and cover logic, sets, functions, and induction (Sections 0.2–0.5) in about one to one and a half weeks. Then we cover the real numbers and completeness in detail, and the construction of the real numbers quickly and with some independent reading. The chapter on sequences (Chapter 2) is covered in detail; there is also the opportunity here for additional exercises and reading on various characterizations of completeness. Chapter 3 contains several new ideas, and we cover most of it, including the Cantor set and several characterizations of compactness; often metric spaces are not covered at the start and only later and as additional reading topics. Chapter 4 on continuous functions is important for developing familiarity with epsilon-delta arguments; the part dealing with metric spaces is usually omitted. Students' familiarity with differentiation is used to cover the first couple of sections in Chapter 5 rather quickly, with emphasis paid to Taylor's theorem. Integration in Chapter 6 is covered in detail, but not improper integrals. Series are also covered rather quickly. Chapter 8 brings together many of the topics that have been studied; emphasis is put on the exchange of limit theorems with applications to power series.

A course with more emphasis on applications could start by reviewing the section on sets and induction. In such a case it is possible to start with Chapter 2 on sequences; one can use the monotone sequence property as the characterization of order completeness. For the reader interested in applications, it is useful to cover metric spaces as there are several applications that can be based on them. The remaining topics are as in the previous course, but there will be more time to spend on Chapter 8.

The book can also be used for a two-semester course that spends a significant amount of time in Chapter 0, where students are introduced to proof-writing techniques. The first part can cover up to limits and continuity, and the second semester could start with differentiation, cover integration, including improper integrals, then series, and finally sequences and series of functions.

For the instructor who wants to include an independent reading component, there are several exercises that complement the material. For example, there is a section on the construction of the integers and rational numbers using equivalence relations, where the student is asked to complete significant parts of the material. The sections on metric spaces could be assigned for independent reading as some parts parallel the theory that has been developed for the real line.

Acknowledgments

This book is based on notes I have used while teaching real analysis at Williams College. I am indebted to all the students in my real analysis classes at Williams who have used versions of this book and made many useful comments. I would like to mention in particular Jared Hallett for his questions and Roshan Sharma and Tarjinder Singh for typo corrections in early versions of the book. I also would like to thank Ran Bi, Josie Maynard, Zane Martin, Jeff Meng, Christina Knapp, Alex Kastner, Rebecca Silva, Ting Da Wang, and Xiwei Yang who read this book at different stages of its development and offered many suggestions and corrections. I would like to thank my former student Kathryn Lindsey, and my colleagues Ed Burger, Steven J. Miller, Frank Morgan, and Stewart Johnson for careful readings of versions of the manuscript. In particular, Stewart Johnson used a manuscript of the book in his classes and offered many comments and suggestions. I have benefitted from discussions with Steve Miller on topics from number theory, and from Matt Foreman for comments and suggestions regarding the sections on set theory. I am also indebted to Joe Auslander, Tom Garrity, Anatoly Preygel, Norton Starr, and Mihai Stoiciu for several suggestions at different stages of this project, as well as to the anonymous reviewers. I thank Chris Marx and May Mei for organizing a Summer Analysis Workshop in 2016 and the participants in the workshop for conversation in real analysis. I would like to thank Emily Silva for her help with the manuscript, and Sergei Gelfand for his support of this project.

Of course, I am indebted to my teachers of real analysis, from whom I learned the subject, and to the books I have read over the years, which are listed in the bibliography.

The book was typeset using LaTeX2e. I received help on the figures from students at different stages of the project. I thank Ran Bi for her work with the early figures, most of which were originally done using Adobe Illustrator. I am also grateful to Zane Martin, who carefully read through the manuscript and worked on the figures in Adobe Illustrator. James Wilcox introduced me to TikZ and typeset many of the figures in TikZ; later Madeleine Elyze also helped with TikZ. More recently, Ting Da Wang worked on the figures, including several new ones. The website for this book is maintained by the publisher and will contain additional material:

www.ams.org/bookpages/amstext-36.

I would like to thank my wife Margaret Oxtoby and daughters Emily and Rebecca for their support in the writing of this book.

The Williams College Science Center and the Hagey Family Chair provided support throughout this project.

Cesar E. Silva
Williamstown, Massachusetts

Chapter 0

Preliminaries: Sets, Functions, and Induction

The language of sets permeates modern mathematics, and we use it in our development of real analysis. Set theory was discovered by Georg Cantor (1845–1918) while studying questions in real analysis. Now sets lie at the foundations of mathematics. The notions we shall study in this book can all be defined in terms of sets. For example, functions are defined in terms of sets.

In this chapter we introduce basic concepts and notation from set theory. Interestingly enough, the notions of a set and set membership remain undefined (they are considered *primitive* notions), but we discuss many properties of sets so that we can develop an intuition for what a set is. At the same time, this chapter covers some basic logic, functions, and mathematical induction.

Readers who are familiar with notions of sets and functions may start with Chapter 1, but they should read Section 0.1 for the notation used in the book and then refer back to sections in this chapter if necessary.

The reader who would like a quick review on sets could work out the proofs in Proposition 0.3.18. Induction is covered in Section 0.5.

Readers who are not familiar with these notions should start with Section 0.2, but should also read Subsection 0.1.4 for some conventions regarding trigonometric functions when used in the exercises.

0.1. Notation on Sets and Functions

Those who are familiar with sets and functions may not need to read this chapter other than this section, where we review the notation on sets and functions used in the book. These notions are defined in detail in later sections of this chapter.

0.1.1. Set Notation. We use upper case letters A, B, etc., to denote sets. If x is an **element** of a set A, we write $x \in A$; we also say that x is a **member** of A. If A is a set whose elements are the numbers 1 and 2, we write $A = \{1, 2\}$. This is the same as the set $\{2, 1\}$ or the set $\{1, 2, 2\}$. To signify that 3 is not a member of A, we write $3 \notin A$. If every element of a set A is an element of another set B, we say A is a **subset** of B and write $A \subseteq B$. Hence $A \subseteq A$ (for us the notation $\subseteq$ means the same as $\subset$). Two sets are equal if they have the same elements. Hence $A = B$ is equivalent to $A \subseteq B$ and $B \subseteq A$. If the sets A and B are not equal, we write $A \neq B$. If $A \subseteq B$ but $A \neq B$, we write $A \subsetneq B$. We let $\varnothing$ denote the **empty set**, i.e., the set with no elements. It follows that for any set A, $\varnothing \subseteq A$.

In a definition, such as "two sets are equal if they have the same elements," when we say "if" we mean "if and only if" since we understand this sentence to also mean that if two sets have the same elements, then they are equal. This usage should always be very clear from the context as "if" and "if and only if" are different logical connectives. Hence in a definition, and only in a definition, we use the convention that "if" stands for "if and only if." The reader is referred to Section 0.2 for a brief introduction to logic, where the logical connectives are introduced (*not, and, or, implication or conditional* (if ..., then ...), *biconditional* (if and only if)), as well as the notions of the converse and contrapositive of a statement.

We assume the reader is familiar with the set of **natural numbers** denoted by $\mathbb{N}$, and which by convention starts at 1, so

$$\mathbb{N} = \{1, 2, 3, \ldots\}.$$

The set of nonnegative integers is

$$\mathbb{N}_0 = \{0, 1, 2, 3, \ldots\}.$$

The set of all **integers** is denoted by $\mathbb{Z}$, and it includes the positive and negative natural numbers and 0. The set of **rational numbers** is denoted by $\mathbb{Q}$, and it consists of all numbers of the form $\frac{m}{n}$, where m, n are in $\mathbb{Z}$ and $n \neq 0$, and where we identify $\frac{m}{n}$ with $\frac{p}{q}$ if $mq = np$.

Given a set A, we can define a new set B by specifying the property that the elements of A have to satisfy to be in B. For example $B = \{x \in \mathbb{N} : 2 < x < 6\}$ defines the set B whose elements are 3, 4, and 5; we can also write $B = \{3, 4, 5\}$.

Given sets A and B, their **union** is denoted $A \cup B$, and the **intersection** of A and B is denoted $A \cap B$. We will also use the **set difference** $A \setminus B$, called "A minus B" and defined by $A \setminus B = \{x \in A : x \text{ is not in } B\}$. If $A \subseteq X$, its **complement** is

$$A^c = \{x \in X : x \notin A\}.$$

(Note that the complement of a set A is always relative to a set X, where A is a subset of X.) We also use the **symmetric difference** of two sets defined by $A \triangle B = (A \setminus B) \cup (B \setminus A)$. Note that $A \triangle B = \varnothing$ if and only if $A = B$.

The **power set** of a set A is the set consisting of all the subsets of A, and it is denoted by $\mathcal{P}(A)$. Thus, for example,

$$\mathcal{P}(\{1, 2\}) = \{\varnothing, \{1\}, \{2\}, \{1, 2\}\}.$$

0.1. Notation on Sets and Functions

Given sets A and B and an element a from A and an element b from B, we can form the **ordered pair** (a, b) which has the property that $(a, b) = (a', b')$ if and only if $a = a'$ and $b = b'$. (There is a formal way, using sets, to define an ordered pair, which is discussed in Subsection 0.4.1.) Now we can define the **Cartesian product** of two sets A and B, denoted by $A \times B$ as the set consisting of all ordered pairs (a, b), where $a \in A$ and $b \in B$. For example, if $A = \{a, b\}$ and $B = \{1, 2, 3\}$, then $A \times B = \{(a, 1), (a, 2), (a, 3), (b, 1), (b, 2), (b, 3)\}$.

0.1.2. Function Notation. If f is a function from a set X to a set Y, we denote this by $f : X \to Y$ and call X the **domain** and Y the **codomain** of the function. Note that while a function f is defined on every element of its domain X, not every element of the codomain Y needs to be attained, i.e., there may be elements y of Y for which there is no x in X such that $f(x) = y$. If $A \subseteq X$, then we define the **image** of the set A by
$$f(A) = \{f(x) : x \in A\}.$$

Given a function $f : X \to Y$ and a set $B \subseteq Y$, the **inverse image** of B under f is defined to be
$$f^{-1}(B) = \{x \in X : f(x) \in B\}.$$

We say that a function $f : X \to Y$ is **surjective** (or **onto**) if for all $y \in Y$ there exists $x \in X$ such that $f(x) = y$. We say $f : X \to Y$ is **injective** (or **one-to-one**) if whenever $f(x_1) = f(x_2)$, for $x_1, x_2 \in X$, then $x_1 = x_2$. If f is both injective and surjective, we say that it is **bijective** or **invertible**.

For example, if we let $A = \{a, b\}$ and $B = \{1, 2, 3\}$, we can define a function f by the relation $f = \{(a, 1), (b, 3)\}$, i.e., f is the function that assigns 1 to a and 3 to b. We write this as $f(a) = 1$, $f(b) = 3$. It is injective but not surjective as the image of the set A is $\{1, 3\}$, which is not B. The inverse image $f^{-1}(\{2\})$ is empty.

We recall the **composition** of two functions. For this we assume that we have a function $f : X \to Y$ and another function $g : Z \to W$ and that $f(X) \subseteq Z$ so that g is defined at every point that is in the image of x under f, i.e., g is defined at every point of the form $f(x)$, where $x \in X$. The composition function $g \circ f$ is a function with domain X and codomain W defined for all $x \in X$ by
$$g \circ f(x) = g(f(x)).$$

To review inverse functions and inverse images, the reader may refer to Subsection 0.4.4.

0.1.3. Basic Prime Number Properties. Recall that a natural number is **prime** if it is only divisible by 1 and itself, and it is not 1. We will use the following two facts about natural numbers. The proofs for these facts can be found in Subsection 0.5.7.

If a prime number p divides a product ab of two natural numbers a and b, then p divides a or p divides b.

Fundamental Theorem of Arithmetic. *Each natural number $n > 1$ can be written as a product of finitely many primes in a unique way up to reordering.*

0.1.4. About Trigonometric and Exponential Functions. The logarithmic function will not be defined until Subsection 6.2.1, and the trigonometric and exponential functions will not be defined until Subsections 8.5.1 and 8.5.2, respectively. However, there are several situations where these functions can be used to give illuminating and interesting examples, and we may use these functions in exercises and examples in the early chapters.

Some exercises will assume your knowledge of sine, cosine, and the exponential and logarithmic functions from calculus, but you may refer to the respective sections to review their properties. After continuity has been defined we will assume that $\sin x$, $\cos x$, and e^x are continuous functions for all real numbers x, and $\ln x$ is continuous for $x > 0$. After differentiation has been defined we will assume they are differentiable and that the derivative of $\sin x$ is $\cos x$, the derivative of $\cos x$ is $-\sin x$, and the derivative of e^x is e^x, for all real numbers x. Also, the derivative of $\ln x$ is $1/x$ for $x > 0$.

0.2. Basic Logic: Statements and Logical Connectives

This can be considered the start of this chapter for the reader who wants to see the complete development of logic, sets, functions, and induction.

All the assertions we shall make about mathematical concepts are statements. We start with a definition.

Definition 0.2.1. A **statement** (also called a *proposition*) is a sentence that has the property of being true or false, and not both; if a statement is not false, then it is true. ◊

Theorems, propositions, lemmas, and corollaries are true statements (well, if they are correct!); logically there is no distinction between them—the different names indicate the relative importance of the statements.

For example, "2 is an odd integer" is a statement that is false. "Williamstown is a small town" is a statement that is true, but "What time is it?" and "The number 8 is the luckiest number" are not statements. (To illustrate new concepts, we will informally use numbers and basic properties of numbers; these are defined or constructed later as the text progresses.) We shall also consider formulas such as $x + 1 = 2$, which contain a free variable, such as x, and which are true or false depending on the value of the free variable. For example, $x + 1 = 2$ is true for $x = 1$ and false for $x = 2$.

Statements can be combined by *logical connectives* to make new statements. We will consider the following **logical connectives**: negation (not), conjunction (and), disjunction (or), implication or conditional (if ..., then ...), and biconditional (if and only if).

Perhaps the simplest logical connective is the negation of a statement.

Definition 0.2.2. If **p** is a statement, "not **p**", also written as

$$\neg \, \mathbf{p},$$

is called the **negation** of **p**. It is true when **p** is false and false when **p** is true. ◊

0.2. Basic Logic: Statements and Logical Connectives

Table 0.1. Truth table for ¬p

p	$\neg p$
T	F
F	T

Table 0.2. Truth table for **p** and **q**

p	q	$p \wedge q$
T	T	T
T	F	F
F	T	F
F	F	F

For example, the negation of $x > 2$ is $x \leq 2$; the negation of "I have a car" is "I do not have a car." The defining properties of the negation are given by its *truth table* and summarized in Table 0.1. This is a table where we list all possible truth value combinations for **p**, **q**, etc., and for each case we show the truth value of the compound statement.

Definition 0.2.3. If **p** and **q** are statements, the **conjunction** of **p** and **q**, written

$$\mathbf{p} \text{ and } \mathbf{q},$$

is a statement that is true only when both **p** and **q** are true. (One may also write **p** ∧ **q**.) ◇

For example,

$$4 > 2 \quad \text{and} \quad 3 < 2$$

is a false statement. "The sky is blue in the morning and there are cows in Williamstown" is a true statement if both "the sky is blue in the morning" and "there are cows in Williamstown" are true. Its truth table is in Table 0.2.

Definition 0.2.4. If **p** and **q** are statements, the **disjunction** of **p** and **q**, written

$$\mathbf{p} \text{ or } \mathbf{q},$$

is a statement that is false only when both **p** and **q** are false. (One may also write **p** ∨ **q**.) ◇

For example, each of the following three statements is a true statement:

$$4 > 2 \quad \text{or} \quad 3 < 2;$$

$$4 > 2 \quad \text{or} \quad 4 > 1;$$

dairy cows eat grass or green is a primary additive color.

The truth table of the disjunction is in Table 0.3.

The negation of (**p** or **q**) is (¬ **p** and ¬ **q**), i.e., these are logically equivalent statements, which we write as

$$\neg(\mathbf{p} \vee \mathbf{q}) \equiv (\neg \mathbf{p} \wedge \neg \mathbf{q}).$$

Table 0.3. Truth table for **p** or **q**

p	q	$p \vee q$
T	T	T
T	F	T
F	T	T
F	F	F

Table 0.4. Truth table for ¬**p** and ¬**q**

p	q	$\neg p$	$\neg q$	$\neg p \wedge \neg q$
T	T	F	F	F
T	F	F	T	F
F	T	T	F	F
F	F	T	T	T

This can be seen by the fact that they have the same truth table; Table 0.4 is the negation of Table 0.3. Similarly, one can see that the negation of (**p** and **q**) is (¬ **p** or ¬ **q**), which we may write as

$$\neg(\mathbf{p} \wedge \mathbf{q}) \equiv (\neg\mathbf{p} \vee \neg\mathbf{q}).$$

Definition 0.2.5. An **implication**, or a **conditional**, is a statement of the form

(0.1)   if **p**, then **q**,

where **p** and **q** stand for statements. We also write this as

$$\mathbf{p} \implies \mathbf{q}$$

and say that **p** implies **q**. Here **p** is called the **hypothesis** and **q** the **conclusion**. We also say that **q** is **necessary** for **p** and that **p** is **sufficient** for **q**. The truth table for the implication is in Table 0.5.   ◊

For example, one can prove that if x is a number and $x > 3$, then $x > 1$; so for $x > 3$ to be true, it is necessary that $x > 1$ be true, and it is sufficient to have $x > 3$ for $x > 1$ to be true. Also, if one wants $x = 0$, it is necessary and sufficient that $x^2 = 0$. The only case when the implication is false is when **p** is true and **q** is false, i.e., if the hypothesis is true and the conclusion is false; in this case it is not a correct implication. For example, if I say "if it rains, then I use an umbrella", this statement is false if it rains and I am not using an umbrella. However, it is important to note that the statement is true if it does not rain; in this case we cannot say that the implication is false, and since it is a statement, if it is not false, it must be true. Another way to think about an implication where the hypothesis is

Table 0.5. Truth table for $\mathbf{p} \implies \mathbf{q}$

p	q	$p \Rightarrow q$
T	T	T
T	F	F
F	T	T
F	F	T

0.2. Basic Logic: Statements and Logical Connectives

false is that the hypothesis does not get tested—in this case we say the implication is *vacuously true*. For example, if in a computer program we have a statement that says "if $0 = 1$, then set $5 = 2$", it will not cause a problem in the program since it never gets executed, since $0 = 1$ is never verified by the program. In summary, the main idea to keep in mind regarding the implication is that it is false only when the hypothesis is true and the conclusion is false, and that when the hypothesis is false, even by correct reasoning we may end up with a false conclusion.

Theorems are typically statements of the form, "if **p**, then **q**".

Definition 0.2.6. The **converse** of the implication "if **p**, then **q**" is the implication "if **q**, then **p**". ◇

The converse of a true statement can be false (or true). For example,

"if x is a positive integer, then $x + 2$ is a positive integer"

is a true statement, but its converse (if $x+2$ is a positive integer, then x is a positive integer) is not true since x could be -1. If I say that "if it rains, then I use an umbrella" its converse, which is "if I use an umbrella, then it rains" may not be true.

Definition 0.2.7. If an implication **p** $\implies$ **q** and its converse **q** $\implies$ **p** are true, then we write

(0.2) $\qquad\qquad$ (**p** $\iff$ **q**) or (**p** if and only if **q**).

This is called the **biconditional**. ◇

To prove a statement of the form (0.2), we need to show that both **p** $\implies$ **q** and **q** $\implies$ **p** are true.

Definition 0.2.8. In contrast to the converse, the **contrapositive** of the implication **p** $\implies$ **q** is

$$\neg\, \mathbf{q} \implies \neg\, \mathbf{p}.$$

◇

For example, the contrapositive of "$x = 2 \implies x^2 = 4$" is "$x^2 \neq 4 \implies x \neq 2$". Upon examination we can see that these two statements say the same thing. In fact, an implication is always equivalent to its contrapositive. We write this as

(0.3) $\qquad\qquad$ (**p** $\implies$ **q**) $\equiv$ ($\neg$ **q** $\implies$ $\neg$ **p**).

This equivalence is proved by showing that $\neg$ **q** $\implies$ $\neg$ **p** has the same truth table as **p** $\implies$ **q**. For another example, the contrapositive of "if it rains, then I use an umbrella" is "if I do not use an umbrella, then it does not rain." If you see me carrying an umbrella it may not be raining, but if you see me not carrying an umbrella, you know it is not raining.

The negation of the implication **p** $\implies$ **q** is "**p** and $\neg$**q**" (see Exercise 0.2.2). For example, to show that the implication $x^2 = 4 \implies x = 2$ is not true (when x is an integer) it suffices find a value of x such that $x^2 = 4$ is true and $x = 2$ is not true; for this we can let $x = -2$, in which case $x^2 = 4$ is true but $x = 2$ is false. Therefore the implication $x^2 = 4 \implies x = 2$ is not true.

We end with a brief discussion of quantifiers. There are two quantifiers, one is **for all**, called the **universal quantifier**, and we use the symbol

$$\forall$$

to denote this quantifier. The other quantifier is **there exists**, called the **existential quantifier**, and we use the symbol

$$\exists$$

for this quantifier. In common language we may use "every" instead of "for all" and sometimes the quantifiers may be implicit. For example, one could make a statement such as, "Every even integer greater than 2 can be written as the sum of two primes". This has two quantifiers: "every" is the universal quantifier, and though hidden there is also an existential quantifier in the claim that there are two primes. This sentence can be written more precisely as, "For all even integers n greater than 2, there exist primes p_1 and p_2 such that $n = p_1 + p_2$." If we let $\mathcal{P}$ denote the set of primes, sometimes it helps to write this sentence more formally as

"$\forall\ n \in \mathbb{N}\ \exists\ p_1, p_2 \in \mathcal{P}$, if n is even, then $n = p_1 + p_2$".

Well, this statement is a famous problem that has been open for over 250 years and is known as *Goldbach's conjecture*. We will not say more about this, except it has served to introduce the use of quantifiers.

We now consider examples where we have only one quantifier at a time. Suppose we have the statement

(0.4)   "there is a room in the math building that has a red chair".

To prove that this statement is true, we only have to find one room in the math building with a red chair; to prove it is false we have to check every room and verify that no room has a red chair. The negation of this statement is

(0.5)   "every room in the math building does not have a red chair".

The statement in (0.4) is of the form "$\exists r\ P(r)$", its negation is "$\forall r\ \neg P(r)$".

Next we note that the order in which quantifiers appear is crucial, since a change in the order can change the meaning of the sentence. For example

(0.6)   "for every ice cream flavor Ic, there is a person who likes that flavor",

is different from

(0.7)   "there is person who likes every ice cream flavor".

Sometimes it helps to write these in symbols. For (0.6) we have

(0.8)   "$\forall$ Ic $\exists P$, P likes Ic".

For (0.7) we have

(0.9)   "$\exists P\ \forall$ Ic, P likes Ic".

Now instead of changing the order we could change the quantifiers, so instead of (0.8) we could have

(0.10)   "$\exists$ Ic $\forall P, P$ likes Ic",

which means that

(0.11)   "there is an ice cream flavor Ic for all persons P, P likes Ic", or
           "there is an ice cream flavor that every person likes".

This sentence again has a different meaning from (0.6) or (0.7). We could also change the quantifiers in (0.9) to obtain

(0.12)   "$\forall P \; \exists$ Ic, P likes Ic",

which means that

(0.13)   for all persons P there exists an ice cream flavor Ic, P likes Ic", or
           "each person likes some flavor".

The negation of (0.6) is, "There exists an ice cream flavor such that every person does not like that flavor"; the negation of (0.7) is, "For every person there is an ice cream flavor so that the person does not like the flavor", or simply "There is no person who likes every ice cream flavor". In general, we have

$$\neg(\forall x \; \phi(x)) \equiv \exists x \; \neg\phi(x),$$
$$\neg(\exists x \; \phi(x)) \equiv \forall x \; \neg\phi(x).$$

The use of quantifiers and their symbols is discussed in more detail in the definitions of limits in Section 2.1 and of continuity in Section 4.1.

Exercises: Basic Logic

0.2.1 Prove the following equivalences:
   (a) $\neg \, (\mathbf{p} \vee \mathbf{q}) \equiv (\neg \, \mathbf{p} \wedge \neg \, \mathbf{q})$.
   (b) $\neg \, (\mathbf{p} \wedge \mathbf{q}) \equiv (\neg \, \mathbf{p} \vee \neg \, \mathbf{q})$.

0.2.2 Prove that $\mathbf{p} \implies \mathbf{q}$ is equivalent to $(\neg \, \mathbf{p}) \vee \mathbf{q}$. Deduce that the negation of $\mathbf{p} \implies \mathbf{q}$ is $\mathbf{p} \wedge (\neg \, \mathbf{q})$. Give a proof using proof tables and a proof using Exercise 0.2.1.

0.2.3 Find the negation of the following statements. Justify your answers.
   (a) $\neg \, (\mathbf{p} \text{ and } \neg \, \mathbf{q})$ or $\mathbf{r}$.
   (b) $\mathbf{p} \implies (\mathbf{q} \text{ or } \mathbf{r})$.
   (c) $\neg \, (\mathbf{p} \text{ or } \mathbf{q}) \implies (\mathbf{r} \text{ or } \mathbf{s})$.

0.2.4 Prove the following equivalences:
   (a) $\mathbf{p} \wedge (\mathbf{q} \vee \mathbf{r}) \equiv (\mathbf{p} \wedge \mathbf{q}) \vee (\mathbf{p} \wedge \mathbf{r})$.
   (b) $\mathbf{p} \vee (\mathbf{q} \wedge \mathbf{r}) \equiv (\mathbf{p} \vee \mathbf{q}) \wedge (\mathbf{p} \vee \mathbf{r})$.

0.2.5 Prove that an implication is equivalent to its contrapositive.

0.2.6 Give an example of an implication whose converse is not true and another example whose converse is true.

0.2.7 Give examples to show that each of the sentences (0.6), (0.7), (0.11), and (0.13) are different.

0.2.8 Write the negation of the following statements.
   (a) For every ice cream flavor there is a pie that goes with that flavor.

(b) For every race car there is a driver who can drive that car.
(c) There exists a race car that every driver can drive.
(d) There exists a driver that can drive every race car.

0.2.9 Write the negation of the following statements.
(a) There is a car that can carry two large suitcases in its trunk.
(b) Every car can carry two large suitcases in its trunk.
(c) For every ice cream flavor there is a student who likes that flavor.
(d) For every car C there is a person P so that if P has the key, then P can drive the car.

0.3. Sets

This section introduces sets and operations on sets. Our discussion is informal but parallels the axiomatic approach. We start with the simplest notions and build our structures from there. We begin with some basic sets and construct new sets by set operations, or by *separation*, also called *comprehension*, as in (0.15). There is an optional discussion of the standard axioms of sets in Subsection 0.6.1.

0.3.1. Elements and Subsets. Sets are specified by their elements. **Set membership**, or being an **element** of the set, is a basic notion; it remains undefined but it is at the heart of our understanding of sets. For example, we can think of a set A consisting of the numbers $0, 2, 4$. The members or elements of this set are 0, 2, and 4. To define this set, we can write

(0.14) $$A = \{0, 2, 4\}.$$

The order is not important: this is the same set as $\{4, 0, 2\}$, and the set $\{3, 3, 3\}$ is the same as the set $\{3\}$. We can also think of the set of all cows in Williamstown; this is a different set from the set of all the houses in Williamstown.

For example, 2 is an element of the set A in (0.14), and we write

$$2 \in A.$$

We also say that 2 **is in** A or **belongs** to A, or is a **member** of A. We may also say that A **contains** 2. However, 3 is not an element of A, and we write

$$3 \notin A.$$

Since sets are determined by their elements, two sets are equal if and only if they have the same elements. If we write $B = \{2\}$, then 2 is an element of B and B has no other elements.

Definition 0.3.1. A set B is defined to be a **subset** of a set A, and we write

$$B \subseteq A$$

if every element of B is an element of A or, equivalently, whenever $x \in B$, then $x \in A$. We also say that A is a **superset** of B. Sometimes we may say that A **includes** B. ◊

0.3. Sets

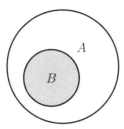

Figure 0.1. Venn diagram illustrating $B \subseteq A$.

For example, the set $\{2,3\}$ is a subset of $\{1,2,3\}$. Figure 0.1 shows the Venn diagram representation of $B \subseteq A$.

We note that some authors use the notation $B \subset A$ to mean $B \subseteq A$.

By abuse of language one may say that A contains B when $B \subseteq A$, but we will try to avoid this: if $A = \{3,4\}$, then A contains 3 and includes $\{3\}$.

We mention an important convention regarding definitions (also remarked upon in Subsection 0.1.1). In Definition 0.3.1 we said "if every element" but what we really meant was "if and only if every element." It is only reasonable that if every element of B is an element of A, then we should understand this to mean that B is a subset of A. In other words, in a definition one often only writes "if" but it only makes sense for this to stand for "if and only if".

Since sets are determined by their elements, two sets A and B are equal, and we write $A = B$, if and only if $A \subseteq B$ and $B \subseteq A$. This is the usual way to show that two sets are equal. If the sets A and B are not equal, we write $A \neq B$. For example, if $A_0 = \{2,3,4\}$ and $B_0 = \{2,3\}$, then $B_0 \neq A_0$ since $4 \notin B_0$ but $4 \in A_0$.

Definition 0.3.2. We write
$$B \subsetneq A$$
when $B \subseteq A$ but $B \neq A$; in this case we say that B is a **proper subset** of A. ◇

As every element of B_0 above is in A_0, we can write $B_0 \subsetneq A_0$. Both $B_0 \subseteq A_0$ and $B_0 \subsetneq A_0$ are true, though the second expression conveys more information.

Perhaps the simplest set is the set with no elements.

Definition 0.3.3. There is a set that has no elements, called the **empty set**, and it is denoted by $\varnothing$. ◇

Since a set is determined by its elements, there is only one set that has no elements, the empty set.

Now we can use the empty set to construct new sets. The set $\{\varnothing\}$ is not $\varnothing$ since it has an element, namely $\varnothing$, and $\varnothing$ has no elements. We can construct the nonnegative numbers in terms of sets. Using sets, one can define 0 to be $\varnothing$,

1 is $\{\varnothing\}$, 2 is $\{\varnothing, \{\varnothing\}\} = \{0,1\}$, 3 is $\{\varnothing, \{\varnothing\}, \{\varnothing, \{\varnothing\}\}\} = \{0,1,2\},\ldots.$

We can continue and set
$$4 = \{0,1,2,3\}, \quad 5 = \{0,1,2,3,4\},\ldots.$$

Note that, corresponding to our intuition, 4 has four elements, etc. The important idea here is that the natural number—for which we have a good intuitive sense about what they are—can be defined in terms of sets, and in a way that agrees with our intuition. Hence we have a model in terms of sets for what we intuitively know as the numbers $0, 1, 2, \ldots$.

It is possible to give a construction of the set of all nonnegative integers, and so also of the set of all natural numbers, from the axioms of set theory. A construction using the notion of *inductive* sets is outlined in Subsection 0.6.1.

From now on we will assume that we have the set of *natural numbers*

$$\mathbb{N} = \{1, 2, \ldots\},$$

and the *nonnegative integers*

$$\mathbb{N}_0 = \{0, 1, 2, \ldots\}.$$

By convention, for us the natural numbers start at 1; other authors start at 0. The basic properties of the natural numbers that we shall use are characterized by the Peano axioms, discussed in Section 0.5.2.

It is interesting that one can define an order based on the definition we have given of nonnegative integers. We will say that $n < m$ when n is a subset of m and is not equal to m. For example, we have $1 < 2$ since $1 = \{\varnothing\}$ is a subset of $2 = \{\varnothing, \{\varnothing\}\}$.

Another way to specify, or construct, a set, and the one used most often, is to give a property that the elements of a given set must satisfy. Here we start with a given set and specify what elements of it will be in our new set. For example, a set E can be defined by saying that the elements of E are those elements of $\mathbb{N}$ (the given set) that are even. In this case we write

(0.15) $$E = \{x \in \mathbb{N} : x \text{ is even}\}.$$

Thus, for example, $4 \in E$ but $11 \notin E$. We say that the set E is defined by *separation* or *comprehension*.

There is an important remark to be made here. One might be tempted to say that it is possible to define a set by specifying the properties that its elements must satisfy (without starting with a given set). That is, one might think that one can define a set C by saying that C consists of all x such that x satisfies some given property P. Bertrand Russell (1872–1970) showed that this leads to a contradiction; this is known as Russell's paradox and is discussed in Subsection 0.3.5. (The difference between this and the definition of the set E above is that in the case of E, we only take those elements of an already existing set, in this case $\mathbb{N}$, such that x satisfies a given property, in this case that of being even.)

0.3.2. Methods of Proof. The first method of proof we consider is the **direct method of proof**. This method consists in assuming the hypothesis and using known facts to deduce the conclusion. We illustrate this method by proving a simple property about subsets. A consequence of the definition of subsets is that for every set A,

(0.16) $$A \subseteq A.$$

0.3. Sets

This statement can be written more precisely as an implication in the form

(0.17) $$\text{if } A \text{ is a set, then } A \subseteq A.$$

In the case of (0.17) the hypothesis is that A is a set. To show the conclusion that $A \subseteq A$, we need to show that every element of A is an element of A. We observe that this is clearly true since if $x \in A$, then $x \in A$. Thus $A \subseteq A$ and this concludes the proof. This is a very simple proof, but it illustrates the structure of direct proofs. From now on, for a simple proof such as this we will typically say that it follows directly from the definition that $A \subseteq A$. We shall see more interesting examples soon.

In our direct proofs we use basic principles of logic:

(1) if we know that **p** implies **q**, and we know that **p** is true, then we can deduce **q**;
(2) if we know that **p** implies **q**, and we know that not **q** is true, then we can deduce not **p**;
(3) if we know **p** or **q**, and that **q** is false, we can deduce **p**;
(4) if we know **p** and **q**, we can deduce **p**.

The second method we consider is **proof by contradiction**. This is a proof technique where we start by assuming that the hypothesis holds and that the conclusion does not hold. If we use known facts to arrive at something that is not true, then it must be the case that the conclusion is true.

As an example we will show that for every set A,

(0.18) $$\varnothing \subseteq A.$$

We prove (0.18) by contradiction. The hypothesis is simply that A is a set. The conclusion is that $\varnothing \subseteq A$. If we suppose that it is not the case that $\varnothing \subseteq A$, then there would exist an element x in $\varnothing$ such that $x \notin A$. But the empty set $\varnothing$ does not contain any elements, so assuming that there exists an element x in $\varnothing$ is a contradiction. Therefore, we can conclude that $\varnothing \subseteq A$.

A third method is **proof by contrapositive**. Here, to prove

$$\mathbf{p} \implies \mathbf{q},$$

we give a direct proof of its contrapositive,

$$\neg \mathbf{q} \implies \neg \mathbf{p}.$$

That is, we start by assuming not **q** and deduce not **p**. For an example, suppose we wanted to show that if the square of a number is different from 4, then the number is different from 2. It would be easier to prove its contrapositive: if a number is 2, then its square is 4, which has a simple direct proof (just square the number).

Remark 0.3.4. It is possible to recast a proof by contradiction as a proof by contrapositive; this is left as an exercise to the interested reader.

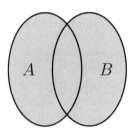

Figure 0.2. Venn diagram for $A \cup B$ (shaded region).

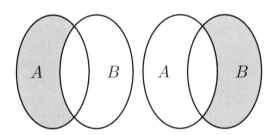

Figure 0.3. Venn diagrams for $A \setminus B$ and $B \setminus A$, respectively (shaded regions).

0.3.3. Constructing New Sets Out of Old. If A and B are sets, there is a new set consisting of those elements that belong to either A or B (recall that the mathematical "or" is inclusive, so the elements may belong to both A and B).

Definition 0.3.5. The **union** of sets A and B, denoted by $A \cup B$, is the set of those elements that are in A or are in B. ◊

This is illustrated in Figure 0.2, where the union is the shaded area. So if $A = \{2, 4, 6\}$ and $B = \{2, 4, 5\}$, then $A \cup B = \{2, 4, 5, 6\}$. (The reader may be worried that there is no superset in this case, but this is a second way of constructing sets where we assert the existence of the union of two sets; we are saying that if we can think of sets A and B, we can think of their union.) Also, we can consider the union of not just two sets but of an arbitrary collection of sets—this is discussed in Subsection 1.3.2.

Question 0.3.6. Prove that $A \cup B = A$ if and only if $B \subseteq A$.[1]

Definition 0.3.7. If A and B are sets, their **set difference** $A \setminus B$, called A **set minus** B or A **minus** B, is defined by

$$A \setminus B = \{x \in A : x \text{ is not in } B\}. \qquad \diamond$$

For example, if $A = \{0, 2, 4\}$ and $B = \{0, 1, 2\}$, then $A \setminus B = \{4\}$ and $B \setminus A = \{1\}$. The Venn diagrams of $A \setminus B$ and $B \setminus A$ are shown in Figure 0.3.

We observe that

(0.19) $\qquad\qquad\qquad A \setminus B = \varnothing$ if and only if $A \subseteq B$.

---

[1] We remind the reader that solutions to all Questions are in Appendix A.

0.3. Sets

In fact, if $A \setminus B = \emptyset$ and $x \in A$, if x were not in B, then x would be in $A \setminus B$, a contradiction. Therefore, $A \subseteq B$. Conversely, if $A \subseteq B$ and x is in $A \setminus B$, then x would be in A but not in B, again a contradiction. Therefore $A \setminus B \subseteq \emptyset$, which implies that $A \setminus B = \emptyset$.

Definition 0.3.8. The **complement** of a set $A \subseteq X$ is
$$A^c = \{x \in X : x \notin A\}. \qquad \diamond$$

That is, the complement of A in X consists of all the elements of X that are not in A. When speaking of the complement of a set (A), we always have a set containing it in mind (X)—we always consider the complement relative to a given set. Note that $A^c = X \setminus A$, and when the set X is not clear from context, we may explicitly write $X \setminus A$ instead of A^c. The Venn diagram of the complement of A is the shaded area in Figure 0.4.

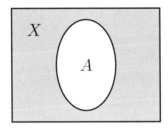

Figure 0.4. Venn diagram for A^c (shaded region).

Question 0.3.9. Prove that
(0.20) $$(A^c)^c = A.$$

Definition 0.3.10. Given two sets A and B, their **intersection** is denoted by $A \cap B$ and defined by
$$A \cap B = \{x \in A \cup B : x \in A \text{ and } x \in B\}.$$
That is, $A \cap B$ consists of all elements of $A \cup B$ that are in A and in B. $\diamond$

For example, if A is the set of all mathematicians and B is the set of all pianists, then $A \cap B$ is the set of all mathematicians who are pianists, which is the same as the set of pianists who are mathematicians. The intersection of two sets is the shaded area in the Venn diagram of Figure 0.5.

The intersection can also be defined in terms of the union and complement. In fact, for any sets A and B, let $C = A \cup B$. Then we have
(0.21) $$A \cap B = C \setminus [(C \setminus A) \cup (C \setminus B)].$$

Question 0.3.11. Prove the equality in (0.21).

Question 0.3.12. Prove that if A is a set, then $A \cap \emptyset = \emptyset$.

Definition 0.3.13. Two sets A and B are said to be **disjoint** if $A \cap B = \emptyset$. $\diamond$

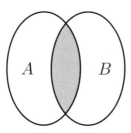

Figure 0.5. Venn diagram for $A \cap B$ (shaded region).

Hence the empty set $\emptyset$ is disjoint from every set A, and a set A is disjoint from itself if and only if $A = \emptyset$.

Definition 0.3.14. Let A be a set. The **power set** of A, denoted by $\mathcal{P}(A)$, is defined to be the set whose elements are all the subsets of A. ◇

In particular, one always has that $\emptyset$ and A are elements of $\mathcal{P}(A)$. For an example, if $A_0 = \{0, 1, 2\}$, then

$$\mathcal{P}(A_0) = \{\emptyset, \{0\}, \{1\}, \{2\}, \{0,1\}, \{0,2\}, \{1,2\}, A_0\}.$$

But note that, for example, $1 \notin \mathcal{P}(A_0)$, while $1 \in A_0$ and $\{1\} \in \mathcal{P}(A_0)$.

Definition 0.3.15. Using set difference we define the **symmetric difference** of two sets by

$$A \triangle B = (A \setminus B) \cup (B \setminus A).$$ ◇

Question 0.3.16. Show that

(0.22) $$A \setminus B = A \cap B^c.$$

Figure 0.6 shows the Venn diagram of the symmetric difference of two sets as the shaded area. We think of $A \triangle B$ as the set of points where A and B differ. In fact we note that

(0.23) $$A \triangle B = \emptyset \text{ if and only if } A = B.$$

To show this, first suppose that $A \triangle B = \emptyset$. Let $x \in A$. If x were not in B, then x would be in $A \setminus B$, which would mean that $x \in A \triangle B$. But this is a contradiction as $A \triangle B$ is empty. Therefore, $x \in B$. In a similar way, one shows that if $x \in B$, then $x \in A$. This shows that $A = B$. For the converse suppose that $A = B$. Then $A \setminus B = \emptyset$ and $B \setminus A = \emptyset$, showing $A \triangle B = \emptyset$. This completes the proof.

Question 0.3.17. Show that $A \triangle B = (A \cup B) \setminus (A \cap B)$.

0.3.4. Basic Properties of Sets. We state a proposition containing many of the properties of sets we shall need in this book. The first two statements are the commutative property of unions and intersections. The next two say that the operations of intersection and union are *associative*: the order in which the operations are performed does not change the result. For this reason one may omit the parentheses when considering the intersection or union of several sets. Properties

0.3. Sets

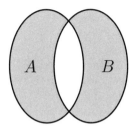

Figure 0.6. Venn diagram for $A \triangle B$ (shaded region).

(5) and (6) of Proposition 0.3.18 are distributive properties of the union and intersection, respectively. Properties (7) and (8) are the well-known De Morgan's laws that relate unions with intersections using complementation.

Proposition 0.3.18. *Let A, B, C be subsets of a set X. Then the following hold.*

(1) $A \cup B = B \cup A$

(2) $A \cap B = B \cap A$

(3) $A \cap (B \cap C) = (A \cap B) \cap C$

(4) $A \cup (B \cup C) = (A \cup B) \cup C$

(5) $A \cap (B \cup C) = (A \cap B) \cup (A \cap C)$

(6) $A \cup (B \cap C) = (A \cup B) \cap (A \cup C)$

(7) (De Morgan's law) $(A \cup B)^c = A^c \cap B^c$

(8) (De Morgan's law) $(A \cap B)^c = A^c \cup B^c$

Proof. The proofs of (1) and (2) follow immediately from the definitions of union and intersection. To prove (3), first let $x \in A \cap (B \cap C)$. Then $x \in A$ and $x \in B \cap C$. Therefore $x \in A$, $x \in B$, and $x \in C$, which we can write as $x \in A \cap B$ and $x \in C$. This means that $x \in (A \cap B) \cap C$. For the converse we let $x \in (A \cap B) \cap C$ and in a similar way show that $x \in A \cap (B \cap C)$.

The proofs of (4) and (6) are left to the reader. To show (5), first let $x \in A \cap (B \cup C)$. Then $x \in A$ and $x \in B \cup C$. Hence, x is in A and x is in B or C. If $x \in B$, then $x \in A \cap B$. If $x \in C$, then $x \in A \cap C$. Hence $x \in (A \cap B) \cup (A \cap C)$, showing that $A \cap (B \cup C) \subseteq (A \cap B) \cup (A \cap C)$. For the reverse inclusion let $x \in (A \cap B) \cup (A \cap C)$. Then x is in $A \cap B$ or $x \in A \cap C$. In the first case x is in A and B, and in the second case x is in A and C. In both cases, x is in A and is in B or C. Hence $x \in A \cap (B \cup C)$. This shows that the sets are equal.

To show (7), let $x \in (A \cup B)^c$. Then x is not in $A \cup B$, which means that x is not in A and is not in B. Hence, $x \in A^c$ and $x \in B^c$, or $x \in A^c \cap B^c$. Therefore, $(A \cup B)^c \subseteq A^c \cap B^c$. Next, let $x \in A^c \cap B^c$. Then $x \in A^c$ and $x \in B^c$. This means that x is not in A and x is not in B, or x is not in their union, i.e., $x \in (A \cup B)^c$. Therefore $A^c \cap B^c \subseteq (A \cup B)^c$, completing the proof. To show (8) apply (7) to the sets A^c and B^c to obtain $(A^c \cup B^c)^c = (A^c)^c \cap (B^c)^c = A \cap B$ by (0.20). Therefore, using (0.20) again, $A^c \cup B^c = ((A^c \cup B^c)^c)^c = (A \cap B)^c$. □

0.3.5. Russell's Paradox. The main notions in analysis involve infinitary processes. The definition of limit involves the notion of infinity. Approximation, which plays a crucial role in analysis, uses infinite processes, so do the notions of differentiation and integration. Compactness, which is a fundamental notion in analysis and which will be studied in Section 3.4, is also defined using infinite operations. These notions lead us to consider infinite sets. Our intuition, however, is not as clear when thinking about infinite sets as it is when thinking about finite sets.

In 1902, Russell wrote a letter to Frege where he stated what is now known as Russell's paradox. Russell found an internal contradiction in the theory of Frege's *The Basic Laws of Arithmetic* that had recently been sent to the publisher. Russell wrote he discovered his paradox in 1901. After this paradox mathematicians developed a new foundation for set theory.

We now describe the paradox. Consider the following property of a set X:

X is not an element of itself, or equivalently, $X \notin X$.

For example, the set $X = \{1\}$ satisfies this property. Suppose we could define sets by specifying a property that its elements satisfy (the reader may want to review the discussion at the end of Subsection 0.3). Then we could define a set S to consist of all sets X such that $X \notin X$, i.e.,

$$S = \{X : X \notin X\}.$$

There are now two possibilities: either $S \in S$ or $S \notin S$. First, suppose that $S \in S$. Then by the definition of S it follows that $S \notin S$, a contradiction. Next suppose that $S \notin S$. Then again by the definition of S we have that $S \in S$, another contradiction. Therefore, neither of the statements $S \in S$ or $S \notin S$ can be true. It follows that sets cannot be defined in this way or, in other words, this broad way of defining sets is not valid.

One of the consequences of Russell's paradox was a carefully formulated collection of axioms for sets. The approach taken to resolve the paradox was to start with very simple assumptions and build sets from there. The standard theory that came out for sets in the early 1900s is an axiomatic formulation called the Zermelo–Fraenkel axioms (this is covered in Subsection 0.6.1, which is optional). These axioms avoid Russell's paradox by not claiming that every property P defines a set. They instead say that if one is given a pre-existing set X, the collection of elements of X with property P is a set. The axioms do state, however, that one can start with basic sets such as $\emptyset$, construct the natural numbers, and do operations on them (such as unions, intersections, complements, power sets, and Cartesian products—defined in 0.4.1).

Exercises: Sets

0.3.1 Let A and B be subsets of a set X. Which of the following statements is equivalent to $A \cup B \neq \emptyset$? Justify your answer.
  (a) $A \neq \emptyset$ and $B \neq \emptyset$
  (b) $A \neq \emptyset$ or $B \neq \emptyset$
  (c) None of the above

0.3.2 Let A and B be subsets of a set X. Which of the following statements is equivalent to $A \cap B = \varnothing$? Justify your answer.
   (a) $A = \varnothing$ and $B = \varnothing$
   (b) $A = \varnothing$ or $B = \varnothing$
   (c) None of the above

0.3.3 Let A and B be sets. Prove that $A = (A \cap B) \cup (A \setminus B)$.

0.3.4 Let A and B be sets. Prove that $A \setminus (A \setminus B) = A \cap B$.

0.3.5 Let A and B be subsets of a set X. Prove that $A \subseteq B$ if and only if $B^c \subseteq A^c$.

0.3.6 Let A, B, and C be sets. Prove that
$$(A \cup B) \setminus C = (A \setminus C) \cup (B \setminus C), \text{ and}$$
$$(A \cap B) \setminus C = (A \setminus C) \cap (B \setminus C).$$

0.3.7 Let A, B, and C be sets. Prove that
$$A \setminus (B \cup C) = (A \setminus B) \cap (A \setminus C), \text{ and}$$
$$A \setminus (B \cap C) = (A \setminus B) \cup (A \setminus C).$$

0.3.8 Let A, B, and C be sets. Prove that
$$(A \setminus B) \setminus C = (A \setminus C) \setminus (B \setminus C), \text{ and}$$
$$A \cap (B \setminus C) = (A \setminus C) \setminus (A \setminus B).$$

0.3.9 Prove Proposition 0.3.18(4).

0.3.10 Prove Proposition 0.3.18(6).

0.3.11 Give a different proof of Proposition 0.3.18(8) without using part (7).

0.3.12 Let A and B be subsets of a set X. Prove that
$$A \triangle B = A^c \triangle B^c.$$

0.3.13 Let A, B, and C be sets. Prove that
$$A \triangle (B \triangle C) = (A \triangle B) \triangle C.$$

0.3.14 Let A and B be sets. Is $(A \triangle B) \triangle A = B$?

0.3.15 Let $A \subseteq I$ and $B \subseteq J$ be sets. Determine whether
$$A \cap B = (I \cap J) \setminus [(I \setminus A) \cup (J \setminus B)].$$

0.3.16 Let A and B be sets. Is $\mathcal{P}(A \cup B) = \mathcal{P}(A) \cup \mathcal{P}(B)$? Does $\mathcal{P}(A \cap B) = \mathcal{P}(A) \cap \mathcal{P}(B)$? Discuss.

0.3.17 Let A and B be two sets, and suppose that $A \subseteq B$. Prove that $\mathcal{P}(A) \subseteq \mathcal{P}(B)$.

0.4. Functions

The reader is familiar with the notion of a function from calculus. A function from a set A to a set B is a correspondence that assigns to each element of A a unique element of B. While this agrees with our intuitive notion of a function, it is not a (formal) definition as the notions of "correspondence" and "assign" have yet to be defined. This section defines the notion of a function. This is done by first introducing Cartesian products. The reader will find that the final definition of a function corresponds to the reader's original intuition. Our definition can be seen as an interesting application of sets. However, there are important new definitions here including those of injective (or one-to-one) and surjective (or onto) functions, and of one-to-one correspondence.

0.4.1. Cartesian Products. To define Cartesian products we use the concept of an ordered pair.

Definition 0.4.1. An **ordered pair** consists of two elements a and b from some set written in a specific order: (a, b). The order here is such that

$$(0.24) \qquad (a,b) = (c,d) \text{ if and only if } a = c \text{ and } b = d. \qquad \Diamond$$

For the reader who is interested in a more precise definition, we note that it is possible to define an ordered pair (a, b) as the set $\{\{a\}, \{a, b\}\}$. This definition was given by Kuratowski (1896–1980) and there are other definitions that also work. In Exercise 0.4.12 the reader is asked to show that under this definition the condition in (0.24) holds. Keep in mind that property (0.24) is what characterizes the concept of an ordered pair.

Definition 0.4.2. Given two sets A and B, their **Cartesian product**, denoted $A \times B$ and read "A cross B", is defined to be the set consisting of all ordered pairs (a, b), where $a \in A$ and $b \in B$. $\qquad \Diamond$

This is an interesting operation that creates a new set from a given pair of sets. For example, if $A_1 = \{a_1, a_2\}$ and $B_1 = \{b_1, b_2, b_3\}$, then

$$A_1 \times B_1 = \{(a_1, b_1), (a_1, b_2), (a_1, b_3), (a_2, b_1), (a_2, b_2), (a_2, b_3)\}.$$

If $A_2 = \{1, 2, 3\}$ and $B_2 = \{2, 4\}$ have representations in the number line, then $A_2 \times B_2$ has a representation in the plane as shown in Figure 0.7.

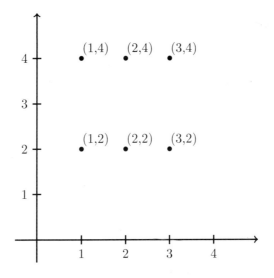

Figure 0.7. Cartesian product example: $\{1, 2, 3\} \times \{2, 4\}$.

0.4. Functions

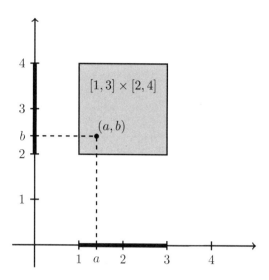

Figure 0.8. Cartesian product example: $[1,3] \times [2,4]$.

Also, if $I = [1,3]$ and $J = [2,4]$ represents intervals in the line, then $I \times J$ is a square in the plane as in Figure 0.8.

The reader who would like to see the Cartesian product as a subset of a specific set should see Exercise 0.4.2.

Question 0.4.3. Prove that for all nonempty sets A and B,
$$A \times B = B \times A \text{ if and only if } A = B.$$

Question 0.4.4. Prove that for every set A,
$$A \times \varnothing = \varnothing.$$

0.4.2. Relations

Definition 0.4.5. A relation between sets A and B is a way to connect elements of A with elements of B, and this is done formally by choosing a subset of $A \times B$. So we define a **relation** on sets A and B to be a subset of $A \times B$. ◇

In many cases of interest we take $B = A$. For example, if A_0 is the set of all cows in Massachusetts, then one can define a relation $R_1 \subseteq A_0 \times A_0$ on A_0 (and A_0) by saying that two cows are related by R_1 if their weight difference is less than 5 pounds; i.e., we say that (a,b) is in R_1 if and only if the difference in the weights of cows a and b is less than 5 pounds. In this case we note that $(a,b) \in R_1$ if and only if $(b,a) \in R_1$. For another relation on $A_0 \times A_0$, we could say that (a,b) is in R_2 if cow b is an ancestor of cow a.

Definition 0.4.6. A relation R on a set A is **symmetric** when
$$(a,b) \in R \text{ if and only if } (b,a) \in R.$$
A relation R on a set A is **reflexive** if
$$(a,a) \in R \text{ for all } a \in A. \qquad \diamond$$

Relation R_1 is symmetric, but relation R_2 is not. We define a third relation R_3 by saying that $(a,b) \in R_3$ if and only if the weight of cow a is greater than the weight of cow b. This relation R_3 is not symmetric; it is not reflexive either as R_3 is defined so that the weight of a is strictly greater than the weight of b. We note, in particular, that for a given a there may be several b such that $(a,b) \in R_3$.

Definition 0.4.7. A relation R is said to be an **equivalence relation** if it is symmetric, reflexive, and **transitive** (i.e., $(a,b) \in R$ and $(b,c) \in R$ imply $(a,c) \in R$). If R is an equivalence relation on A and if $a \in A$, the **equivalence class** of a under R, denoted $[a]$, is defined by
$$[a] = \{b \in A : (a,b) \in R\}.$$
◊

The relation R_3 is transitive but not an equivalence relation.

The set of all equivalence classes on a set A partitions the set in the following way: every element of A is in some equivalence class, and any two distinct equivalence classes do not intersect, or in other words each element of the set is in one and only one equivalence class (Exercise 0.4.21). For an example of an equivalence relation, define a relation R_4 on the set on nonnegative integers $\mathbb{N}_0$ by saying that the element (n,m) is in $R_4 \subseteq \mathbb{N}_0 \times \mathbb{N}_0$ if and only if $m - n$ is divisible by 2. Note that if $m - n$ is divisible by 2, then so is $n - m$, thus R_4 is symmetric. Also, $m - m = 0$ is divisible by 2, so R_4 is reflexive. In a similar way one can verify that R_4 is transitive. It follows that R_4 is an equivalence relation. Under this relation we note that $0, 2, 4$ are equivalent to each other, and furthermore the equivalence class of 0 consists of the set of all even numbers; similarly, one can see that the equivalence class of 1 consists of all odd numbers. In this case we have only two equivalence classes, the even and the odd numbers, they are disjoint and their union is $\mathbb{N}_0$.

0.4.3. Functions.

Definition 0.4.8. A **function** f from a set X to a set Y is a relation on X and Y; i.e., it is a subset $f \subseteq X \times Y$ that satisfies the special property that

for all $x \in X$ there exists exactly one $y \in Y$ such that $(x,y) \in f$.

In this case we write $f(x) = y$ instead of $(x,y) \in f$. We use the notation $f: X \to Y$ to represent a function f from X to Y. We call X the **domain** of the function f and Y the **codomain** of the function f. ◊

Note that a function f is defined on every element of its domain X, but not every element of the codomain Y needs to be attained.

For an example, let X be the set of towns in Massachusetts, and let Y be the set of cows in Massachusetts. Let $R_3 = \{(a,b) \in X \times Y : \text{cow } b \text{ lives in town } a\}$. Then R_3 is not a function since there are at least two cows in Williamstown. On the other hand, if Z is the set of possible weights and $R_4 = \{(b,z) \in Y \times Z : \text{cow } b \text{ weighs } z\}$, then R_4 is a function.

Definition 0.4.9. We define the **image** of X under f to be the set
$$f(X) = \{f(x) : x \in X\}.$$

0.4. Functions

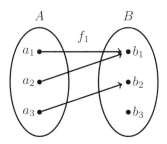

Figure 0.9. Function example f_1 that is not injective and not surjective.

The notion of image under f may be defined for every set $C \subseteq X$ in a similar way:
$$f(C) = \{f(c) : c \in C\}.$$
(In the literature, the word "range" is sometimes used for the codomain and sometimes for the image.) ◇

For example, let $A = \{a_1, a_2, a_3\}$ and $B = \{b_1, b_2, b_3\}$, and define a relation $f_1 \subseteq A \times B$ by
$$f_1 = \{(a_1, b_1), (a_2, b_1), (a_3, b_2)\}.$$
This relation is a function since for every element x in A there is exactly one $y \in B$ such that $(x, y) \in f_1$. The function is $f_1(a_1) = f_1(a_2) = b_1, f_1(a_3) = b_2$. While f_1 is defined on all of A, not every element of B is attained, as there is no $x \in A$ such that $f_1(x) = b_3$. The image of A under f_1 is $f_1(A) = \{b_1, b_2\}$. Figure 0.9 illustrates this function.

Regarding notation, when defining a function, instead of saying, for example, let f be the function defined by $f = \{(n, n^2) : n \in \mathbb{N}\}$, we will say let $f : \mathbb{N} \to \mathbb{N}$ be the function defined by $f(n) = n^2$. However, defining a function as a list or table is useful when defining functions in a computer program.

Definition 0.4.10. A function $f : X \to Y$ is a **surjective** (or **onto**) function if for all $y \in Y$ there exists $x \in X$ such that $f(x) = y$. ◇

Clearly, the function f_1 above is not surjective. A function $f : X \to Y$ is surjective if and only if $f(X) = Y$.

Definition 0.4.11. A function $f : X \to Y$ is said to be **injective** (or **one-to-one**) if whenever $f(x_1) = f(x_2)$, for $x_1, x_2 \in X$, then $x_1 = x_2$. This is equivalent to saying that
$$x_1 \neq x_2 \implies f(x_1) \neq f(x_2).$$
◇

The function f_1 is not injective either, since $f_1(a_1) = f_1(a_2)$ and $a_1 \neq a_2$.

Definition 0.4.12. A function that is injective and surjective is called a **bijection**. A bijection is also called a **one-to-one correspondence**. ◇

Figure 0.10 shows examples of a function that is injective but not surjective, a function that is surjective but not injective, and a bijection.

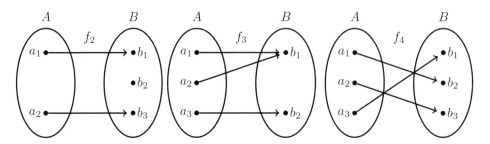

Figure 0.10. Function examples: f_2 is injective but not surjective, f_3 is surjective but not injective, f_4 is a bijection.

Definition 0.4.13. Given a function $f : X \to Y$ and a set $C \subseteq X$, one can define a new function $g : C \to Y$ by setting

$$g(x) = f(x) \text{ for every } x \in C.$$

This function is called the **restriction** of f to C and is denoted by $f|_C$. ◇

Subsection 0.5.4 gives an interesting and powerful use of the notion of one-to-one correspondence.

0.4.4. Inverse Function. Inverse Image. Composition of Functions. Let $f : X \to Y$ be a bijection. As f is surjective, for each $y \in Y$ there exists $x \in X$ such that $f(x) = y$, and since f is injective, this x is unique. Then we can define a function $g : Y \to X$ by letting $g(y)$ be the unique $x \in X$ such that $f(x) = y$. Hence this means that

$$f(g(y)) = f(x) = y \text{ and } g(f(x)) = g(y) = x.$$

Also, if f is a bijection and there is a function $h : Y \to X$ that satisfies $h(f(x)) = x$ for all $x \in X$, it follows that h must be g. This is so since each $y \in Y$ can be written as $y = f(x)$ for $x \in X$, and we have $h(y) = h(f(x)) = x = g(f(x)) = g(y)$, so $h = g$. This means that given a bijection f there is a unique function g such that $f(g(y)) = y$ and $g(f(x)) = g(y)$ for all $x \in X$ and $y \in Y$.

Definition 0.4.14. Let $f : X \to Y$ be a bijection. The unique function g that satisfies $f(g(y)) = y$ and $g(f(x)) = x$ for all $x \in X$ and $y \in Y$ is called the **inverse function** of f. We write f^{-1} for the inverse of f, and in this case we say f is **invertible**. ◇

Thus the inverse function of f satisfies

$$f(f^{-1}(y)) = y \text{ for all } y \in Y, \text{ and}$$
$$f^{-1}(f(x)) = x \text{ for all } x \in X.$$

For example, the function f_4 of Figure 0.10 is a bijection, and for its inverse we have $f^{-1}(b_1) = a_3$, $f^{-1}(b_2) = a_1$, $f^{-1}(b_3) = a_2$. We remark that if a function f has a function g such that $f(g(y)) = y$ for all $y \in Y$, then f must be surjective (Exercise 0.4.14), and if there is a function g such that $g(f(x)) = x$ for all $x \in X$, then f must be injective (Exercise 0.4.15), but $f^{-1}(y)$ makes sense only when f is invertible.

0.4. Functions

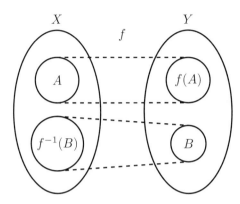

Figure 0.11. Image and inverse image of a function.

A notion related to the inverse function is that of the **inverse image of a set**. However, the definition of the inverse image does not need the function to be invertible; moreover, it is not necessary that f be injective or surjective.

Definition 0.4.15. Let $f : X \to Y$ be a function, and let $B \subseteq Y$. The **inverse image** of B, denoted $f^{-1}(B)$, is defined to be

$$f^{-1}(B) = \{x \in X : f(x) \in B\};$$

i.e., we think of $f^{-1}(B)$ as all the points in X that are sent to B under f (see Figure 0.11). ◇

We note again that no assumptions regarding injectivity or surjectivity are made on the function f, and $f^{-1}(B)$ is well-defined for any function $f : X \to Y$ and any subset B of Y (note that the definition only depends on f). It is perhaps an abuse of notation that we also use f^{-1} here, but when applied to a set, it does not need the existence of the inverse of f. However, if f happens to be invertible, then it is the case that $f^{-1}(B) = \{f^{-1}(b) : b \in B\}$. It should be clear from the context whether we mean the inverse function or the inverse image; whenever f is applied to a set, the reader should always interpret it as an inverse image. For example, $f^{-1}(\{y\})$ is always defined and consists of the set of x in X such that $f(x) = y$ (which may be empty).

Definition 0.4.16. Let X, Y, Z, W be sets, and let $f : X \to Y$ and $g : Z \to W$ be two functions. If the image of f is a subset of the domain of g, i.e.,

$$f(X) \subseteq Z,$$

then we can define a new function, called the **composition** of g and f, denoted by $g \circ f : X \to W$, and defined by

$$g \circ f(x) = g(f(x)) \text{ for each } x \in X.$$

The domain of $g \circ f$ is X and its codomain is W. The image of $g \circ f$ is $g(f(X))$; see Figure 0.12. ◇

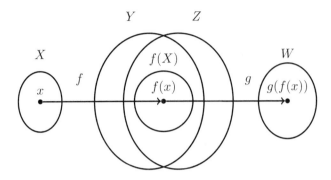

Figure 0.12. Function composition: $g \circ f$.

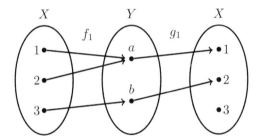

Figure 0.13. Example: $g_1 \circ f_1$.

For example, let $X = \{1, 2, 3\}$, $Y = \{a, b\}$, $Z = Y$, $W = X$ and define $f_1 : X \to Y$ by $f_1(1) = f_1(2) = a$, $f_1(3) = b$, and $g_1 : Y \to X$ by $g_1(a) = 1, g_1(b) = 2$. Then $g_1 \circ f_1(1) = 1$, $g_1 \circ f_1(2) = 1$, and $g_1 \circ f_1(3) = 2$. This is illustrated in Figure 0.13.

In the special case when the codomain of a function is the same as its domain (i.e., when for a function $f : X \to X$), then the composition $f \circ f$ is defined, since in this case $f(X)$ is certainly a subset of X.

Question 0.4.17. Let $f : X \to Y$ be a function. Show that if f is a surjection and there is a function $g : Y \to X$ such that $g(f(x)) = x$ for all $x \in X$, then f is a bijection and g is the inverse of f. Show that if f is an injection and there is a function $g : Y \to X$ such that $f(g(y)) = y$ for all $y \in Y$, then f is a bijection and g is the inverse of f.

Exercises: Functions

0.4.1 Let A, B be sets. Prove that $A \times B = \varnothing$ if and only if $A = \varnothing$ or $B = \varnothing$.

0.4.2 Let A, B be sets. Prove that
$$A \times B \subseteq \mathcal{P}(\mathcal{P}(A \cup B)).$$
(*Hint*: Use the definition of ordered pairs as sets.)

0.4.3 Let A, B, and C be sets. Prove that

(a) $A \times (B \cup C) = (A \times B) \cup (A \times C)$.
(b) $A \times (B \cap C) = (A \times B) \cap (A \times C)$.

0.4.4 Let A, B, C, and D be sets. Prove that
$$(A \times B) \cap (C \times D) = (A \cap C) \times (B \cap D).$$

0.4.5 Let A, B, C, and D be sets. Prove that
$$(A \times B) \cup (C \times D) = (A \cup C) \times (B \cup D).$$

0.4.6 Let A, B, C, and D be sets. Find a formula for $(A \times B) \setminus (C \times D)$ and prove your claim.

0.4.7 Let $X = \{a, b, c, d\}$ and $Y = \{1, 2, 3\}$. For each of the following sets in $X \times Y$, determine whether they are functions or not, and in the case that they are functions determine whether they are surjective, injective, or bijections.
  (a) $R_1 = \{(a, 1), (b, 3), (c, 2)\}$
  (b) $R_2 = \{(a, 2), (b, 3), (b, 2), (c, 1)\}$
  (c) $R_3 = \{(a, 2), (b, 1), (c, 3), (d, 1)\}$
  (d) $R_4 = \{(a, 1), (b, 2), (c, 1), (d, 2)\}$

0.4.8 Let $f : X \to X$ be a function, and let $A \subseteq X$. Prove that $A \subseteq f^{-1}(A)$ if and only if $f(A) \subseteq A$.

0.4.9 Let $f : X \to Y$ be a function.
  (a) Prove that $A \subseteq f^{-1}(f(A))$ for every $A \subseteq X$.
  (b) Prove that $A = f^{-1}(f(A))$ when f is injective.

0.4.10 Let $f : X \to Y$ be a function.
  (a) Prove that $f(f^{-1}(B)) \subseteq B$ for every $B \subseteq Y$.
  (b) Prove that $f(f^{-1}(B)) = B$ when f is surjective.

0.4.11 Let X, Y, Z be sets, and let $f : X \to Y$ and $g : Y \to W$ be two functions. Prove that if f and g are bijective, then $g \circ f$ is bijective and its inverse is given by $(g \circ f)^{-1} = f^{-1} \circ g^{-1}$.

* 0.4.12 Prove that property (0.24) holds for ordered pairs under the definition $(a, b) = \{\{a\}, \{a, b\}\}$. Show that (0.24) does not hold if we were to define (a, b) as $\{a, \{b\}\}$.

0.4.13 Let $f : X \to Y$ and $g : Y \to Z$ be two functions.
  (a) Prove that if f and g are surjective, then $g \circ f$ is surjective.
  (b) Prove that if f and g are injective, then $g \circ f$ is injective.

0.4.14 Let $f : X \to Y$ be a function. Prove that if there exists a function $g : Y \to X$ such that $f(g(y)) = y$ for all $y \in Y$, then f is surjective. We call g the **right inverse** of f.

0.4.15 Let $f : X \to Y$ be a function. Prove that if there exists a function $g : Y \to X$ such that $g(f(x)) = x$ for all $x \in X$, then f is injective. We call g the **left inverse** of f.

0.4.16 Let $f : X \to Y$ be a function. Prove that if there exists a function $g : Y \to X$ such that $g(f(x)) = x$ and $f(g(y)) = y$ for all $x \in X$ and $y \in Y$, then f is a bijection and g is its inverse function.

0.4.17 Give an example of a function with a left inverse but no right inverse, and a function with right inverse but no left inverse.

0.4.18 Let $f : X \to Y$ be a function. Prove that
$$f(A \cup B) = f(A) \cup f(B) \text{ for all sets } A, B \subseteq X.$$

0.4.19 Let $f : X \to Y$ be a function.
  (a) Prove that
  $$f(A \cap B) \subseteq f(A) \cap f(B)$$
  for all sets $A, B \subseteq X$, but that the equality does not hold in general (i.e., prove that there exists a function $f : X \to Y$ and sets A, B such that $f(A \cap B) \subsetneq f(A) \cap f(B)$).
  (b) Prove that if f is injective,
  $$f(A \cap B) = f(A) \cap f(B).$$

0.4.20 Let $f : X \to Y$ be a function, and let $A, B \subseteq Y$. Show that
$$f^{-1}(A \cup B) = f^{-1}(A) \cup f^{-1}(B),$$
$$f^{-1}(A \cap B) = f^{-1}(A) \cap f^{-1}(B),$$
$$f^{-1}(A \setminus B) = f^{-1}(A) \setminus f^{-1}(B).$$

0.4.21 Prove that the set of all equivalence classes forms a (disjoint) partition of A, i.e., prove that every element of A is in some equivalence class and that any two distinct equivalence classes do not intersect.

0.5. Mathematical Induction

Often we will want to prove that a statement that depends on a natural number n is true for all natural numbers n. An example of such a statement is:

> Let n be a natural number. If A is a set with n elements, then the power set $\mathcal{P}(A)$ is a set with 2^n elements.

Denote this statement by $S(n)$ (this is a statement that depends on n). We would like to show that $S(n)$ is true for each natural number n. We can verify that, for example, $S(1)$ is true: If A is a set with one element, we can assume it has the form $A = \{a\}$. Then we find that $\mathcal{P}(A) = \{\varnothing, \{a\}\}$ has two elements, so $S(1)$ is true. Similarly, if A is a set with two elements, we can assume A can be written as $A = \{a, b\}$. Then $\mathcal{P}(A) = \{\varnothing, \{a\}, \{b\}, \{a, b\}\}$ has $4 = 2^2$ elements, so $S(2)$ is true. In a similar way one can verify $S(3)$, for example, is true. But we cannot use this technique to prove that $S(n)$ holds for every single n, as we would need infinitely many proofs. The most useful technique to prove these kinds of statements is *mathematical induction*, which we state in Subsection 0.5.1.

Mathematical induction is at the heart of our understanding of the natural numbers. Mathematical induction is equivalent to another principle that is simpler to state and which says that every nonempty subset of natural numbers has a smallest element. The equivalence of these two principles is shown in Theorem 0.5.8, which also shows an equivalence with a second version of induction that is useful in some cases. We take mathematical induction as a property basic to our understanding of the natural numbers. It can be shown, using the axioms of set theory, that the construction of the natural numbers we have outlined satisfies this property (Subsection 0.6.2).

0.5.1. Principle of Mathematical Induction

Definition 0.5.1. *Mathematical Induction.* Let $S(n)$ be a statement about the natural number n. To show that $S(n)$ is true for all $n \in \mathbb{N}$, it suffices to prove the following two statements.

(1) *Base step*: $S(1)$ holds.
(2) *Inductive step*: For $k \geq 1$, if $S(k)$ holds, then $S(k+1)$ holds. ◊

To prove the inductive step, we have to prove is that the implication is true, i.e., that the truth of $S(k)$ implies the truth of $S(k+1)$. Thus we are allowed to assume that $S(k)$ is true; this assumption is called the *inductive hypothesis*.

Remark 0.5.2. One can formulate mathematical induction in terms of sets in the following way. Suppose P is a subset of $\mathbb{N}$ such that

(1) *Base Step*: $1 \in P$.
(2) *Inductive Step*: If $k \in P$, then $k+1 \in P$.

Then $P = \mathbb{N}$.

To see the equivalence, given a statement $S(n)$, one can let P be the set of all n such that $S(n)$ holds. And given a set P, one can let $S(n)$ be all the natural numbers n that are in P.

We illustrate the use of mathematical induction with the following example.

Example 0.5.3. Let $n \in \mathbb{N}$. We prove that if A is a set with n elements, then $\mathcal{P}(A)$ is a set with 2^n elements. (We are using the notion of number of elements informally; it is defined in Subsection 0.5.4.) We proceed by induction. Let $S(n)$ denote the statement, "If A is a set with n elements, then $\mathcal{P}(A)$ is a set with 2^n elements." To show the base step $S(1)$, let $A = \{a\}$. Hence $\mathcal{P}(A) = \{\varnothing, \{a\}\}$, which has $2 = 2^1$ elements. Therefore $S(1)$ holds.

To show the inductive step, we have to show that if $S(k)$ is true, then $S(k+1)$ is true. Note that $S(k)$ asserts that if A is a set with k elements, then $\mathcal{P}(A)$ has 2^k elements. We need to show that if B is a set with $k+1$ elements, then $\mathcal{P}(B)$ is a set with 2^{k+1} elements. Hence, let B be a set with $k+1$ elements. Then B is nonempty (as $k \geq 1$), and we can let b be an element of B. Let $A = B \setminus \{b\}$. Then A is a set with k elements. By the inductive hypothesis $\mathcal{P}(A)$ has 2^k elements. Let $A_1, \ldots, A_{2^k}$ be all the 2^k subsets of A. These are all different sets. Then

$$A_1 \cup \{b\}, \ldots, A_{2^k} \cup \{b\}$$

are 2^k subsets of B that are different from $A_1, \ldots, A_{2^k}$ (they differ precisely in the element b). These two groups of sets together give 2^{k+1} subsets of B. It is also clear that there cannot be any other subset of B, as the first list has the subsets of B that do not contain b, and the second list has the subsets of B that contain b. Therefore $\mathcal{P}(B)$ has 2^{k+1} elements, completing the proof of the inductive step. Induction now implies that $S(n)$ is true for all $n \in \mathbb{N}$.

Example 0.5.4. Here we try to prove that all cows have the same color. We will try to prove by induction that any set of n cows has the same color. If the set has one cow, then clearly they all have the same color. Now assume any set of n cows has the same color, and let C be a set of $n+1$ cows. Divide C into two sets of n cows, A and B, whose union is C. By the induction hypothesis, all cows in A have the same color, and all cows in B have the same color. Since there is a cow in common, then all cows in C have the same color. This seems to prove that all cows have the same color! What happened? Let us be more precise with the inductive step. Let a be a cow in C and set $A = C \setminus \{a\}$. Let b be another cow in C (we need $b \neq a$), and set $B = C \setminus \{b\}$. For A and B to have a cow in common, we need C to have at least three cows, or $n \geq 2$. So the base step cannot start at $n = 1$. But if the base step is $n = 2$, we note that any two cows do not necessarily have the same color, so the base step for $n = 2$ does not work. Thus induction fails in this case.

0.5.2. The Natural Numbers $\mathbb{N}$. As we have said, we view mathematical induction as part of our understanding of what the natural numbers are, and it is standard practice to take it as an axiom of the natural numbers.

There are basically three ways one can introduce the natural numbers. The first one is an informal way in which one appeals to our basic intuition and understanding of the natural numbers, but one points out that there is one important property of the natural numbers that may not be as obvious, and that property is mathematical induction.

A second approach is an axiomatic presentation of the natural numbers, where one gives a set of axioms and postulates the existence of a set (namely $\mathbb{N}$) satisfying these axioms. This is close to the informal presentation, and as the most difficult axiom in the axiomatic presentation is the induction axiom, we briefly mention the axiomatic presentation. The axioms are called the Peano axioms. They were introduced by Giuseppe Peano in 1889.

Definition 0.5.5. The **Peano axioms** for the natural numbers are the following.

(1) There is an element in $\mathbb{N}$ denoted by 1 (this establishes the existence of an element in $\mathbb{N}$);

(2) there exists a function, called the *successor* function, that assigns to each natural number n another natural number n^+ (the successor of 1 is 2, etc.);

(3) 1 is not the successor of any natural number (this guarantees that 1 is the "first" element);

(4) if $n^+ = m^+$ for any natural numbers n and m, then $n = m$ (this establishes uniqueness of the successors);

(5) mathematical induction. ◇

One can see that our informal notion of the natural numbers satisfies the Peano axioms if we define the successor function n^+ by $n^+ = n+1$. Conversely, assuming the Peano axioms, we can construct the natural numbers in the following way. First, addition by 1 is defined by the successor function ($n+1 = n^+$). Using this function also, we define 2 as the successor of 1, 3 as the successor of 2, etc. Addition by an arbitrary natural number $k > 1$ is defined by induction: $n + k = (n + (k-1))^+$.

Once addition has been defined, multiplication can also be defined by induction: $n \cdot 1 = n$ and $n \cdot k = n \cdot (k-1) + n$ for $k > 1$.

A third approach is a constructive presentation from the set axioms, where one constructs the set of natural numbers using set theory and proves from this construction the main properties of the natural numbers, such as the Peano axioms. This approach starts with more basic axioms than the Peano axioms (those of set theory). While a detailed construction would take us too much time, we have outlined the ideas in Subsection 0.6.2; here natural numbers are sets, and the successor of n is $n+ = n \cup \{n\}$. We mention that in this case it is possible to prove induction from a set theoretic axiom called the *axiom of infinity*.

0.5.3. The Well-Ordering Principle for $\mathbb{N}$**.** There is another property of the natural numbers, called the well-ordering principle for $\mathbb{N}$, which is equivalent to induction and may be more intuitive to some readers than induction. We need a definition before stating this principle.

Definition 0.5.6. Given a nonempty set A of natural numbers we say that A has a **least element** if there is an integer n such that
$$n \in A \text{ and } n \leq k \text{ for all } k \in A.$$
In this case we write $n = \min A$. $\Diamond$

This notation is justified as there cannot be more than one least element. For example, $\min \mathbb{N} = 1$. (Here the inequality is defined by $n < m$ if $m \notin \{1, \ldots, n\}$, and of course $\leq$ means $<$ or $=$.)

Definition 0.5.7. The **well-ordering principle for** $\mathbb{N}$ states that if A is a nonempty subset of $\mathbb{N}$, then A has a least element. $\Diamond$

Theorem 0.5.8 shows that the well-ordering principle is equivalent to induction. The theorem also gives another equivalent form of induction that is used in some problems, which is called *strong induction*.

Theorem 0.5.8. *The following statements are equivalent.*

(1) Mathematical induction.
(2) Strong mathematical induction: *Let $S(n)$ be a statement about the natural number n. To show that $S(n)$ is true for all $n \in \mathbb{N}$ it suffices to show the following two statements.*
   (a) Base step. $S(1)$ holds.
   (b) Strong inductive step: *If $S(1), S(2), \ldots, S(k)$ hold for $k \geq 1$, then $S(k+1)$ holds.*
(3) Well-ordering principle for $\mathbb{N}$: *If A is a nonempty subset of $\mathbb{N}$, then A has a least element.*

Proof. To prove such an equivalence, it suffices to show that (1) implies (2), that (2) implies (3), and that (3) implies (1).

To prove that (1) implies (2), let $S(n)$ be a statement for each $n \in \mathbb{N}$ and suppose that the statements $S(n)$ satisfy conditions (a) and (b) of strong induction. We want to show that $S(n)$ holds for all $n \in \mathbb{N}$, and we may assume induction.

The idea is to define a new statement $T(n)$, for each $n \in \mathbb{N}$, that groups all the statements $S(1)$ and $S(2), \ldots, S(n)$ together. (More formally, we show how to define $T(n)$ using induction—recursive definitions are discussed in more detail in Subsection 0.5.5. Define $T(1)$ to be $S(1)$ and, assuming that $T(n-1)$ has been defined, define $T(n)$ to be $T(n-1)$ with $S(n)$; then induction can be used to justify that $T(n)$ is defined for all $n \in \mathbb{N}$.) Now apply induction to the statements $T(n)$ for $n \in \mathbb{N}$. We first observe that $T(1)$ is true since it is the same as $S(1)$, and $S(1)$ is assumed true by (a) of strong induction. Also, if $T(k)$ holds, then $S(1), S(2), \ldots, S(k)$ hold, so using step (b) of strong induction we obtain that $S(k+1)$ holds. This means that $T(k+1)$ is true. Then induction gives that $T(n)$ holds for all n, and so $S(n)$ holds for all n, completing the proof.

We show that (2) implies (3). Let A be a nonempty set in $\mathbb{N}$. We proceed by contradiction. Suppose that A does not have a least element. Let $S(n)$ be the statement $n \in A^c$. It suffices to show that $S(n)$ is true for all $n \geq 1$, as this would show that $A^c = \mathbb{N}$ and A is empty, a contradiction. Clearly, $1 \in A^c$ (otherwise, 1 would be the least element of A). Next assume that $1, 2, \ldots, k$ are in A^c. If $k+1$ were an element of A, then $k+1$ would be the least element of A. As A does not have a least element, $k+1 \in A^c$. This shows that the strong inductive step is true. Therefore $S(n)$ is true for all $n \in \mathbb{N}$, completing the proof.

To show that (3) implies (1), suppose that every nonempty set in $\mathbb{N}$ has a least element, and let $S(n)$ be a statement about the natural number n. Assume that the two steps of induction hold for $S(n)$. We want to show that $S(n)$ holds for all $n \in \mathbb{N}$. Define a set A as

$$A = \{n \in \mathbb{N} : S(n) \text{ holds}\}.$$

To show that $S(n)$ holds for all n, it suffices to show that $A = \mathbb{N}$. We proceed by contradiction. Suppose that $A \neq \mathbb{N}$. Then A^c is a nonempty set in $\mathbb{N}$ and thus must have a least element, call it k. We know that $k \neq 1$ ($S(1)$ is true) and that $k-1 \in A$. Hence, $S(k-1)$ is true, and this implies that $S(k)$ is true, a contradiction. Therefore, $A = \mathbb{N}$, completing the proof of this part. $\square$

We remark that it is easier to give a direct proof that strong induction implies induction.

Example 0.5.9. Suppose we wish to prove that for all natural numbers n,

$$\text{(0.25)} \qquad \sum_{i=1}^{n} 2^i = 2^{n+1} - 2.$$

We can start by letting A be the set of all $n \in \mathbb{N}$ that satisfy (0.25). We wish to show $A = \mathbb{N}$. Clearly, $1 \in A$. Now suppose k is in A. To show that $k+1$ is in A, we write

$$\text{(0.26)} \qquad \sum_{i=1}^{k+1} 2^i = \sum_{i=1}^{k} 2^i + 2^{k+1}.$$

Since $k \in A$, we know that $\sum_{i=1}^{k} 2^i = 2^{k+1} - 2$ (this is the induction hypothesis). Replacing this in (0.26), we have

$$\sum_{i=1}^{k+1} 2^i = 2^{k+1} - 2 + 2^{k+1} = 2^{k+2} - 2,$$

showing that $k+1 \in A$. By the induction principle we conclude then that $A = \mathbb{N}$.

We can also prove this using the well-ordering principle for $\mathbb{N}$. If (0.25) is not true for all natural numbers, then there must be a smallest $k \in \mathbb{N}$ for which it is not true. (More formally, we let B be the set of n such that (0.25) is not true for n. We wish to show that B is empty. Proceeding by contradiction, we assume that B is nonempty; by well-ordering it must have a smallest element k. This shows that if (0.25) is not true for all natural numbers, then there is a smallest k for which (0.25) is not true.) Next we observe that k cannot be 1, as one can verify the equation for $k = 1$. Since k is the smallest for which it is not true and $k > 1$, then (0.25) must be true for $k-1$ (equivalently $k-1 \notin B$). Now a similar argument to the one we have just completed shows that the equation holds for k, giving a contradiction. Therefore (0.25) holds for all $n \in \mathbb{N}$ (equivalently, $B = \varnothing$).

0.5.4. Finite Sets. The reader has a good intuition of what a finite set should be (in fact we have used this informally in Section 0.5), and will agree that the following definition is reasonable. For each $n \in \mathbb{N}$, let J_n denote the set

$$J_n = \{i \in \mathbb{N} : i \leq n\} = \{1, \ldots, n\}.$$

We think of J_n as the "canonical set of n elements". (The reader may recall that the definition we gave of $0, 1, \ldots$, in terms of sets, defines n as the set $\{0, 1, \ldots, n-1\}$.)

Definition 0.5.10. A set A is defined to be a **finite set** if it is empty or if there is a natural number n and a bijection $f : J_n \to A$. If a set is finite, then there is only one natural number n such that there is a bijection $f : J_n \to A$ (this follows from Exercise 0.5.3); in this case we say that n is the **cardinality** of A, and we write

$$\text{card}(A) = n.$$

If the set is empty, we say it has cardinality 0. A set A is said to be **infinite** if it is not finite. ◇

Clearly, $\text{card}(J_n) = n$, since the identity function (i.e., $\phi(n) = n$) is a bijection from J_n to itself.

At the moment the notion of cardinality is defined only for finite sets. While all the obvious properties for finite sets can be proved by induction (for example, if A is finite and $B \subseteq A$, then B is finite), we will not emphasize this, and instead we will assume the reader is familiar with the basic properties of finite sets (another such property, for example, is that the union of two finite sets is finite); the exception being when we explicitly ask in the exercises for a particular property to be proved.

0.5.5. Recursive Definitions. Mathematical induction can be used to define expressions for each natural number n. We illustrate this with examples. Suppose we desire to define the exponential a^n where a is a rational number (or a real

number once we have defined real numbers) and n is an integer. First we define it when $n \in \mathbb{N}$ by induction in the following way. Set
$$a^1 = a,$$
$$a^{n+1} = a^n \cdot a.$$
Induction can be used to show that the set of natural numbers n, such that a^n has been defined, is all of $\mathbb{N}$. The case of other integers can be completed by setting
$$a^0 = 1,$$
$$a^{-n} = \frac{1}{a^n} \text{ for each } n \in \mathbb{N}.$$

In a similar way we define the summation sign $\sum$. Suppose that for each $i \in \mathbb{N}$ we have a number a_i. Define
$$\sum_{i=1}^{1} a_i = a_1,$$
$$\sum_{i=1}^{n+1} a_i = (\sum_{i=1}^{n} a_i) + a_{n+1}.$$
The product $\prod_{i=1}^{n} a_i$ of finitely many numbers is defined analogously.

One can also define the factorial function by recursion: set $0! = 1$ and $(n+1)! = n!(n+1)$. We note that, as in this example, sometimes we may start at 0 and sometimes at an integer greater than 1.

We make a remark regarding recursive definitions. Let's look again at the definition of the factorial function. We are defining a function f on the nonnegative integers by $f(0) = 1$ and $f(n+1) = f(n)n$. We note that in this last part we are using the notation of a function while the function has not been completely defined yet. One can indeed use induction to show that this indeed works and defines a unique function f; we will not cover the details but refer the interested reader to the Bibliographical Notes.

Definition 0.5.11. Another example of a recursive definition is that of the **Fibonacci sequence** (F_n). The sequence is defined by
$$F_0 = 0, F_1 = 1,$$
$$F_n = F_{n-1} + F_{n-2}.$$
◇

0.5.6. The Integers $\mathbb{Z}$ and Rational Numbers $\mathbb{Q}$. The set of integers $\mathbb{Z}$ and the set of rational numbers $\mathbb{Q}$ can be defined from the natural numbers. The integers consist of the natural numbers, the number 0, and the negative natural numbers:
$$\mathbb{Z} = \{\ldots, -2, -1, 0, 1, 2, \ldots\}.$$
The set of rational numbers $\mathbb{Q}$ consists of numbers of the form p/q where p and q are integers and $q \neq 0$, with an identification, where we regard p/q the same as p'/q' provided $pq' = qp'$. This is necessary as we consider, for example, the rational number $4/6$ to be the same as $2/3$. Formally, this identification is done using the notion of equivalence relations, and each rational number is a particular equivalence class. This is discussed in more detail in Subsection 0.6.2; however, it is all right

0.5. Mathematical Induction

for the reader to use the usual intuitive notion of rational numbers with the usual identification of fractions.

We shall assume familiarity with the basic arithmetic properties of the natural numbers, the integers, and the rationals. Basic properties of the natural numbers that we shall need are proved in Subsection 0.5.7.

0.5.7. Some Properties of the Natural Numbers. We prove some properties about the natural numbers that will be useful in later sections.

Definition 0.5.12. A natural number p is **prime** if $p > 1$ and its only divisors are 1 and itself. ◊

Example 0.5.13. We show that there exist infinitely many primes. This argument goes back to Euclid (323–283 BCE). We first note that a natural number other than 1 is either a prime or is divisible by a prime (see Exercise 0.5.4). Suppose to the contrary that there existed finitely many primes. Then we could list them all as $p_1, \ldots, p_n$, for some fixed natural number n. Let
$$a = (p_1 p_2 \cdots p_n) + 1.$$
Then a cannot be a prime as it is greater than each of $p_1, \ldots, p_n$. Hence it must be divisible by a prime. Thus there exists some $i \in \{1, \ldots, n\}$ such that p_i divides a. But then p_i would divide 1, a contradiction. Therefore, there exist infinitely many primes. (There are now many proofs of the infinitude of the primes, and the reader is encouraged to find others.)

Definition 0.5.14. If a and b are in $\mathbb{N}$, the **greatest common divisor** of a and b, written $\gcd(a, b)$, is a number d that is the maximum among all common divisors of a and b (this exists as it is the maximum of a finite set; see Exercise 0.5.15). ◊

Example 0.5.15. Assume that a and b are natural numbers that do not have any divisor in common other than 1, i.e., $\gcd(a, b) = 1$. We show that
$$x_0 a + y_0 b = 1$$
for some $x_0, y_0 \in \mathbb{Z}$. First set
$$A = \{xa + yb : x, y \in \mathbb{Z}\} \cap \mathbb{N},$$
and let
$$k = \min A.$$
(The set A is nonempty, as a and b are in A, since $a = 1 \cdot a + 0 \cdot b$ and $b = 0 \cdot a + 1 \cdot b$.) Then k can be written as $k = x_0 a + y_0 b$ for some $x_0, y_0 \in \mathbb{Z}$. We now show that $k = 1$. For this we first show that k divides every element of A. Suppose that this were not the case. Then there would exist a smallest element of A that k does not divide. Denote this element by $x_1 a + y_1 b$. From the definition of k we have $k < x_1 a + y_1 b$. Let q be the largest natural number such that $qk < x_1 a + y_1 b < (q+1)k$. It must be the case that
$$0 < (x_1 a + y_1 b) - qk < k.$$
But this gives an element of A that is smaller than k, a contradiction. Therefore, k divides every element of A, in particular it divides a and b. Thus $k = 1$.

Example 0.5.16. We show that if a prime number p divides a product ab of two natural numbers a and b, then p divides a or p divides b. We start by supposing that p does not divide a. Then $\gcd(p,a) = 1$. By Example 0.5.15, there exist integers x, y such that $xp + ya = 1$. Hence $b = xpb + yab$. As p divides ab, it divides $xpb + yab$, so it divides b.

Example 0.5.17. We show that each natural number $n > 1$ can be written as a product of finitely many primes in a unique way (up to the order of the prime factors). This fact is known as the *Fundamental Theorem of Arithmetic*. First we use induction to show that each $n > 1$ can be written as a finite product of primes (where, to simplify the statement, we understand a prime as being a product of a single prime). The statement is clearly true for $n = 2$. Now suppose the statement holds for $n = 2, 3, \ldots, k$. If $k + 1$ is prime, we are done. If not, it has a divisor that is not itself, so $k + 1$ can be written as $k + 1 = \ell m$, for some natural numbers ℓ and m bigger than 1. The inductive hypothesis implies that each ℓ and m is a product of finitely many primes. Hence $k + 1$ is a finite product of primes.

Finally, to show uniqueness, we proceed by contradiction. Suppose n is the smallest natural number that can be written in two different ways as

$$n = p_1^{i_1} \cdots p_a^{i_a} = q_1^{j_1} \cdots q_b^{j_b},$$

for some primes $p_1 < p_2 < \cdots < p_a$ and $q_1 < q_2 < \cdots < q_b$, and natural numbers $i_1, \ldots, i_a, j_1, \ldots, j_b$. By Example 0.5.16, p_1 divides $q_1^{j_1}$ or it divides $q_2^{j_2} \cdots q_b^{j_b}$. In the first case $p_1 = q_1$. In the second case, as $q_2^{j_2} \cdots q_b^{j_b}$ is smaller than n, it has a unique factorization into primes (up to the order of the primes), so $p_1 = q_j$ for some $j \in \{2, \ldots, b\}$. In either case we can remove p_1 from both sides of the equation and obtain two different factorizations of n/p_1, a number smaller than n, yielding a contradiction.

0.5.8. Cantor–Schröder–Bernstein Theorem. The Cantor–Schröder–Bernstein theorem, also known as the Schröder–Bernstein theorem, basically states that if the cardinality of a set A is less than or equal to the cardinality of a set B, and the cardinality of B is less than or equal to the cardinality of A, then the cardinalities are equal. This is not surprising when the sets A and B are finite, and this fact can be shown by induction. When the sets are not finite, we have not formally defined the notion of cardinality as it is rather technical and would take us too far afield, but we think of it intuitively as a notion of size. One can say that A has cardinality less than or equal to B when there is an injection from A to B. Also, the notion of two sets having the same cardinality can easily be defined by saying that there is a bijection between the two sets. (All this clearly makes sense in the case of finite sets.) The Cantor–Schröder–Bernstein theorem is very useful when one wants to show that there is a bijection between two sets.

Theorem 0.5.18 (Cantor, Schröder, and Bernstein)**.** *Let A and B be two nonempty sets. If there exists a function $f : A \to B$ that is injective and a function $g : B \to A$ that is injective, then there exists a bijection $h : A \to B$.*

Proof. If f (or g) were surjective, we would be done. The idea for the proof will be to partition A into two disjoint parts so that to define h we can use f in one part and use g^{-1} in the other part.

To find this partition of the set A, we introduce some notation. Consider a point a in A. Either it has a preimage under g (precisely when a is in $g(B)$), or it does not. If a has a preimage under g, as g is injective, this preimage is unique and can be written as $g^{-1}(a)$; call it the 1-preimage of a. Now, it may happen that $g^{-1}(a)$ has a preimage under f (precisely when $g^{-1}(a)$ is in $f(A)$). If it does, call it the 2-preimage of a and note that it is $f^{-1}(g^{-1}(a))$. In this way we can define for each $n \in \mathbb{N}$ the n-preimage of a, if it exists. The 0-preimage of a is defined to be a. Then, each element a of A either has infinitely many preimages (i.e., it has an n-preimage for each n) or finitely many (i.e., there is an n so that a has an n-preimage but not an $(n+1)$-preimage). Let A_0 consist of all elements of A that have infinitely many preimages, let A_1 consist of those elements of A that have an odd number of preimages, and let A_2 be those that have an even number of preimages (a 0-preimage is considered an even preimage). It is clear that A is the disjoint union of A_0, A_1, and A_2. In a similar way define B_0 to consist of all elements of B that have infinitely many preimages (first under f, then under g, etc.), and analogously define B_1 and B_2.

We observe that f maps A_0 bijectively onto B_0. (This is because if $a \in A_0$, $f(a)$ is in B_0, as a is a preimage of $f(a)$, so $f(a)$ has infinitely many preimages and is in B_0. Also, if $z \in B_0$, it has infinitely many preimages, so its preimage $f^{-1}(z)$ has infinitely many preimages, thus it is in A_0. This shows f maps A_0 onto B_0 and we already know it is injective.) Similarly, f maps A_2 bijectively onto B_1. Also, elements of A_1 have a unique inverse under g, so g^{-1} maps A_1 bijectively onto B_2. Then we define a function $h : A \to B$ by

$$h(x) = \begin{cases} f(x), & \text{if } x \in A_0 \sqcup A_2, \\ g^{-1}(x), & \text{if } x \in A_1. \end{cases}$$

We first show h is injective. Suppose $h(a_1) = h(a_2)$ for $a_1, a_2 \in A$. If a_1 and a_2 are both in $A_0 \sqcup A_2$ or in A_1, then it is clear that $a_1 = a_2$ as f and g^{-1} are injective on these sets, respectively. So suppose $a_1 \in A_0 \sqcup A_2$ and $a_2 \in A_1$. Then $h(a_1) \in B_0 \sqcup B_1$ and $h(a_2) \in B_2$, a contradiction as the sets are disjoint, so it follows that h is injective. To show h is surjective, let $b \in B$. If $b \in B_0$, we have already seen that there is a in A_0 such that $f(a) = b$; so $h(a) = b$. If $b \in B_1$, then there is an $a \in A$ such that $f(a) = b$, and as f is injective, a must be in A_2, so again $h(a) = b$. If $b \in B_2$, then $g(b) \in A_1$, so $h(g(b)) = b$. This shows that h is surjective and therefore a bijection. $\square$

Example 0.5.19. We use the Cantor–Schröder–Bernstein theorem to show that $\mathbb{Z}$ and $\mathbb{Z} \times \mathbb{Z}$ have the same cardinality. There is a natural injective map from $\mathbb{Z}$ to $\mathbb{Z} \times \mathbb{Z}$ that sends n in $\mathbb{Z}$ to $(n,0)$ in $\mathbb{Z} \times \mathbb{Z}$. To apply the Cantor–Schröder–Bernstein theorem, it remains to show that there is an injection from $\mathbb{Z} \times \mathbb{Z}$ to $\mathbb{Z}$. Let $(m, n) \in \mathbb{Z} \times Z$. Define $h(n, m) = 2^n 3^m$ for $m, n \geq 0$, and for $m, n < 0$ set $h(n, m) = 5^{-n} 7^{-m}$, and $h(n, 0) = 5^{-n}$, $h(0, m) = 7^{-m}$. By the unique factorization of natural numbers into primes it follows that h is injective. Theorem 0.5.18 then implies that there is a bijection between $\mathbb{Z}$ and $\mathbb{Z} \times \mathbb{Z}$, they have the same cardinality.

Exercises: Mathematical Induction

0.5.1 Give a direct proof, without using Theorem 0.5.8, that strong mathematical induction implies mathematical induction.

0.5.2 Formulate and prove an extended strong mathematical induction principle.

0.5.3 Let $n, m \in \mathbb{N}$. Prove that if there exists a bijection $f : J_n \to J_m$, then $n = m$.

0.5.4 Prove that a natural number greater than 1 is either a prime or is divisible by a prime. (*Hint*: If a number is not a prime, then by definition it must be divisible by another number. Show that one can choose this other number to be a prime.)

0.5.5 *Bézout's identity*: Let $a, b \in \mathbb{N}$, and let $d = \gcd(a, b)$. Prove that $d = xa + yb$ for some $x, y \in \mathbb{Z}$.

0.5.6 Prove that for each integer $n \geq 1$, $n < 2^n$.

0.5.7 Prove that for each integer $n \geq 4$, $n^2 \leq 2^n$.

0.5.8 Let $\{F_n\}$ be the Fibonacci sequence. Prove that for each $n \in \mathbb{N}$,
$$\sum_{i=0}^{n} F_i = F_{n+2} - 1.$$

0.5.9 Prove that for each integer $n \geq 1$,
$$\sum_{i=1}^{n} \frac{1}{2^i} = 1 - \frac{1}{2^n}.$$

0.5.10 Prove that for each integer $n \geq 1$,
$$\sum_{i=1}^{n} i = \frac{n(n+1)}{2}.$$

0.5.11 Prove that for each integer $n \geq 1$,
$$\sum_{i=1}^{n} i^2 = \frac{n(n+1)(2n+1)}{6}.$$

0.5.12 Prove that for each integer $n \geq 1$,
$$\sum_{i=1}^{n} i^3 = \left(\frac{n(n+1)}{2}\right)^2.$$

0.5.13 Let $\{F_n\}$ be the Fibonacci sequence. Prove the Cassini identity: for each $n \in \mathbb{N}$,
$$F_n^2 - F_{n+1}F_{n-1} = (-1)^{n-1}.$$

0.5.14 For each $n \in \mathbb{N} \cup \{0\}$, define the Fermat number $F_n = 2^{2^n} + 1$. Prove by induction that for all $n \geq 1$,
$$\prod_{i=0}^{n-1} F_i = F_n - 2.$$

0.6. More on Sets: Axioms and Constructions

Use this equality to prove there exist infinitely many primes (*Hint*: For $n \neq m$, F_n and F_m do not have prime divisors in common.) Here $\prod_{i=0}^{n-1} F_i$ is the product $F_0 F_1 \cdots F_{n-1}$.

0.5.15 Let $n \in \mathbb{N}$, and let $A \subseteq \{1, \ldots, n\}$ be nonempty. Prove that A has a **maximum** element, i.e., an element $d \in A$ such that $i \leq d$ for all $i \in A$, and a **minimum** element, i.e., an element $e \in A$ such that $i \geq e$ for all $i \in A$. Write $\max A$ for d and $\min A$ for e.

0.5.16 Let A and B be nonempty sets of natural numbers. Prove that if $A \subseteq B$, then $\min A \geq \min B$.

0.5.17 Let A and B be finite sets. Show that if there is a surjection $f : A \to B$, then $\mathrm{card}(A) \geq \mathrm{card}(B)$, and show that if there is an injection $g : A \to B$, then $\mathrm{card}(A) \leq \mathrm{card}(B)$.

0.6. More on Sets: Axioms and Constructions

This section is optional; it is not used later in the book and could also be used as an independent project. The first part covers the standard axioms of set theory, called the Zermelo–Fraenkel axioms, and also the axiom of choice. The second part covers the construction of the natural numbers, the integers, and the rationals using these axioms.

0.6.1. The Axiom of Choice and Other Axioms. We describe in an informal way the Zermelo–Fraenkel axioms as well as the axiom of choice. The axiom of choice is an interesting axiom that has a lot of nontrivial equivalent formulations. It has some nonintuitive consequences, and also there are times when it is used without knowing since it is considered very intuitive. This axiom does not follow from the other standard axioms of set theory.

The standard axioms of set theory are the Zermelo–Fraenkel axioms, and while we have not stated them, we have used most of them informally in our presentation of sets.

The first axiom is the **axiom of extension**. This says that two sets are equal if and only if they have the same elements. It connects the notions of set membership and set, and it is what we use when we prove that two sets are equal.

The **empty set axiom** states that the empty set exists. It stipulates the existence of the simplest possible set.

Next we have the **axiom of pairing**. This says that given sets A and B there is a set consisting of exactly A and B; we write this set as $\{A, B\}$. A consequence of this is that there is a set $\{\varnothing, \varnothing\}$, which is the same as $\{\varnothing\}$. Then we also have the set $\{\varnothing, \{\varnothing\}\}$. We note that in the Zermelo–Fraenkel theory, all elements of sets are sets themselves; natural numbers, for example, are defined as sets.

Then we have the **axiom of separation or comprehension**. The idea of the axiom is that given a set X and a property ϕ there is a set consisting of all $x \in X$ such that $\phi(x)$ is satisfied. This axiom says that given a set X and a formula, we can separate the elements of X and create a new set consisting of those elements of X that satisfy the formula. For example, this is used to define the set of even numbers as the set $\{n \in \mathbb{N} : n \text{ is even}\}$ once we have the set of natural numbers $\mathbb{N}$. We can also use this to define the relative complement of a set by saying that $A \setminus B$ consists of all elements of A that are not in B.

The **union axiom** says that given a set C there is a set U such that for all x, we have $x \in U$ if and only if there exists A such that $A \in C$ and $x \in A$. For example, given sets A and B, we can use the axiom of pairing to construct the set $C = \{A, B\}$. Then the elements of U are precisely the elements of A or of B; this is what we have called the union of A and B. We note that this axiom also allows for the union of an arbitrary set of sets, not just a set of two sets. Now that we have unions and relative complements, one can define intersections as in (0.21).

The **power set axiom** says that for every set A there is a set P such that $x \in P$ if and only if $x \subseteq A$. Hence this states the existence of the power set of every set.

The **axiom of infinity** says that there exists an "infinite" set. This is an important axiom in the construction of the set of natural numbers. At this moment of the axiomatic presentation we cannot assume that the notion of an infinite set has been defined; we use the notion of an *inductive* set to capture the intuitive idea of an infinite set. A set A is said to be **inductive** if $\varnothing \in A$, and when $x \in A$, then $x \cup \{x\} \in A$. This axiom states the existence of an inductive set.

Recall that 0 stands for $\varnothing$, and 1 is $\{\varnothing, \{\varnothing\}\}$, which is $0 \cup \{0\}$. Hence $0, 1, 2, \ldots$ are in any inductive set. An inductive set is our model for an infinite set. We can now define the nonnegative integers as the intersection of all inductive sets, hence we have the definition of the set of nonnegative integers and also of the set of all natural numbers. It can be shown that this intersection is itself an inductive set, and this can be used to prove the Peano axioms for the natural numbers; we discuss this in more detail in Subsection 0.6.2.

The next axiom is more technical and is called the **axiom of foundation**. It states that for all nonempty sets A there is a set B such that $B \in A$ and $B \cap A = \varnothing$. It implies, in particular, that there is no set A with $A \in A$.

The axioms we have seen so far were all in Zermelo's original formulation. Fraenkel added one axiom called the **axiom of replacement**. This is also a more technical axiom and asserts a different way of constructing sets. We have seen only a few ways of constructing sets, and we need to be cautious not to assert the existence of too big a set, or we are in danger of running into Russell's paradox. We note that if we have a function $f : X \to Y$, then the image of X under f is no bigger than X is. Informally, the axiom of replacement asserts that if we can define a mapping, then the image of a set under that mapping is also a set. More formally, if we have a definition $\phi(x, y)$ such that for a set X we can prove that for all $a \in X$ there is at most one y such that $\phi(x, y)$ (this says that ϕ defines a function on X), then there is a set Y such that for all b we have b is in Y if and only if there is an $a \in X$ with $\phi(a, b)$ (this says that Y is the image of X under ϕ).

All the axioms we have seen so far are in some form constructive. The next axiom, called the **axiom of choice** (AC), is purely an existence axiom. It is not part of the standard Zermelo–Fraenkel axioms, and when added to them the axioms

0.6. More on Sets: Axioms and Constructions

are referred to as ZFC. The axiom of choice has some startling consequences and has sometimes been rejected by some philosophers and mathematicians. It is however, extremely useful and as such it is accepted for its usefulness.

It was shown by Kurt Gödel (1906–1978) in 1938 that if the Zermelo–Fraenkel axioms are consistent, then so is ZFC. So adding choice does not risk adding any inconsistencies. It was shown by Paul Cohen (1934–2007) in 1963 that the axiom of choice cannot be proven from the Zermelo–Fraenkel axioms. Since axiom of choice is consistent with the Zermelo–Fraenkel axioms and is not provable from them, axiom of choice is independent of Zermelo–Fraenkel.

If A is a nonempty set, we can choose an element a in A; this follows from the definition of "there is an element of A", and no axiom of choice is needed. If A and B are nonempty sets, we can choose an element a in A and an element b in B (again the axiom of choice is not needed). One can use induction to show that, for each $n \in \mathbb{N}$, if $A_1, \ldots, A_n$ are nonempty, then there exist elements a_1 in A_1, a_2 in $A_2, \ldots, a_n$ in A_n, respectively. If we are given, however, an infinite collection of sets A_α, where α ranges over an arbitrary set Γ, it does not follow from mathematical induction and the standard Zermelo–Fraenkel axioms that for each $\alpha \in \Gamma$ we can choose an element a_α in A_α. The axiom of choice states that we can do this. More precisely, if $\{A_\alpha\}_{\alpha \in \Gamma}$ is a family of sets such that each A_α is nonempty, then it states that there exists a function $C : \Gamma \to \bigcup_{\alpha \in \Gamma} A_\alpha$ such that $C(\alpha) \in A_\alpha$ (here $\bigcup_{\alpha \in \Gamma} A_\alpha$ denotes the union of all the sets A_α). This function gives the choice of an element in each A_α. The axiom of choice is equivalent to many other statements, such as Zorn's lemma (which is useful in algebra and topology), and it is used to prove many theorems in mathematics (for example, to prove that an arbitrary vector space—without assuming it is finitely generated—has a basis).

We also note that the axiom of choice, in the case where the set Γ is merely countably infinite, is called the countable axiom of choice and is used extensively in analysis without making its use explicit. (Countable and countably infinite sets are defined in Section 1.3.1.) For example, the countable axiom of choice is used implicitly in the proof that a countable union of countable sets is countable (though it is not immediately obvious where it is needed). The interested reader may refer to [20, Chapter 10] where it is shown that countable axiom of choice is necessary for this.

0.6.2. Construction of the Sets $\mathbb{N}$, $\mathbb{Z}$, and $\mathbb{Q}$. This is again an optional section where we outline a construction of the natural numbers, the integers, and the rational numbers.

Construction of the Natural Numbers. In fact, we do start with the nonnegative integers $\mathbb{N}_0$ and then obtain the natural numbers $\mathbb{N}$ from $\mathbb{N}_0$. We remind the reader that our convention is that the natural numbers start at 1. However, it is more "natural" to first construct the nonnegative integers $\mathbb{N}_0 = \mathbb{N} \cup \{0\}$ when using the axioms.

By the axiom of infinity, there exists a set S that is inductive. Define $\mathbb{N}_0$ be the intersection of all inductive subsets of S. Then $\mathbb{N}_0$ is the smallest inductive set (in the sense that it is a subset of every inductive set), and in particular it is a subset of S. Also, $0 = \varnothing$ is in $\mathbb{N}_0$, and if n is in $\mathbb{N}_0$, then $n + 1 = n \cup \{n\}$ is in $\mathbb{N}_0$. So, in particular, 1 is in $\mathbb{N}_0$. We define $\mathbb{N}$ to be the elements of $\mathbb{N}_0$ except for 0.

We now show that $\mathbb{N}_0$ satisfies the induction principle (or the modified version of this principle that starts at 0). Let P be a subset of $\mathbb{N}_0$ such that 0 is in P, and whenever n is in P, then $n + 1$ is in P. This means that P is an inductive set, therefore $\mathbb{N}_0$ is a subset of P, which means that $P = \mathbb{N}_0$. From here it follows that the natural numbers also satisfy the induction principle: let $P \subseteq \mathbb{N}$ such that $1 \in P$ and $n \in P$ implies $n + 1 \in P$. Set $P' = P \cup \{0\}$. From what we have shown it follows that $P' = \mathbb{N}_0$, so $P = \mathbb{N}$. The other Peano axioms are more straightforward and are left as an exercise.

Remark 0.6.1. After constructing $\mathbb{N}$ one can define finite and infinite sets as in Subsection 0.5.4. It can then be shown that $\mathbb{N}$ is infinite (a proof can be found in Subsection 1.3.1). There is another interesting definition of an infinite set. A set A is said to be **Dedekind infinite** if there is a proper subset $B \subsetneq A$ and a bijection $f : A \to B$. The set $\mathbb{N}$ is Dedekind infinite since one can define a bijection with a proper subset by setting $f(n) = n + 1$ for $n \in \mathbb{N}$. Using the bijection in the definition of Dedekind infinite, it can be shown that a Dedekind infinite set includes a "copy" of $\mathbb{N}$, i.e., it includes a set bijective with $\mathbb{N}$, hence it is an infinite set. The converse, that an infinite set is Dedekind infinite, is also true, though surprisingly uses the axiom of choice.

Construction of the Integers. Now that we have the nonnegative integers, one may ask how exactly it is that we define -1, -2, etc. There is an interesting construction of the integers from the nonnegative integers (or from the natural numbers) that we now outline (the details are left to the Exercises).

A motivation for this construction is to look at equations of the form $a + x = b$ where a and b are nonnegative integers. For example, if we consider the equation $1 + x = 0$, we note that it has no solution in $\mathbb{N}_0$, since $1 + x > 0$ for every nonnegative integer x. The solution should be what we intuitively know as -1, but we do not have this number yet. However, we can think of this number as being specified by the equation $1 + x = 0$ or just simply by the pair $(1, 0)$. Next we note that this x also satisfies the equation $2 + x = 1$, so the pair $(2, 1)$ should represent the same solution, and similarly, x also satisfies the equation $(n + 1) + x = n$ for each nonnegative integer n. Hence we can identify the solution of this equation with all the pairs that have the same solution as the pair $(1, 0)$. What we do now is form the set of all pairs that specify the same solution as the pair $(1, 0)$. This is a set depending on $(1, 0)$, and we denote it by $[(1, 0)]$. It is

$$[(1, 0)] = \{(n + 1, n) : n \in \mathbb{N}_0\}.$$

This set will turn out to be the equivalence class of $(1, 0)$, i.e., the set of all pairs that are equivalent to $(1, 0)$, after we define an equivalence relation on pairs of elements from $\mathbb{N}_0$. Note that the definition of $[(1, 0)]$ depends only on elements of $\mathbb{N}_0$. Now, similarly, we can think of the solution of $2 + x = 0$, which is also the solution of $3 + x = 1$, and in general of $a + 2 + x = a$. Hence we can identify the set of all pairs of the form $(n + 2, n)$, where n is a nonnegative integer. In general, the solution of $a + x = 0$, for $a \in \mathbb{N}$, is identified with the set

$$[(a, 0)] = \{(n + a, n) : n \in \mathbb{N}\},$$

which will be the equivalence class of $(a, 0)$, the same as the equivalence class of $(a+1, 1)$.

Now define an equivalence relation R_1 on elements of $\mathbb{N}_0 \times \mathbb{N}_0$ by

(0.27) $\qquad\qquad (m, n) R_1 (p, q)$ if and only if $n + p = q + m$.

(This notation is simpler here than the equivalent notation $((m, n), (p, q)) \in R_1$.) Then, for example, $(2, 1)$ is related to $(3, 2)$ since $2 + 2 = 3 + 1$. We now define the set of integers $\mathbb{Z}$ as the set of equivalence classes under the relation R_1.

Equivalence classes have the important property that each pair $(m, n), n, m \in \mathbb{N}_0$ belongs to one and only one equivalence class. For example, $(7, 5)$ is in the equivalence class of $(3, 1)$, or of $(2, 0)$, and all other equivalence classes are disjoint from this one. Hence we can let $[(m, n)]$ denote the unique equivalence class containing the pair (m, n). For example, the class $[(4, 2)]$ is the same as $[(3, 1)]$.

We have that each element of $\mathbb{Z}$ is an equivalence class of some ordered pair of nonnegative integers. The difficult part is that we need to think of the set integers as the set of all such equivalence classes. However, the interesting and surprising fact is that on this collection (set) of equivalence classes, one can define the operation of addition, and multiplication, so that the resulting set has all the algebraic and arithmetic properties of the set of integers that we expect. To define addition of equivalence classes, we first define addition of two ordered pairs by

$$(a, b) + (c, d) = (a + c, b + d).$$

This extends in a natural way to addition of equivalence classes (one can verify that the addition of equivalence classes does not depend on the particular representative of the class one uses). For example, one has that $(1, 2) + (2, 1) = (3, 3)$. Note in particular that $[(m, n)] + [(1, 1)] = [(m+1, n+1)] = [(m, n)]$, so $[(1, 1)]$ (which is the same as $[(0, 0)]$) acts like the zero element of $\mathbb{Z}$. Since $[(1, 2)] + [(2, 1)] = [(3, 3)] = [(1, 1)]$, one can see that $[(1, 2)]$ and $[(2, 1)]$ act as additive inverses of each other.

Multiplication is defined along similar lines, and the properties are developed in the Exercises. Finally, we identify each a in $\mathbb{N}_0$ with $[(0, a)]$, and in this way we can view $\mathbb{N}_0$ (in fact, a copy of it) as a subset of $\mathbb{Z}$.

In summary, in this construction $\mathbb{Z}$ consists of equivalence classes of pairs of nonnegative integers, where we can define an operation of addition and multiplication. Elements of the form $[(0, a)]$ can be seen to satisfy the Peano axioms and to form a copy of $\mathbb{N}_0$. Each element $[(0, a)]$ has an additive inverse, which is $[(a, 0)]$, since $[(a, 0)] + [(0, a)] = [(a, a)] = [(0, 0)]$.

Construction of the Rationals. After constructing the integers $\mathbb{Z}$, the rational numbers can be constructed using similar ideas. To motivate this construction, we consider solutions of equations of the form

$$qx = p,$$

where p and q are integers and $q \neq 0$. We think of the pair (p, q) as representing the solution of this equation. Since (ap, aq) should represent the same solution, we want to identify (p, q) with (ap, aq) for all nonzero integers a. So a relation has to be defined to make them part of the same equivalence class. Then an equivalent class of such pairs is thought of as a rational number. For example, the equivalence class $[(1,2)]$ represents what we otherwise write as the rational number $\frac{1}{2}$; note that this equivalence class also represents $\frac{2}{4}$. (This makes clear the fact that we regard $\frac{1}{2}$ and $\frac{2}{4}$ as representing the same rational number.) The operations of addition and multiplication can be defined and all the expected properties of the rational numbers can be proved; this is done in the Exercises.

Exercises: More on Sets

0.6.1 Use the construction of $\mathbb{N}$ from Subsection 0.6.2 and prove it satisfies the Peano axioms.

0.6.2 Let R_1 be the relation defined in (0.27). Prove that R_1 is an equivalence relation.

0.6.3 Let R_1 be the equivalence relation of Exercise 0.6.2. Define an addition of equivalence classes by
$$[(a,b)] + [(c,d)] = [(a+c, b+d)].$$

(a) Prove that this addition is well-defined, i.e., that it does not depend on the choice of element of the equivalent class: if $((a,b),(a',b')) \in R_1$ and $((c,d),(c',d')) \in R_1$, then
$$[(a,b)] + [(c,d)] = [(a',b')] + [(c',d')].$$

(b) Prove that $[(c,c)]$ acts as a zero: $[(a,b)] + [(c,c)] = [(a,b)]$.
(c) Prove that addition is commutative: $[(a,b)] + [(c,d)] = [(c,d)] + [(a,b)]$.
(d) Prove that that $[(b,a)]$ acts as the additive inverse of $[(a,b)]$: $[(a,b)] + [(b,a)] = [(c,c)]$.

0.6.4 Let R_1 be the equivalence relation of Exercise 0.6.2. Define a product of equivalence classes by
$$[(a,b)] \cdot [(c,d)] = [(bc+ad, ac+bd)].$$

(a) Prove that this product is well-defined, i.e., that it does not depend on the choice of element of the equivalent class.
(b) Prove that $[(a,b)] \cdot [(1,2)] = [(a,b)]$.
(c) Prove the distributive property $[(a,b)] \cdot ([(c,d)] + [(e,f)]) = [(a,b)] \cdot [(c,d)] + [(a,b)] \cdot [(e,f)]$.
(d) Give an argument for why this set of equivalence classes of pairs of natural numbers, just defined with the operations of multiplication and addition, can be considered a construction of the integers $\mathbb{Z}$.

* 0.6.5 In this exercise we construct the rational numbers as certain equivalence classes of pairs of integers. Define a relation R_2 on subsets of $\mathbb{Z} \times \mathbb{Z} \setminus \{0\}$ by
$$((a,b),(c,d)) \in R_2 \text{ if and only if } ad = bc.$$

(a) Prove that R_2 is an equivalence relation.
(b) Define an addition of equivalence classes by
$$[(a,b)] + [(c,d)] = [(ac, bc+ad)],$$
and prove similar properties for this operation as in Exercise 0.6.3.

(c) Define a product on equivalence classes by
$$[(a,b)] \cdot [(c,d)] = [(ac, bd)],$$
and prove similar properties for this operation as in Exercise 0.6.4.

(d) Why can this new set of equivalence classes be considered a construction of the rational numbers?

* 0.6.6 Prove that an infinite set includes a subset that is bijective with $\mathbb{N}$. (A detailed proof of this needs the axiom of choice for countable families.)

0.6.7 Prove that $\mathbb{N}_0$ is a subset of every inductive set.

0.6.8 Use induction to prove the axiom of choice for the case of finite families of sets: for each $n \in \mathbb{N}$, if $A_1, \ldots, A_n$, are nonempty sets, then for all $i \in \{1, \ldots, n\}$ there exists $a_i \in A_i$.

* 0.6.9 A set A is **Dedekind finite** if every injection from A to A is a surjection. Prove that a set is finite if and only if it is Dedekind finite. (One direction uses the axiom of choice.)

0.6.10 Use the axiom of foundation to show that for every set A is it the case that $A \notin A$.

Chapter 1

The Real Numbers and the Completeness Property

This chapter defines the real numbers and studies their basic properties. The real numbers are characterized by three types of properties. The first type is algebraic properties (having to do with addition and multiplication). These properties are called the field axioms and are shared by other sets of numbers such as the rational numbers and the complex numbers, but not, for example, by the integers. The second consists of the order properties, which introduce an order structure (an inequality), and are also shared by the rational numbers, but not, for example, by the complex numbers. The third type is completeness, also called order completeness, which is probably new to the reader as it is not satisfied by the rational numbers and is the more difficult to state. There are several formulations of the completeness property; we defined completeness in terms of the supremum and discuss equivalent formulations in later chapters.

We define the real numbers by the properties they possess. The **real numbers** are defined to be an ordered field that is complete under that order. Ordered fields are defined in Section 1.1 and completeness is defined in Section 1.2.

There are two natural questions that arise from this definition. The first one is of the existence of a set that is a complete ordered field; i.e., is there a set satisfying the properties of the real numbers? The second question is whether there is only one such set; i.e., in what sense is the set satisfying these properties unique? There are now several constructions of the real numbers, and we outline the construction due to Dedekind in Section 1.4; this establishes the existence of a complete ordered field. The construction of the real numbers is optional as all the properties that we need of the real numbers can be deduced from the axioms of field, order, and completeness that we cover in this chapter. Regarding the question of uniqueness, we note that it can be shown that in some reasonable sense a set satisfying these properties is unique; this is outlined in one of the exercises.

1.1. Field and Order Properties of $\mathbb{R}$

1.1.1. Field Properties. The field properties consist of the axioms of addition, multiplication, and the distributive axiom. We state them in the setting of a field F (defined after the statement of the distributive axiom), but the case we are most interested in is when F is the set of real numbers.

Definition 1.1.1. Axioms for Addition. A set F is said to satisfy the axioms of addition if it has a function from $F \times F$ to F, called an operation and denoted by $+$, so that for any pair of elements x and y in F there is an element denoted $x + y$ in F satisfying the following properties.

(A1) *Commutativity of addition:* $x + y = y + x$ for all x and y in F.

(A2) *Associativity of addition:* $(x + y) + z = x + (y + z)$ for all x, y and z in F.

(A3) *Existence of a zero:* There is an element of F denoted by 0 that satisfies $x + 0 = x$ for all $x \in F$.

(A4) *Existence of additive inverse:* For every $x \in F$ there exists an element of F denoted $-x$, called its **additive inverse**, such that $x + (-x) = 0$. ◊

We note that the zero element, 0, is unique. What we mean is that there is no other element of the set F that satisfies the characteristic or defining property of 0, i.e., property (A3). To show this, we start by supposing that z is an arbitrary element of F satisfying the same property as 0, i.e., that $x + z = x$ for all $x \in F$. It suffices to show that z must be 0. In fact, by letting $x = 0$ in the property of z and then using the commutativity of addition, we have

$$0 = 0 + z = z + 0 = z.$$

Also, the additive inverse (defined by property (A4)) is unique. If for a given x in F there is a w in F such that $x + w = 0$, then

$$w = 0 + w = (x + (-x)) + w = (-x) + (x + w) = -x.$$

We write $x - y$ instead of $x + (-y)$. It follows from here that $-(-x) = x$.

Definition 1.1.2. Axioms for Multiplication. A set F, which already satisfies the axioms for addition, is said to satisfy the axioms of multiplication if it has an operation denoted $\cdot$ so that for any pair of elements x and y in F there is an element in F denoted $x \cdot y$ satisfying the following properties.

(M1) *Commutativity of multiplication:* $x \cdot y = y \cdot x$ for all x and y in F.

(M2) *Associativity of multiplication:* $(x \cdot y) \cdot z = x \cdot (y \cdot z)$ for all x, y and z in F.

(M3) *Existence of a unit:* There is an element of F denoted by 1 and different from 0 that satisfies $1 \cdot x = x$ for all $x \in F$.

(M4) *Existence of multiplicative inverse:* For every $x \neq 0$ in F there exists an element of F denoted $\frac{1}{x}$ (or $1/x$), called its **multiplicative inverse**, such that $x \cdot \frac{1}{x} = 1$. ◊

1.1. Field and Order Properties of $\mathbb{R}$

The element 1 is also unique. If z satisfies $z \cdot x = x$ for all $x \in F$, then
$$z = 1 \cdot z = z \cdot 1 = 1.$$
Similarly, if for $x \neq 0$ there exists w such that $x \cdot w = 1$, then
$$w = 1 \cdot w = \left(\frac{1}{x} \cdot x\right) \cdot w = \frac{1}{x} \cdot (x \cdot w) = \frac{1}{x} \cdot 1 = \frac{1}{x}.$$
We often write x^{-1} instead of $\frac{1}{x}$ and write xy instead of $x \cdot y$.

Definition 1.1.3. Distributive Axiom. A set F with operations $+$ and $\cdot$ satisfies the distributive axiom if

(D) $\qquad\qquad x(y+z) = xy + xz$ for all x, y and z in F. $\qquad\qquad\diamond$

Definition 1.1.4. A **field** consists of a set F with two operations $+$ and $\cdot$ that satisfies the addition, multiplication, and distributive axioms. $\qquad\diamond$

For example, the set of rational numbers $\mathbb{Q}$ with $+$ and $\cdot$ is a field. The set of integers $\mathbb{Z}$ with $+$ and $\cdot$ is not a field since elements of $\mathbb{Z}$, other than 1 and -1, do not have multiplicative inverses.

It is possible to derive all the algebraic properties we use for the real numbers from the axioms of a field. The following proposition is included as an example of some of the many algebraic properties of the real numbers that can be deduced from the field axioms; additional properties are in the Exercises. We will use algebraic properties of the reals even if they are not listed here, though the reader should know how to derive them from the axioms.

Proposition 1.1.5. *The elements of a field F satisfy the following properties.*

(1) *If $x + w = y$, then $w = y - x$.*
(2) $0 \cdot x = 0$.
(3) *If $xw = y$ and $x \neq 0$, then $w = x^{-1}y$.*
(4) $-1 \cdot x = -x$.
(5) $(-x)y = x(-y) = -xy$.

Proof. We show (1) one step at a time. We start with $x + w = y$. Then
$$(x + w) - x = y - x, \text{ by (A4) and the property of equality},$$
$$(w + x) - x = y - x, \text{ by (A1)},$$
$$w + (x - x) = y - x, \text{ by (A2)},$$
$$w + 0 = y - x, \text{ by (A4)},$$
$$w = y - x, \text{ by (A3)}.$$

For the remaining properties we give an outline of the proof, but the reader should know how to justify each step. To show (2) note that
$$0x = (0 + 0)x = 0x + 0x.$$
Hence $0x = 0x - 0x = 0$. For (3), use that x^{-1} exists and multiply both sides of $xw = y$ by x^{-1}. For part (4) note that
$$x + (-1 \cdot x) = 1 \cdot x + (-1) \cdot x = (1 + (-1)) \cdot x = 0x = 0.$$

By the uniqueness of the additive inverse, $-1 \cdot x = -x$. For part (5) we use (4) and associativity to write
$$(-x)y = ((-1)x)y = (-1)(xy) = -(xy).$$
A similar argument shows that $x(-y) = -xy$. □

As a final example we mention that division by 0 is not consistent with our properties. Suppose w was the multiplicative inverse of 0; then it would satisfy $0w = 1$, but this contradicts Proposition 1.1.5(2).

Definition 1.1.6. For x in F and each positive integer n, we can define x^n by induction. Let $x^1 = x$ and $x^{n+1} = x \cdot x^n$. We extend this to all integers by setting $x^0 = 1$ and $x^{-n} = 1/x^n$ when $x \neq 0$. Also, for $y \neq 0$, write
$$\frac{x}{y} = x \cdot \frac{1}{y}. \qquad \diamond$$

Example 1.1.7. We show that if F is a field, then for all $x \in F$ such that $x \neq 0$ we have that $-(\frac{1}{x}) = \frac{-1}{x}$. The defining property of $1/x$ is that it is the multiplicative inverse of x, so $x \cdot (1/x) = 1$. Multiplying both sides by -1 and using Proposition 1.1.5, $x \cdot [-(1/x)] = -1$. Next we multiply both sides by $1/x$ and use Proposition 1.1.5 to obtain $-(1/x) = (-1)(1/x) = \frac{-1}{x}$.

1.1.2. Order Axioms

Definition 1.1.8. A field F with operations $+$ and $\cdot$ is said to be an **ordered field** if there is a subset of F denoted by F^+ and called the **positive set** satisfying the following properties.

(O1) *Closure of F^+ under $+$ and $\cdot$* : $x + y$ and xy are in F^+ for all x, y in F^+.

(O2) *Trichotomy property*: For every x in F, exactly one of the following is true: $x = 0$, or $x \in F^+$, or $-x \in F^+$. $\qquad \diamond$

The idea is to think of F^+ as the set of positive elements of the field. Then the closure property is simply saying that the "sum and product of positive elements are positive". The trichotomy property then states that a nonzero element is either positive or its negative is positive.

We typically do not use the set F^+ but instead use the $>$ notation. Write

- $a > 0$ if and only if $a \in F^+$;
- $b > a$ if and only if $b - a > 0$ (i.e., $b - a \in F^+$).

Also, write $a < b$ when $b > a$. Hence $a < 0$ if and only if $-a > 0$, and $a < b$ is equivalent to $a - b < 0$. Further, write $x \geq y$ if $x > y$ or $x = y$, and similarly for $x \leq y$. This establishes an "order" in F, i.e., any two elements x and y in F can be compared: if $x \neq y$, then either $x > y$ or $y > x$.

We claim that $1 \in F^+$ (equivalently, $1 > 0$). From the trichotomy property we know that, as $1 \neq 0$, either $1 \in F^+$ or $-1 \in F^+$. If we had that $-1 \in F^+$, by the closure property $1 = (-1) \cdot (-1) \in F^+$, contradicting the trichotomy property. Thus $-1 \notin F^+$ and $1 \in F^+$. In particular, F^+ is nonempty.

Question 1.1.9. Let F be an ordered field. Prove that $-3 < -2$.

1.1. Field and Order Properties of $\mathbb{R}$

We note that $\mathbb{Q}$ is an ordered field (Exercise 1.1.13). We now mention the important fact that one can regard the natural numbers $\mathbb{N}$ as a subset of any ordered field F. The identification of elements of $\mathbb{N}$ with elements of F is the natural one. We identify 1 in $\mathbb{N}$ with the 1 in F, 2 in $\mathbb{N}$ with $1+1$ in F, completing the process by induction. (There is something that needs to be clarified. Since $1 > 0$, using induction one shows that $1 + \cdots + 1 (n+1 \text{ times}) > 1 + \cdots + 1 (n \text{ times})$.) Technically, we would say that F contains a copy of $\mathbb{N}$, but in practice we consider $\mathbb{N}$ a subset of F. Since F is a field, if it contains n it must contain $-n$. It follows that an ordered field includes the set of integers $\mathbb{Z}$, and thus it includes the rational numbers $\mathbb{Q}$ (again, technically a copy of $\mathbb{Q}$). A consequence of this is that an ordered field must be infinite.

The following proposition shows some properties of the order we have introduced.

Proposition 1.1.10. *The following properties hold in an ordered field.*

(1) *For each x one and only one of the following hold: $x = 0$, $x < 0$, or $0 < x$.*
(2) *If $x < y$ and $y < z$, then $x < z$.*
(3) *If $x < y$, then $x + z < y + z$ for all z.*
(4) *If $x < y$ and $z > 0$, then $xz < yz$.*

Proof. Part (1) follows from the definition and the trichotomy property. For (2), assume $x < y$ and $y < z$, which means that $y - x \in F^+$ and $z - y \in F^+$. Then $y - x + z - y \in F^+$, so $x < z$. For (3), $y - x > 0$, so for every z, $(y+z) - (x+z) > 0$. For (4), $y - x > 0$, so for every $z > 0$, $(y-x)z > 0$. Thus $yz - xz > 0$, or $yz > xz$. □

Corollary 1.1.11. *The following properties hold in an ordered field.*

(1) *If $x < y$, then $-x > -y$.*
(2) *If $x < y$ and $z < 0$, then $xz > yz$.*
(3) *If $x > 0$, then $-x < 0$ and $1/x > 0$.*

Proof. If $x < y$, then $x - y < 0$, so $-(x-y) > 0$ or $(-x) - (-y) > 0$, which means $-x > -y$. For part (2) first note that we have $y - x > 0$ and $-z > 0$, so $(y-x)(-z) > 0$. Thus, $-yz + xz > 0$ or $xz > yz$. For the last part, let $x > 0$. By (2), $0 > -x$, so $-x < 0$. Now, if $1/x < 0$, then $-(1/x) > 0$, so $-(1/x)x > 0$, which would imply $-1 > 0$, a contradiction. □

Question 1.1.12. Let F be an ordered field. Prove that if $xy > 0$, then either $x > 0$ and $y > 0$ or $x < 0$ and $y < 0$, and that if $xy < 0$, then either $x > 0$ and $y < 0$ or $x < 0$ and $y > 0$.

Remark 1.1.13. We have seen that from a positive set F^+ in a field F we can define an order on F; namely, we can define $a < b$ if and only if $b - a \in F^+$. There is a parallel theory which starts with the idea of an abstract order. While we do not need this, we discuss it now as it is interesting to see how one can define an abstract order. From this, one can define a positive set F^+.

We can define an **order** on a set S as a relation on S (i.e., a subset of $S \times S$) that we denote by $<$, satisfying two properties. First, regarding notation, rather than writing $(x,y) \in <$ (i.e., that (x,y) belongs to the order $<$), we write $x < y$. Then we require the following properties.

(1) For each $x, y \in S$, one and only one of the following hold: $x = y$, $x < y$, or $y < x$.

(2) If $x < y$ and $y < z$, then $x < z$.

If now F is an ordered field, we can define a relation $<$ on F so that $a < b$ if and only if $b - a \in F^+$. Then by Proposition 1.1.10, $<$ is an order on the set F. In addition, this order satisfies the following:

(3) $x < y$, then $x + z < y + z$ for all z;

(4) if $x < y$ and $z > 0$, then $xz < yz$.

Exercise 1.1.18 shows that one could alternatively define an ordered field using the notion of order.

Exercises: Field and Order Properties of $\mathbb{R}$

1.1.1 Give complete details in the proof of Proposition 1.1.5(5).

1.1.2 Let F be a field, and let $x, y \in F$. Prove that $(-x)(-y) = xy$.

1.1.3 Let F be a field, and let $x, y \in F$. Prove that
$$x^3 - y^3 = (x - y)(x^2 + xy + y^2).$$

1.1.4 Let F be an ordered field. Prove that for every $x \in F$, $x^2 \geq 0$.

1.1.5 Prove that for all x in a field F, $x \neq 0$, $-(1/x) = 1/(-x)$.

1.1.6 Let F be an ordered field. Let $x \geq 0$ be in F. Prove that if $x < \varepsilon$ for all $\varepsilon > 0$, $\varepsilon \in F$, then $x = 0$.

1.1.7 Let F be an ordered field. Suppose that x and y are in F and satisfy $x < y + \varepsilon$ for all $\varepsilon > 0$, $\varepsilon \in F$. Prove that then $x \leq y$.

1.1.8 Let F be an ordered field, and let $\varepsilon \in F$. Show that if $0 < \varepsilon < 1$, then $\varepsilon^2 < \varepsilon$.

1.1.9 Let F be an ordered field, and let $a, b \in F$. Prove that if $0 < a < b$, then $1/b < 1/a$.

1.1.10 *Binomial formula*: Let F be a field. Prove that for all a and b in F, and for all $n \in \mathbb{N}$,
$$(a+b)^n = \sum_{i=0}^{n} \binom{n}{i} a^i b^{n-i},$$
where
$$\binom{n}{i} = \frac{n!}{i!(n-i)!}$$
is the **binomial coefficient**.

1.1.11 *Bernoulli's inequality*: Let F be an ordered field, and let $x \in F$. Prove that
$$(1+x)^n \geq 1 + nx$$
for all $n \in \mathbb{N}$ and $x > -1$.

1.1.12 *Geometric sum*: Let F be a field. Use induction to prove that for all $r \in F$ with $r \neq 1$ and all $n \in \mathbb{N}$,
$$\sum_{i=1}^{n} r^{i-1} = \frac{1-r^n}{1-r}.$$
Write and prove a formula for $\sum_{i=k}^{n} r^{i-1}$ for $k \in \mathbb{N}$.

1.1.13 Verify that the rational numbers $\mathbb{Q}$ satisfy the properties of an ordered field.

1.1.14 Let $\mathbb{Z}_2$ denote the set $\{0,1\}$ with addition defined by $0+0 = 0$, $0+1 = +0 = 1$, $1+1 = 0$ (this is called addition mod 2) and multiplication defined by $0 \cdot 0 = 0 \cdot 1 = 1 \cdot 0 = 0$, $1 \cdot 1 = 1$. Prove that this is a field. Is it ordered?

1.1.15 Construct a set with three elements and two operations so that it satisfies the axioms of a field.

1.1.16 Let F be an ordered field, and let $x \in F$. Prove that if $x > 1$, then $x^n \geq x$ for all $n \in N$.

1.1.17 Let F be a field, and let $a, b \in F^+$. Prove that if $a^2 > b^2$, then $a > b$.

1.1.18 Let F be a field. Let $<$ be an order on F (see Remark 1.1.13), which in addition satisfies (a) if $x < y$, then $x + z < y + z$ for all z; and (b) if $x < y$ and $z > 0$, then $xz < yz$. Define a set F^+ by $F^+ = \{x \in F : x > 0\}$. Prove that F^+ is a positive set for F.

1.1.19 Let F be a field, and let $F[x]$ consist of the set of polynomials with coefficients in F, i.e., the set of functions of the form $f(x) = a_n x^n + \cdots + a_1 x + a_0$ where $a_i \in F$ (call a_n the leading coefficient). Let $F(x)$ denote the set of rational functions with coefficients in F, i.e., functions of the form $f(x)/g(x)$ where $f(x), g(x) \in F[x]$, $g(x)$ is not identically zero, and $f(x)$ and $g(x)$ do not have any common factors ($f(x)/g(x)$ is defined at the points where g does not vanish). Prove that $F[x]$ is not a field but $F(x)$ is a field with the usual addition and multiplication, i.e.,
$$\frac{f(x)}{g(x)} + \frac{h(x)}{k(x)} = \frac{k(x)f(x) + g(x)h(x)}{g(x)k(x)} \quad \text{and} \quad \frac{f(x)}{g(x)} \cdot \frac{h(x)}{k(x)} = \frac{f(x)h(x)}{g(x)k(x)}.$$

1.1.20 Let $F(x)$ be the field defined in Exercise 1.1.19. Define the set $F(x)^+$ to consist of all elements $f(x)/g(x)$ where the leading coefficient of f and g have the same sign. Prove that with this definition $F(x)$ is an ordered field.

1.2. Completeness Property of $\mathbb{R}$

The completeness property, also called order completeness, is one of the most important properties of the real numbers. Together with the field and order axioms, the completeness property finishes the definition of the real numbers.

1.2.1. Numbers that Are Not in the Field $\mathbb{Q}$.

There are some important numbers that are missing from the set of rational numbers $\mathbb{Q}$ and that will be added by the completeness property (namely the *irrational numbers*). It turns out that numbers such as $\sqrt{2}$ are not rational. What we mean by $\sqrt{2}$ is a positive number α such that $\alpha^2 = 2$. It follows from Pythagoras' theorem that if α is the hypotenuse of a right-angle isosceles triangle with sides of length 1, then α must satisfy $\alpha^2 = 2$. Thus the number α is needed to measure simple geometric figures. To their amazement, the ancient Greeks discovered that α is not a rational number. Before studying the completeness property, we prove that α is missing from $\mathbb{Q}$. As we shall see, a consequence of this fact is that the ordered field $\mathbb{Q}$ is not complete.

The following is a classic theorem; a proof of this theorem can already be found in Euclid's *Elements*, written around 300 BCE.

Theorem 1.2.1. *There is no rational number α such that $\alpha^2 = 2$ (i.e., $\sqrt{2} \notin \mathbb{Q}$).*

Proof. We show by contradiction that there is no positive rational number α satisfying $\alpha^2 = 2$ (this also implies there is no such negative rational number α). Consider the set A defined by

$$A = \{q \in \mathbb{N} : \alpha = \frac{p}{q} \text{ for some } p \in \mathbb{N}\}.$$

If α is a rational number, then the set A is a nonempty subset of $\mathbb{N}$. By the well-ordering principle, it has a least element, which we denote by b. Then

$$\alpha = \frac{a}{b} \text{ for some } a \in \mathbb{N}.$$

Therefore

$$(1.1) \qquad 2b^2 = a^2.$$

This implies that 2 is a factor of a^2, and as 2 is prime, 2 is a factor of a. Then we can write a as $a = 2c$, for some positive integer c. From (1.1) we obtain

$$(1.2) \qquad 2b^2 = 4c^2,$$
$$(1.3) \qquad b^2 = 2c^2.$$

By the same argument as before, this implies that 2 divides b, so $b = 2d$ for some $d \in \mathbb{N}$. Again, replacing b in (1.3), we obtain

$$(1.4) \qquad 2d^2 = c^2 \text{ or}$$
$$(1.5) \qquad \alpha = \frac{c}{d} \text{ for } c, d \in \mathbb{N}.$$

Thus $d \in A$, but $d < b$, contradicting that b is the least element of A. Therefore α is not a rational number. $\square$

There are now many different proofs of this theorem, and our proof can be extended to show that many other numbers are not rational. We mention one modification of the argument. We could start the proof by contradiction by supposing that α can be written in the form $\alpha = \frac{a}{b}$, where a and b have been simplified so that they do not both have 2 as a factor. Then, following as in the proof of Theorem 1.2.1, one can obtain a contradiction by showing that 2 must divide both a and b.

1.2. Completeness Property of $\mathbb{R}$

There are numbers, however, that are still not known to be rational. For example, while it is believed that these numbers are not rational, it is not known whether either of 2^e or $\pi + e$ is rational (though it is known that either $\pi + e$ or $e\pi$ is not rational).

1.2.2. Infimum and Supremum. As we shall see, there are many equivalent formulations of the completeness property. We start our development in terms of the notions of infimum and supremum. We state these definitions in the context of ordered fields to emphasize that the main property of the real numbers used here is that of being an ordered field.

Definition 1.2.2. Let S be a nonempty subset of an ordered field F. An element $b \in F$ is said to be an **upper bound** for S if
$$x \leq b \text{ for all } x \in S.$$
Similarly, $c \in F$ is said to be a **lower bound** for S if
$$c \leq x \text{ for all } x \in S. \qquad \diamond$$

In the case of the empty set, every element of F is both an upper bound and a lower bound for $\varnothing$.

Definition 1.2.3. A set S in F is said to be **bounded below** if it has a lower bound and it is **bounded above** if it has an upper bound. A set $S \subseteq F$ is said to be **bounded** if it is bounded above and bounded below. $\qquad \diamond$

For example, $\mathbb{N}$ is not bounded above and is bounded below by 0 (and also by $1/2$, for example). The empty set is a bounded set.

Question 1.2.4. Let F be an ordered field, and let $S \subseteq F$. Prove that $b \in F$ is an upper bound of S if and only if $-b$ is a lower bound of $-S$, where $-S = \{-x : x \in S\}$.

Definition 1.2.5. The **supremum** or **least upper bound** of S, denoted $\sup S$, is defined to be an element β of F such that

- β is an upper bound for S;
- if c is any other upper bound for S, then $c \geq \beta$ (i.e., β is less than any other upper bound). $\qquad \diamond$

The second property in the definition readily implies that the supremum, when it exists, is unique.

Question 1.2.6. Let F be an ordered field, and let A and B be subsets of F such that $\sup A$ and $\sup B$ exist (and so are elements of F). Prove that $\sup A \cup B$ exists and $\sup A \cup B = \max\{\sup A, \sup B\}$. Conclude that if $A \subseteq B$, then $\sup A \leq \sup B$.

Definition 1.2.7. We extend the definition of supremum to a nonempty set S that does not have an upper bound by writing $\sup S = \infty$. Also, for the empty set we write $\sup \varnothing = -\infty$. $\qquad \diamond$

We justify the definition above by noting that every element x of an ordered field F is an upper bound for $\varnothing$ (otherwise, there would be an element of $\varnothing$ greater than x); thus every such element x should be greater than or equal to the supremum

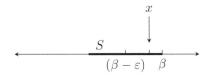

Figure 1.1. For each $\epsilon > 0$ there exists $x \in S$ with $\beta - \epsilon < x$.

of $\varnothing$. Thus it makes sense to set $\sup \varnothing = -\infty$. A consequence of this is that for all sets B, one has $\sup \varnothing \leq B$. So, in particular, when $A \subseteq B$, then $\sup A \leq \sup B$ even when A is empty. We note, however, that when we say $\sup S$ exists, we mean that it exists as an element of F.

For example, if $S = \{x \in F : 0 < x < 1\}$, then $\sup S = 1$. Clearly, 1 is an upper bound for S and, in addition, if c is any upper bound for S, then $c \geq 1$.

Question 1.2.8. Let F be an ordered field, and let S be a subset of F. Prove that if β is an upper bound of S and $\beta \in S$, then $\sup S = \beta$.

Definition 1.2.9. The **infimum** or **greatest lower bound** of a set $S \subseteq F$, denoted $\inf S$, is defined to be the element α of F such that

- α is a lower bound for S;
- if b is any lower bound for S, then $b \leq \alpha$ (i.e., α is greater than any other lower bound). ◇

The infimum of the empty set is ∞, and we write $\inf \varnothing = \infty$. If S is nonempty and does not have a lower bound, we write $\inf S = -\infty$.

The following lemma gives a useful characterization of the supremum of a set; a similar statement holds for the infimum.

Lemma 1.2.10. *Let S be a nonempty subset of an ordered field F. An element $\beta \in F$ satisfies $\beta = \sup S$ if and only if β is an upper bound for S and for every $\varepsilon > 0$, there exists $x \in S$ such that $x > \beta - \varepsilon$.*

Proof. Suppose that $\beta = \sup S$. Clearly, β is an upper bound for S. Let $\varepsilon > 0$. Suppose to the contrary that every $x \in S$ satisfies $x \leq \beta - \varepsilon$. Then $\beta - \varepsilon$ is also an upper bound for S, but this contradicts that β is the least upper bound. Therefore, there exists $x \in S$ with

$$x > \beta - \varepsilon.$$

To show the converse assume that β is an upper bound and that for every $\varepsilon > 0$ there is $x \in S$ with $x > \beta - \varepsilon$; see Figure 1.1.

Let c be an upper bound for S. Assume to the contrary that $c < \beta$. Then we can take $\varepsilon = \beta - c > 0$. Therefore, there exists $x \in S$ with

$$x > \beta - \varepsilon = \beta - (\beta - c) = c,$$

contradicting that c is an upper bound for S. It follows that $c \geq \beta$, and as β is an upper bound, $\beta = \sup S$. □

1.2. Completeness Property of $\mathbb{R}$

1.2.3. Completeness Property of the Reals. Now we state the completeness property, which together with the field and order properties, completes the characterization of the real numbers.

Definition 1.2.11. An ordered field F satisfies the **completeness property**, or is **order complete**, if every nonempty subset of F that has an upper bound has a supremum in F. ◇

We have already seen that the supremum of the empty set is $-\infty$. Also, when a set is not bounded above, its supremum is ∞. So if we allow $\pm\infty$ as a possible value, we can say that for any set S, its supremum $\sup S$ is defined. We know that when the set is nonempty and bounded above, this supremum is in the ordered field, but it will be convenient to know that we can write $\sup S$ for any set $S \subseteq F$. The completeness property has an equivalent formulation in terms of the infimum (see Exercise 1.2.3).

The following theorem will be proved in Section 1.4.

Theorem 1.2.12. *There exists a field that is an ordered field and satisfies the completeness property.*

There is a natural way in which one can claim there is a unique complete ordered field: it can be shown (see Exercise 1.4.5) that given any two complete ordered fields F_1 and F_2 there is a bijection between them that preserves the field and order operations.

Definition 1.2.13. We choose a complete ordered field (for example, the one constructed in Section 1.4), call it the field of **real numbers**, and denote it by $\mathbb{R}$. Whenever we use the real numbers, we will only use the properties of a complete ordered field. ◇

We have already seen that any ordered field, hence $\mathbb{R}$, includes the set of rational numbers $\mathbb{Q}$.

Definition 1.2.14. Define an **irrational** number to be an element of $\mathbb{R} \setminus \mathbb{Q}$. ◇

We can think of the completeness property as "completing" the rational numbers by adding the irrational numbers. For example, we will see in Proposition 1.2.19 that $\sqrt{2}$ is the supremum of a certain set of rational numbers; therefore, it is an element of $\mathbb{R}$. (It will also follow from this that $\mathbb{Q}$ is not complete.)

The completeness property has many applications, and we start with an important property that is named after Archimedes of Syracuse (an equivalent property appears as an axiom in his *On the Sphere and Cylinder* treatise). While we obtain the Archimedean property of the real numbers as a consequence of the completeness property, one can verify that the rational numbers also satisfy this property.

Theorem 1.2.15 (Archimedean property). *If x is a real number, then there exists a natural number n such that $n > x$.*

Proof. If this were not the case, there would exist $x \in \mathbb{R}$ such that $n \leq x$ for all $n \in \mathbb{N}$. This would mean that the set $\mathbb{N}$ is bounded above. Let $\beta = \sup \mathbb{N} \in \mathbb{R}$. By Lemma 1.2.10, for $\varepsilon = 1$ there exists $n \in \mathbb{N}$ such that

$$\beta - 1 < n.$$

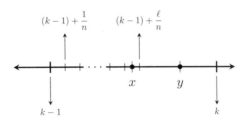

Figure 1.2. Construction of q such that $x < q < y$ (when $y - x < 1$).

Then $\beta < n+1 \in \mathbb{N}$, a contradiction to the fact that β is an upper bound for $\mathbb{N}$. □

This property can also be stated in the following equivalent way (see Exercise 1.2.6).

Corollary 1.2.16. *For each $x \in \mathbb{R}$, $x > 0$, there exists $n \in \mathbb{N}$ such that $\frac{1}{n} < x$.*

The following is an important consequence, which shows that given any real number there is a rational number that is arbitrarily close to it.

Corollary 1.2.17. *Let x and y be real numbers. If $x < y$, then there exists a rational number q such that*
$$x < q < y.$$

Proof. We may assume that $0 \leq x < y$ (if $x < 0 < y$, let $q = 0$; if $x < y \leq 0$, consider $0 \leq -y < -x$). Let k be the smallest natural number greater than x, i.e.,
$$k = \min\{i \in \mathbb{N} : x < i\}.$$
Then $k - 1 \leq x < k$. If $k < y$, then we are done by letting $q = k$, so we may assume that $y \leq k$, the case illustrated by Figure 1.2.

Since $y - x > 0$, by the Archimedean property (Corollary 1.2.16) there exists $n \in \mathbb{N}$ such that
$$\frac{1}{n} < y - x.$$
Starting at $k - 1$, we want to move forward by multiples of $1/n$ until we have a rational number that lands inside the interval (x, y). This is possible since $1/n$ is less than the distance from x to y.

Hence, let
$$\ell = \min\{i \in \mathbb{N} : x < k - 1 + \frac{i}{n}\},$$
and set
$$q = k - 1 + \frac{\ell}{n}.$$
Figure 1.2 shows the construction of q (it shows the case when $y - x < 1$).

We claim that $x < q < y$. First, from the definition of q we have that $x < q$. Now, if it were the case that $q \geq y$, then
$$k - 1 + \frac{\ell - 1}{n} = q - \frac{1}{n} > q + x - y \geq x,$$
contradicting the definition of ℓ. Therefore $q < y$, completing the proof. □

1.2. Completeness Property of $\mathbb{R}$

Definition 1.2.18. A set D in $\mathbb{R}$ is said to be **dense** (in $\mathbb{R}$) if for any $x \in \mathbb{R}$ and every $\varepsilon > 0$ there exists $q \in D$ such that $q \in (x - \varepsilon, x + \varepsilon)$. $\diamond$

Corollary 1.2.17 states that the set of rational numbers is dense in the set of real numbers. The set of integers $\mathbb{Z}$ is not dense in $\mathbb{R}$ as there are no integers that are arbitrarily close to $1/2$, for example. But the set $\mathbb{R} \setminus \mathbb{Z}$ is dense in $\mathbb{R}$.

Now we are in a position to prove that there is a positive number β in $\mathbb{R}$, and only one such element of $\mathbb{R}$, such that $\beta^2 = 2$. We denote this β by $\sqrt{2}$ and recall that we have already shown it is not in $\mathbb{Q}$.

Proposition 1.2.19. *There is a unique positive number $\beta \in \mathbb{R}$ such that $\beta^2 = 2$.*

Proof. Let $S = \{x \in \mathbb{R} : x^2 < 2\}$. Then $S \neq \varnothing$ (as $0 \in S$) and is bounded above by 2 (if $x \in S$, then $x^2 < 2 < 4$, so $x < 2$). Hence by the completeness property, S has a supremum in $\mathbb{R}$, which we denote by β. It is clear that $\beta > 0$ as $1 \in S$. We will show that assuming $\beta^2 < 2$ and assuming $\beta^2 > 2$ each lead to a contradiction. Therefore, it must be that $\beta^2 = 2$.

Now, if $\beta^2 < 2$, we will show there exists $\varepsilon > 0$ such that $(\beta + \varepsilon)^2 < 2$. This will imply $\beta + \varepsilon \in S$, contradicting that β is an upper bound of S as $\beta + \varepsilon > \beta$. For this we need ε to satisfy

(1.6) $$\beta^2 + 2\beta\varepsilon + \varepsilon^2 < 2 \text{ or}$$

(1.7) $$2\beta\varepsilon + \varepsilon^2 < 2 - \beta^2.$$

If we choose $\varepsilon < 1$, then $\varepsilon^2 < \varepsilon$, so if ε satisfies

(1.8) $$2\beta\varepsilon + \varepsilon < 2 - \beta^2 \text{ or}$$

(1.9) $$\varepsilon(2\beta + 1) < 2 - \beta^2,$$

we would have the desired inequality (1.6). This suggests that we choose

$$\varepsilon = \frac{1}{2} \cdot \frac{2 - \beta^2}{2\beta + 1}.$$

Then clearly ε satisfies (1.9). Also, as $\beta > 1$, $0 < \varepsilon < 1$, so ε satisfies (1.6). This shows that assuming $\beta^2 < 2$ leads to a contradiction.

Finally, we assume $\beta^2 > 2$. Now we show that there exists $\varepsilon > 0$ with $\varepsilon < \beta$ such that $(\beta - \varepsilon)^2 > 2$. This would contradict that β is the smallest upper bound of S. Hence we need ε to satisfy

(1.10) $$\beta^2 - 2\beta\varepsilon + \varepsilon^2 > 2 \text{ or}$$

(1.11) $$2\beta\varepsilon < \beta^2 - 2 + \varepsilon^2.$$

It suffices to have

$$2\beta\varepsilon < \beta^2 - 2.$$

Thus choose ε such that $0 < \varepsilon < \beta$ and

$$\varepsilon < \frac{\beta^2 - 2}{2\beta}.$$

This shows that assuming $\beta^2 > 2$ leads to a contradiction. We conclude that $\beta^2 = 2$. It is clear that $\beta \in \mathbb{R}$.

To see uniqueness, let y be any positive element of $\mathbb{R}$ such that $y \neq \beta$. If $y < \beta$, then $y^2 < \beta^2 = 2$, so $y^2 \neq 2$; similarly, if $y > \beta$, then $y^2 \neq 2$. □

Remark 1.2.20. The proof of Proposition 1.2.19 only uses the properties of a complete ordered field. Hence it also follows from the proof that the ordered field $\mathbb{Q}$ is not complete, since if it were complete it would have to have a number whose square is 2, but we have shown no such number exists in $\mathbb{Q}$.

1.2.4. Absolute Value and Intervals

Definition 1.2.21. We define the **absolute value** of a real number x denoted $|x|$. For $x \in \mathbb{R}$ set

$$(1.12) \qquad |x| = \begin{cases} x, & \text{if } x \geq 0; \\ -x, & \text{if } x < 0. \end{cases}$$
◇

Proposition 1.2.22. *The absolute value satisfies the following properties.*

(1) $|x| \geq 0$, *with equality holding if and only if $x = 0$;*

(2) $|xy| = |x||y|$;

(3) *Triangle inequality:* $|x + y| \leq |x| + |y|$.

Proof. We show the triangle inequality. The other properties are immediate from the definition. First note that $x \leq |x|$ and $-x \leq |x|$ for every $x \in \mathbb{R}$. Let $x, y \in \mathbb{R}$. If $x + y \geq 0$, then
$$|x + y| = x + y \leq |x| + |y|.$$
If $x + y < 0$, then
$$|x + y| = -x - y \leq |x| + |y|.$$
Hence, $|x + y| \leq |x| + |y|$. □

We conclude with the notation and definition for **intervals** in $\mathbb{R}$.

Definition 1.2.23. For $a \leq b$, write

$$(a, b) = \{x \in \mathbb{R} : a < x < b\}, \text{ this is called an \textbf{open interval}},$$
$$[a, b] = \{x \in \mathbb{R} : a \leq x \leq b\}, \text{ this is called a \textbf{closed interval}},$$
$$(a, b] = \{x \in \mathbb{R} : a < x \leq b\},$$
$$[a, b) = \{x \in \mathbb{R} : a \leq x < b\}.$$
◇

Note that when $b = a$, the interval (a, a) is the empty set. (Some authors require intervals to be nonempty.)

Definition 1.2.24. We define the **infinite intervals**:

$$(a, \infty) = \{x \in \mathbb{R} : x > a\},$$
$$[a, \infty) = \{x \in \mathbb{R} : x \geq a\},$$
$$(-\infty, b) = \{x \in \mathbb{R} : x < b\},$$
$$(-\infty, b] = \{x \in \mathbb{R} : x \leq b\},$$
$$(-\infty, \infty) = \mathbb{R}.$$
◇

Exercises: Completeness Property of $\mathbb{R}$

1.2.1 Prove that $\sqrt{p} \notin \mathbb{Q}$ when p is prime.

1.2.2 Prove that $\sqrt{a} \notin \mathbb{Q}$ when a is not a perfect square (i.e., a is an integer not of the form b^2 for some integer b).

1.2.3 State the analogue of the completeness property using the infimum instead of the supremum. Prove that the completeness property for the infimum is equivalent to the completeness property for the supremum. (*Hint*: Note that for any set $S \subseteq \mathbb{R}$, $\sup S = -\inf(-S)$, where $-S = \{-x : x \in S\}$.)

1.2.4 Let S be a nonempty set of real numbers. Show that a real number α satisfies $\alpha = \inf S$ if and only if α is a lower bound for S and for every $\varepsilon > 0$ there exists $x \in S$ such that $x < \alpha + \varepsilon$.

1.2.5 Prove that the rational numbers satisfy the Archimedean property: for each $r \in \mathbb{Q}$ there exists $n \in \mathbb{N}$ such that $n > r$. Prove this without using the completeness property of the real numbers.

1.2.6 Prove that the following property is equivalent to the Archimedean property: for each $x \in \mathbb{R}$, $x > 0$, there exists $n \in \mathbb{N}$ such that $\frac{1}{n} < x$.

1.2.7 (Gauss) Let $p(x)$ be a polynomial with integer coefficients of the form $p(x) = x^n + a_{n-1}x^{n-1} + \cdots + a_1 x + a_0$, where all $a_{n-1}, \ldots, a_0 \in \mathbb{Z}$. Show that if $a \in \mathbb{R}$ is a **root** of $p(x)$, i.e., $p(a) = 0$, and a is not an integer, then a is irrational.

1.2.8 Prove that for each $n \in \mathbb{N}$, if $\sqrt{n}$ is not an integer, then it is irrational.

1.2.9 Extend the proof of Proposition 1.2.19 to show that for every real number $x > 0$, the square root of x is a real number.

* 1.2.10 Prove that for each $a \in \mathbb{R}^+$ and each $q \in \mathbb{N}$ there exists a unique $s \in \mathbb{R}^+$ such that $s^q = a$; this number s defines $a^{1/q}$.

1.2.11 Let $a \in \mathbb{R}^+$. Use Exercise 1.2.10 to define a^r for each rational number r.
(a) Prove that $(a^{1/q})^p = (a^p)^{1/q}$ for $p, q \in \mathbb{N}$.
(b) Prove that if $r, s \in \mathbb{Q}$, then $a^r a^s = a^{r+s}$ and $a^{-r} = 1/a^r$.

1.2.12 Let $a \in \mathbb{R}^+$. For $x \in \mathbb{R}$ define
$$a^x = \sup\{a^r : r \in \mathbb{Q}, r < x\}.$$
(a) Prove that when $x \in \mathbb{Q}$ this gives the same definition as Exercise 1.2.11.
(b) Prove that $a^{x+y} = a^x a^y$ for $x, y \in \mathbb{R}$.

1.2.13 Prove that for all positive real numbers x and y there exists $n \in \mathbb{N}$ such that $y < nx$.

1.2.14 Find:
(a) $\sup(0, 1)$;
(b) $\inf(0, 1)$.
Prove your answer in each case.

1.2.15 Find:
  (a) $\sup[(1,2) \cap \mathbb{Q}]$;
  (b) $\inf[(1,2) \cap \mathbb{Q}]$ in the ordered field $\mathbb{Q}$.
  Prove your answer in each case.

1.2.16 Complete the details in the proof of Corollary 1.2.17.

1.2.17 Let F be an ordered field and let $A \subseteq B \subseteq F$.
  (a) Prove that $\inf A \geq \inf B$.
  (b) Prove that if A is nonempty, $\inf A \leq \sup A$.

1.2.18 Give another proof of the triangle inequality by expanding the expression $(x+y)^2$.

1.2.19 Let $x, y \in \mathbb{R}$, $y > 0$. Prove that $|x| < y$ if and only if $-y < x < y$.

1.2.20 Prove that $|x - a| < \varepsilon$ if and only if $a - \varepsilon < x < a + \varepsilon$ for every $x, a, \varepsilon \in \mathbb{R}, \varepsilon > 0$.

1.2.21 Prove that for all real numbers x and all $n \in \mathbb{N}$, $|x^n| = |x|^n$.

1.2.22 Prove that for all real numbers x and y, $||x| - |y|| \leq |x - y|$.

1.2.23 Let F be an ordered field. Prove that F satisfies the Archimedean property if and only if the natural numbers in F are not bounded above.

1.2.24 Prove that the ordered field $F(x)$ of Exercise 1.1.20 does not satisfy the Archimedean property (i.e., the set $\mathbb{N}$ is bounded in $F(x)$).

1.3. Countable and Uncountable Sets

While the definition of finite (and hence infinite) sets does not hold any surprises, we shall see that there are different "sizes" of infinite sets. This is a fundamental and nontrivial observation that we owe to Cantor. The simplest (i.e., smallest in some reasonable sense) kind of infinite sets are the countably infinite sets. To distinguish the subtle differences between different kinds of infinity, we use the notion of a function. The reader who has not seen finite sets should read Subsection 0.5.4.

1.3.1. Countable Sets

Definition 1.3.1. A set A is **countable** if it is finite or if there is a bijection $f : \mathbb{N} \to A$ (in this case we can see A as the set $\{f(1), f(2), \ldots\}$). A set is **countably infinite** if it is countable and not finite. ◊

In particular, the set of natural numbers $\mathbb{N}$ is a countable set. We observe now that $\mathbb{N}$ is not finite. In fact, if for some $n \in \mathbb{N}$ we have a function $f_n : J_n \to \mathbb{N}$, we can define
$$a = f_n(1) + \cdots + f_n(n) + 1.$$
Then a is in $\mathbb{N}$ and, from its definition, it is greater than $f_n(i)$ for all $i = 1, \ldots, n$. It follows that f_n cannot be a surjective function. Thus $\mathbb{N}$ is not finite, and so it is countably infinite. From this argument it also follows that if a set is bijective with $\mathbb{N}$, then it is not finite, so is countably infinite.

1.3. Countable and Uncountable Sets

We make a remark regarding notation. There is no universal agreement in the literature on the use of the term "countable"; sometimes countable is used to mean what we call "countably infinite". A countably infinite set is also called "denumerable".

Definition 1.3.2. A set is **uncountable** if it is not countable. ◇

Clearly, an uncountable set has to be infinite, but it must be an infinity "greater" than the infinity of sets like $\mathbb{N}$. The existence of uncountable sets is a surprising result which is due to Cantor. Before proving this, we show that many familiar sets are countable. We start with a lemma that will be used later.

Lemma 1.3.3. *If $K \subseteq \mathbb{N}$, then K is countable.*

Proof. If K is finite, we are done, so suppose K is infinite. We wish to define a function $h : \mathbb{N} \to K$ that is a bijection. We define $h(i)$ recursively. Define first
$$h(1) = \min K,$$
which exists as K is nonempty. Assuming that $h(1), \ldots, h(n-1)$ have been defined, set
$$h(n) = \min(K \setminus \{h(1), \ldots, h(n-1)\}).$$
The induction principle is what justifies that h is well-defined. By definition $h(n) \neq h(i)$ for all $i < n, i \in \mathbb{N}$, so h is injective. If h were not surjective, there would exist $k \in K$ such that $h(i) \neq k$ for all $i \in \mathbb{N}$. If $k < h(\ell)$ for some $\ell \in \mathbb{N}$, then k should have been chosen as a value of h before the ℓth step. Thus $k > h(i)$ for all $i \in \mathbb{N}$, a contradiction since h takes infinitely many values (see Exercise 1.3.4). Therefore, h is a bijection. □

Proposition 1.3.4. *Let A be a nonempty set. The following statements are equivalent.*

(1) A is countable.

(2) There exists a function $f : \mathbb{N} \to A$ that is surjective.

(3) There exists a function $g : A \to \mathbb{N}$ that is injective.

Proof. We start by showing that if A is countable, then (2) holds. If A is finite, there exists a bijection $h : J_n \to A$ for some integer n. Define $f : \mathbb{N} \to A$ by

(1.13) $$f(i) = \begin{cases} h(i) & \text{for } 1 \leq i \leq n, \\ h(1) & \text{for } i > n. \end{cases}$$

(There is a lot of freedom in defining f.) It is clear that f is surjective. Next, when A is infinite, it is countably infinite, so there is already a bijection $\mathbb{N} \to A$, which is, of course, surjective.

Now we show that (2) implies (3). Using f, we define g for $a \in A$ by
$$g(a) = \min f^{-1}(\{a\}).$$
Recall that $f^{-1}(\{a\})$ is the set of all points n in $\mathbb{N}$ whose image under f is a, i.e., $f(n) = a$. Since f is surjective, this set is nonempty for each $a \in A$. As it is a subset of $\mathbb{N}$, it has a least element, or minimum, so g is defined for all $a \in A$. Note

also that by its definition, g satisfies the property that $f(g(a)) = a$ for all $a \in A$. Then, by Exercise 0.4.15, g is injective.

Finally, assume that (3) holds. We first note that the function g defines a bijection from A to $g(A)$. Therefore, it follows that if $g(A)$ is countable, then A is countable. Next we observe that $g(A)$ is a subset of $\mathbb{N}$, and so by Lemma 1.3.3, $g(A)$ is countable, completing the proof. □

Question 1.3.5. Let A and B be two sets. Prove that if A is countable and $B \subseteq A$, then B is countable.

It follows that the set of even integers $2\mathbb{N}$, for example, is countably infinite. Something unexpected occurs with infinite sets that does not happen with finite sets: a proper subset of an infinite set, such as $2\mathbb{N}$, may be bijective with the bigger set. We can express this in terms of the notion of "cardinality" of sets.

Definition 1.3.6. We extend the notion of having the same cardinality to arbitrary sets by saying that two sets A and B have the **same cardinality** if they are in one-to-one correspondence, i.e., there is bijection between them. We may write $\text{card}(A) = \text{card}(B)$. The set $\mathbb{N}$ is said to have **cardinality** $\aleph_0$ (pronounced "aleph naught"), where $\aleph$ is the first letter of the Hebrew alphabet. ◊

Thus a countably infinite set is said to have cardinality $\aleph_0$.

We can express the remark above by saying that both $2\mathbb{N}$ and $\mathbb{N}$ have cardinality $\aleph_0$, while $2\mathbb{N}$ is a proper subset of $\mathbb{N}$. (In contrast to this, no proper subset of a finite set A can have the same cardinality as A; see Exercise 1.3.4.)

In the proofs that follow it will be useful to have the notion of a sequence. The reader may be familiar with the basic notion of a sequence from a calculus course, and we will study sequences in greater depth in Chapter 2, but here we introduce some useful notation.

Definition 1.3.7. A **sequence** in a (nonempty) set A is a function from $\mathbb{N}$ to A. We denote this function with $a : \mathbb{N} \to A$. Rather than writing the values of the function as $a(1), a(2), \ldots$, we write $a_1, a_2, \ldots$. We call a_n a **term** of the sequence. In particular a_n is the nth term; we write the sequence as (a_n) or $(a_n)_{n \in \mathbb{N}}$. (Some authors may also denote the sequence (a_n) by $\{a_n\}$ or $\{a_n\}_{n \in \mathbb{N}}$.) ◊

For example, we can define a sequence by setting $a_n = 2n$, for $n \in \mathbb{N}$; this is the sequence of even integers $2, 4, 6, \ldots$.

Proposition 1.3.8. *Let A and B be two sets. If A and B are countable sets, then $A \times B$ is a countable set.*

Proof. We give two proofs as they illustrate different ideas.

Let $f : A \to \mathbb{N}$ and $g : B \to \mathbb{N}$ be injective functions. Let p and q be two distinct primes in $\mathbb{N}$. Define a function $h : A \times B \to \mathbb{N}$ by

$$h(a, b) = p^{f(a)} q^{g(b)}.$$

As integers have a unique decomposition as a product of primes p and q and f, g are injective, it follows that h is injective. Proposition 1.3.4 implies that $A \times B$ is countable.

1.3. Countable and Uncountable Sets

For the second proof assume that there are surjective functions $a : \mathbb{N} \to A$ and $b : \mathbb{N} \to B$. It suffices to define a surjective function from $\mathbb{N}$ to $A \times B$. We order the elements of $A \times B$ in an array in the following way. First think of a row of the form

$$(a_1, b_1), (a_1, b_2), (a_1, b_3), \ldots, (a_1, b_j), \ldots.$$

(The row is finite if B is finite.) For the second row use a_2 wherever a_1 was used in the first row to obtain

$$(a_2, b_1), (a_2, b_2), (a_2, b_3), \ldots, (a_2, b_j), \ldots.$$

Continue in this way so that the ith row is

$$(a_i, b_1), (a_i, b_2), (a_i, b_3), \ldots, (a_i, b_j), \ldots.$$

If A is finite, there are finitely many rows. Now we define $f : \mathbb{N} \to A \times B$, which is surjective. First set $f(1) = (a_1, b_1)$. Then move down in the array so that each diagonal is covered completely one at a time from top to bottom; this is illustrated in Figure 1.3. (Another equally valid construction could go from bottom to top). The reader should make the appropriate changes in case either A or B is finite. Hence, to cover the second diagonal, we set $f(2) = (a_1, b_2)$ and $f(3) = (a_2, b_1)$. Then start with the third diagonal by setting $f(4) = (a_1, b_3)$, etc. In this way we have a surjective function f. □

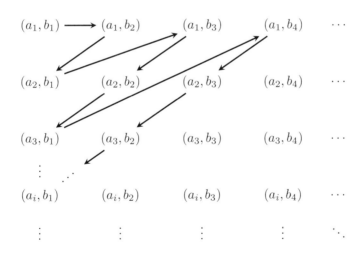

Figure 1.3. Construction of the function f in Proposition 2.3.2.

Corollary 1.3.9. *The set of rational numbers $\mathbb{Q}$ is countable.*

Proof. Let $\mathbb{Q}^+$ stand for the set of positive rational numbers. Define a function $f : \mathbb{N} \times \mathbb{N} \to \mathbb{Q}^+$ by

$$f(m, n) = \frac{m}{n}.$$

Then by definition f is surjective. As $\mathbb{N} \times \mathbb{N}$ is countable (by Proposition 1.3.8), it follows that $\mathbb{Q}^+$ is countable. So there is a surjective function $g : \mathbb{N} \to \mathbb{Q}^+$. Now, using g, define a function $h : \mathbb{N} \to \mathbb{Q}$ by
$$h(1) = 0,$$
$$h(2i) = g(i) \text{ for } i \in \mathbb{N},$$
$$h(2i+1) = -g(i) \text{ for } i \in \mathbb{N}.$$
The reader can verify that h is surjective, and therefore $\mathbb{Q}$ is countable. □

The reader is asked to give another proof that $\mathbb{Q}$ is countable along the lines of the second proof in Proposition 1.3.8 (Exercise 1.3.10).

1.3.2. Unions and Intersections of Families of Sets. The definition of union and intersection of two sets can be extended to an arbitrary collection of sets. We use "collection of sets" or "family of sets" instead of "set of sets".

Definition 1.3.10. Let Γ be an arbitrary nonempty set. If for each $\alpha \in \Gamma$ there is a set denoted A_α, we say that $\{A_\alpha : \alpha \in \Gamma\}$ is a **family of sets indexed by** Γ. When Γ is a countable set, we say it is a **countable family**. ◇

For example, if $\Gamma = \mathbb{N}$, then we have a family denoted by $\{A_n : n \in \mathbb{N}\}$. It is possible to have a countable family of sets $\{A_n : n \in \mathbb{N}\}$ where each set A_n is uncountable, as well as an infinite family of sets $\{A_\alpha : \alpha \in \Gamma\}$ (Γ is infinite) where each set A_α is finite.

Definition 1.3.11. Given a family of sets $\{A_\alpha : \alpha \in \Gamma\}$, we define their **union** or **intersection** as follows:
$$\bigcup_{\alpha \in \Gamma} A_\alpha = \{x : x \in A_\alpha \text{ for some } \alpha \in \Gamma\},$$
$$\bigcap_{\alpha \in \Gamma} A_\alpha = \{x : x \in A_\alpha \text{ for all } \alpha \in \Gamma\}.$$
◇

In the case of a countable collection we usually write $\bigcup_{n=1}^\infty A_n$ and $\bigcap_{n=1}^\infty A_n$ for $\bigcup_{n \in \mathbb{N}} A_n$ and $\bigcap_{n \in \mathbb{N}} A_n$, respectively.

Question 1.3.12. Show that $\bigcap_{n \in \mathbb{N}} \{i \in \mathbb{N} : i > n\} = \varnothing$.

Definition 1.3.13. A family of sets $\{A_\alpha\}_{\alpha \in \Gamma}$ is said to be **disjoint**, sometimes also called *pairwise disjoint*, if $A_\alpha \cap A_\beta = \varnothing$ for all $\alpha, \beta \in \Gamma$ with $\alpha \neq \beta$. ◇

When $\{A_\alpha : \alpha \in \Gamma\}$ is a family of disjoint sets, we may write
$$\bigsqcup_{\alpha \in \Gamma} A_\alpha$$
for the union $\bigcup_{\alpha \in \Gamma} A_\alpha$ to make explicit in the notation that the union is over a pairwise disjoint family. If two sets A and B are disjoint, we may write $A \sqcup B$ for the union $A \cup B$.

Theorem 1.3.14. *A countable union of countable sets is countable: if $\{A_n : n \in \mathbb{N}\}$ is a family such that each set A_n is countable, then their union $\bigcup_{n=1}^\infty A_n$ is a countable set.*

1.3. Countable and Uncountable Sets

Proof. First assume that the sets A_n are disjoint. For each $n \in \mathbb{N}$, let $f_n : A_n \to \mathbb{N}$ be injective, and let p_n be the nth prime (we use that there exist infinitely many primes). Define a function $f : \bigsqcup_{n=1}^{\infty} A_n \to \mathbb{N}$ by

$$f(x) = p_n^{f_n(x)} \text{ if } x \in A_n,$$

where f is a function since the sets are disjoint. We claim that f is injective. Indeed, if $f(x) = f(y)$, then

$$p_n^{f_n(x)} = p_m^{f_m(y)},$$

which implies that $p_n = p_m$ and $f_n(x) = f_m(y)$. Then $n = m$ and $x = y$. Then by Proposition 1.3.4, $\bigsqcup_{n=1}^{\infty} A_n$ is countable.

If the sets A_n are not disjoint, we define a new sequence of sets by

$$B_n = A_n \times \{n\}.$$

The sets B_n are disjoint since if B_n and B_m have a common element (x, y), then $y = n$ and $y = m$, so $n = m$ and they are the same set. Also, $\operatorname{card}(A_n) = \operatorname{card}(B_n)$ for all $n \in \mathbb{N}$, since there is a natural bijection that sends $x \in A_n$ to (x, n) in B_n. By the first part, the union $\bigsqcup_{n=1}^{\infty} B_n$ is countable. The function

$$g : \bigsqcup_{n=1}^{\infty} B_n \to \bigcup_{n=1}^{\infty} A_n,$$

defined by $g(x, n) = x$, is surjective. Proposition 1.3.4 implies that $\bigcup_{n=1}^{\infty} A_n$ is countable. $\square$

Question 1.3.15. Show that if Γ is a countable set and for each $\alpha \in \Gamma$ the set A_α is countable, then $\bigcup_{\alpha \in \Gamma} A_\alpha$ is countable.

1.3.3. An Uncountable Set: $\mathcal{P}(\mathbb{N})$. Let $\{0, 1\}^{\mathbb{N}}$ denote the set of all functions from $\mathbb{N}$ to $\{0, 1\}$. According to our notation, this is the same as the set of all sequences with values in $\{0, 1\}$. For example, $b : \mathbb{N} \to \{0, 1\}$ defined by

(1.14) $$b_i = \begin{cases} 0 & \text{if } i \text{ is even,} \\ 1 & \text{if } i \text{ is odd,} \end{cases}$$

is an element of $\{0, 1\}^{\mathbb{N}}$. This is the sequence whose terms are $0, 1, 0, 1, \ldots$.

The set $\{0, 1\}^{\mathbb{N}}$ can be identified with the set of all subsets of $\mathbb{N}$, namely $\mathcal{P}(\mathbb{N})$. To see this, we will define a one-to-one correspondence that will associate to each subset of $\mathbb{N}$ a unique sequence with values in $\{0, 1\}$. Given a set $A \subseteq \mathbb{N}$, we can define a sequence (a_n) by setting

(1.15) $$a_n = \begin{cases} 0 & \text{if } n \notin A, \\ 1 & \text{if } n \in A. \end{cases}$$

The idea is that we could think of tagging each natural number with 1 if that number is in A and with 0 if the number is not in A. This gives a sequence of 0's and 1's, so in this way each subset of $\mathbb{N}$ is associated with an element of $\{0, 1\}^{\mathbb{N}}$. Conversely, given a sequence $a \in \{0, 1\}^{\mathbb{N}}$, one can define a set $A \subseteq \mathbb{N}$ by setting $A = \{n \in \mathbb{N} : a_n = 1\}$. For example, the sequence $0, 1, 0, 1, \ldots$ corresponds to the even integers. This identification defines a one-to-one correspondence between sequences with values in $\{0,1\}$ ($\{0,1\}^{\mathbb{N}}$) and $\mathcal{P}(N)$. Therefore $\mathcal{P}(\mathbb{N})$ and $\{0,1\}^{\mathbb{N}}$ have the same cardinality.

Theorem 1.3.16 (Cantor). *The set of subsets of $\mathbb{N}$, $\mathcal{P}(N)$, is uncountable.*

Proof. We give two proofs, both due to Cantor. The first one proves that $\mathcal{P}(N)$ is uncountable. The second one uses the identification of $\mathcal{P}(\mathbb{N})$ with $\{0,1\}^{\mathbb{N}}$ and shows that $\{0,1\}^{\mathbb{N}}$ is uncountable.

We proceed by contradiction and suppose that the set $\mathcal{P}(\mathbb{N})$ is countable. By Theorem 1.3.4 there exists a surjection $g : \mathbb{N} \to \mathcal{P}(\mathbb{N})$. Note that $g(i)$ is a subset of $\mathbb{N}$ for each $i \in \mathbb{N}$. Define the set S by

$$S = \{i \in \mathbb{N} : i \notin g(i)\}.$$

By definition S is a subset of $\mathbb{N}$, and since g is surjective there must exist $n \in \mathbb{N}$ such that $g(n) = S$. Now, either n is an element of S or not. But if $n \in S$, by the definition of n and S we obtain that $n \notin S$. And if $n \notin S$, again by the definition of S we have that $n \in S$. In both cases we have a contradiction; therefore, the function g cannot be surjective. It follows that $\mathcal{P}(\mathbb{N})$ is not countable.

The second proof is a well-known proof that introduces what is known as Cantor's *diagonalization argument*. We show that the set $\{0,1\}^{\mathbb{N}}$ is uncountable.

We again proceed by contradiction and suppose that the set $\{0,1\}^{\mathbb{N}}$ is countable. Then there exists a surjection $f : \mathbb{N} \to \{0,1\}^{\mathbb{N}}$. Each $f(n)$ is a sequence whose ith element we denote by $f(n)_i$ (where $n, i \in \mathbb{N}$). Now we construct a sequence that is not in the image of f, showing that f could not be surjective, a contradiction. Define a new sequence $\alpha \in \{0,1\}^{\mathbb{N}}$ by

$$(1.16) \qquad \alpha_n = 1 - f(n)_n = \begin{cases} 1 & \text{if } f(n)_n = 0, \\ 0 & \text{if } f(n)_n = 1. \end{cases}$$

(The sequence α has been constructed so that at the nth place it differs from the value of the sequence $f(n)$ at the nth place.) As $\alpha \in \{0,1\}^{\mathbb{N}}$ and f is surjective, there must exist $k \in \mathbb{N}$ so that $f(k) = \alpha$. But, from the construction of α, $\alpha_k \neq f(k)_k$, a contradiction. Therefore, f is not surjective and $\{0,1\}^{\mathbb{N}}$ is uncountable. $\square$

It is interesting to note that the argument in the first proof of Theorem 1.3.16 is similar to the argument in Russell's paradox, which appeared a couple of decades after Cantor's theorem. Theorem 1.3.16 will be used to show that the set of real numbers $\mathbb{R}$ is uncountable. This will follow from the representation of the real numbers in binary. The details (addressing the fact that some real numbers may have two representations) will be discussed in Section 2.2 after we cover additional properties of the real numbers that are a consequence of the completeness property.

Exercises: Countable and Uncountable Sets

1.3.1 Write $\mathbb{N}$ as a union of infinitely many disjoint sets each of which is infinite.

1.3.2 Let A and B be sets. Prove that if A is countable and there exists a function $f : A \to B$ that is surjective, then B is countable.

1.3.3 Let A and B be sets. Prove that if B is countable and there exists a function $f : A \to B$ that is injective, then A is countable.

1.3.4 Let A be a finite set. Prove that if $B \subseteq A$, then B is finite. Furthermore, prove that if in addition $B \neq A$, then $\operatorname{card}(B) < \operatorname{card}(A)$.

1.3.5 Let K be an infinite subset of $\mathbb{N}$. Prove that there does not exist $n \in \mathbb{N}$ such that $n > k$ for all $k \in K$.

1.3.6 Let A be a finite nonempty set. A *word* in A is a finite sequence of symbols from A. For example, if $A = \{a, b, c\}$, then an example of a word in A is *abca*. We consider *ab* and *ba* to be different words. Prove that the set of all words in A is countable.

1.3.7 Give a definition of the Cartesian product of k sets $A_1 \times A_2 \times \cdots \times A_k$, and show that if the sets $A_1, \ldots, A_k$ are countable, then their Cartesian product is countable.

1.3.8 Let A be a finite nonempty set. Prove that the set of all functions from A to $\mathbb{N}$ is countable.

1.3.9 Give another proof of the part that (2) implies (3) in Proposition 1.3.4 by using the following construction. Assume that A is infinite. Using f, define a new function $\tilde{f} : \mathbb{N} \to A$ by

$\tilde{f}(1) = f(1)$,

$\tilde{f}(n) = f(\tilde{k})$ where $\tilde{k} = \min\{k : f(k) \notin \{\tilde{f}(1), \ldots, \tilde{f}(n-1)\}\}$, for $n > 1$.

We use induction to justify in detail that $\tilde{f}$ is defined for every element of $\mathbb{N}$. Prove that $\tilde{f}$ is a bijection, and use this to complete the proof.

1.3.10 Give a second proof that $\mathbb{Q}$ is countable using the diagonal construction of the second proof in Proposition 1.3.8.

1.3.11 Give another proof of Theorem 1.3.14 using part (2) in Proposition 1.3.4.

1.3.12 Let A be a countably infinite set. Is the set of all functions from A to $\mathbb{N}$ a countably infinite set? Prove your answer.

1.3.13 Prove that every collection of disjoint intervals (of positive length) is countable.

1.3.14 A number a is said to be an **algebraic number** if it is a root of a polynomial with integer coefficients; i.e., there exists a polynomial p of the form $p(x) = a_n x^n + a_{n-1} x^{n-1} + \cdots + a_1 x + a_0$, where $a_n, \ldots, a_0$ are integers, and $p(a) = 0$. (For example, $\sqrt{2}$ is a root of the polynomial $x^2 - 2 = 0$, so it is algebraic.) Prove that the set of algebraic numbers is countable. (You may use that a polynomial of degree n has at most n roots.)

1.3.15 Let A be a set, and let $\{B_\alpha\}$ be a collection of sets indexed by some set Γ. Show the following distributive properties.
   (a) $A \cap (\bigcup_{\alpha \in \Gamma} B_\alpha) = \bigcup_{\alpha \in \Gamma} (A \cap B_\alpha)$
   (b) $A \cup (\bigcap_{\alpha \in \Gamma} B_\alpha) = \bigcap_{\alpha \in \Gamma} (A \cup B_\alpha)$

1.3.16 Let A be a set, and let $\{B_\alpha\}$ be a collection of sets indexed by some set Γ. Determine whether the following equality holds. If it does not hold, determine whether one is a subset of the other.
$$A \triangle (\bigcup_{\alpha \in \Gamma} B_\alpha) = \bigcup_{\alpha \in \Gamma} A \triangle B_\alpha.$$

1.3.17 (Inclusion-Exclusion) Let A, B be finite sets. Prove that
$$\text{card}(A \cup B) = \text{card}(A) + \text{card}(B) - \text{card}(A \cap B).$$
Formulate a formula for the cardinality of the union of n finite sets.

1.3.18 (De Morgan's laws) Let Γ be a set, and let $\{G_\alpha\}$ be a collection of sets in some set X indexed by Γ. Show that
$$(\bigcup_{\alpha \in \Gamma} G_\alpha)^c = \bigcap_{\alpha \in \Gamma} G_\alpha^c$$
and
$$(\bigcap_{\alpha \in \Gamma} G_\alpha)^c = \bigcup_{\alpha \in \Gamma} G_\alpha^c.$$

1.3.19 Define the **limit supremum** and **limit infimum** of the sequence of sets $(A_n), n \geq 1$ in a set X by
$$\liminf_{n \to \infty} A_n = \bigcup_{m=1}^{\infty} \bigcap_{n=m}^{\infty} A_n,$$
$$\limsup_{n \to \infty} A_n = \bigcap_{m=1}^{\infty} \bigcup_{n=m}^{\infty} A_n.$$
   (a) Show that $\liminf_{n \to \infty} A_n$ is precisely the set of points $x \in X$ such that there exists $k \geq 1$ with $x \in A_n$ for all $n \geq k$. (In this case we say that this is the set of points that are eventually in A_n, i.e., in A_n for all large n.)
   (b) Show that $\limsup_{n \to \infty} A_n$ consists of the sets of points $x \in X$ that are in infinitely many A_n.
   (c) When does $\liminf_{n \to \infty} A_n = \limsup_{n \to \infty} A_n$? Give a condition that guarantees equality and an example when equality does not hold.

1.3.20 Show that $\{0,1\}^{\mathbb{N}}$ and $\{0,1\}^{\mathbb{N}} \times \{0,1\}^{\mathbb{N}}$ have the same cardinality.

* 1.3.21 Prove that a set A is finite if and only if whenever $f : A \to A$ is injective, then f is bijective.

1.3.22 Let A be a finite nonempty set. Prove that $A^{\mathbb{N}}$, the set of sequences in A (i.e., functions from $\mathbb{N}$ to A), is uncountable if and only if A has more than one element.

1.3.23 Let A be a set. Prove that there is no surjective function from A to $\mathcal{P}(A)$.

1.4. Construction of the Real Numbers

In this section we outline a construction of the real numbers. It is not necessary for later chapters, and the reader may omit this section. At the same time, many details are left for the interested reader and completing all the details of this section could be viewed as a project.

We follow Dedekind's construction which is based on Dedekind cuts. We will construct an ordered field and prove it has all the properties that we expect: it is a field, it is ordered, and it is order complete. This construction was published by Richard Dedekind in 1872; another construction due to Cantor was published at about the same time (it uses Cauchy sequences to construct what is now called a *metric completion*; see Subsection 4.5.1).

The idea of the construction is to start with the field of rational numbers (which we have already seen is not order complete) and then construct a new set which is an order-complete ordered field. Each element of this new field will be a particular collection of rational numbers, so each real number will be a set of rational numbers with some special properties.

We want this new set (which we will call the set of real numbers) to contain numbers such as $\sqrt{2}$, for example (where as before, by $\sqrt{2}$ we understand a positive number whose square is 2). The idea is to represent $\sqrt{2}$ by the set ρ of all rational numbers that approximate $\sqrt{2}$ from below; we wish to think of ρ as an infinite "ray" to the left of the real number we want to represent. Then we can define ρ as consisting of all negative rational numbers plus the nonnegative rational numbers whose square is less than 2. (If we just say that ρ consists of rational numbers x such that $x^2 < 2$, then this set would not include numbers such as -4, for example, so we had to be more careful with the definition.) Note that we only need properties of $\mathbb{Q}$ to define this set. Now we note two properties of this set ρ. The first is that if x is in ρ and $y < x$, then y is also in ρ. (The idea is that ρ is an infinite ray to the left.) The second property is that if x is in ρ, there is also another rational z in ρ such that $x < z$. This is crucial to the idea that ρ "approximates" $\sqrt{2}$ from below; it says we can get closer and closer to $\sqrt{2}$. These two properties characterize what we call a Dedekind cut, which we define next.

Definition 1.4.1. A subset α of $\mathbb{Q}$ is said to be a **Dedekind cut** (sometimes simply called a **cut**) if it satisfies the following three properties.

(1) Both α and its complement α^c in $\mathbb{Q}$ are nonempty.

(2) If $x \in \alpha$ and $y < x$, then $y \in \alpha$.

(3) If $x \in \alpha$, then there is some $z \in \alpha$ such that $x < z$. ◇

It follows from these properties that if α is a cut, then there must be a rational number r in its complement, and every element of α must be less than r. For an example of a cut, the reader should verify that the set $\rho = \{r \in \mathbb{Q} : r < 0, \text{ or } r \geq 0 \text{ and } r^2 < 2\}$ is a Dedekind cut.

Definition 1.4.2. We define the set of **real numbers** $\mathbb{R}$ to be the set of all Dedekind cuts. Hence, $\mathbb{R}$ is a certain subset of the power set of $\mathbb{Q}$. ◇

There is a natural way in which we can think of $\mathbb{R}$ as containing $\mathbb{Q}$, which will be a consequence of the following definition.

Definition 1.4.3. For each rational number r, define the **cut corresponding to** r by
$$[r] = \{x \in \mathbb{Q} : x < r\}.$$
◇

It follows that each set $[r]$ is a cut. By identifying each rational number r with the cut $[r]$, we can think of $\mathbb{R}$ as containing a copy of $\mathbb{Q}$.

Our next task is to define operations of addition "+" and multiplication "·" on $\mathbb{R}$ with respect to which $\mathbb{R}$ becomes a field, and then to define an order on $\mathbb{R}$ so that $\mathbb{R}$ is order complete with respect to this order.

We first define addition.

Definition 1.4.4. Given two cuts α and β in $\mathbb{R}$, let
$$\alpha + \beta = \{x + y : x \in \alpha \text{ and } y \in \beta\}.$$
◇

One needs verify that $\alpha+\beta$ is indeed a cut (Exercise 1.4.1). From the definitions one can see that $\alpha + \beta = \beta + \alpha$ and $\alpha + [0] = \alpha$. The verification of associativity is also straightforward. The definition of multiplication needs some care, and we first define the additive inverse.

Definition 1.4.5. For each α in $\mathbb{R}$ define
$$-\alpha = \{y \in \mathbb{Q} : -y - r \notin \alpha \text{ for some } r \in \mathbb{Q}, r > 0\}.$$
◇

For example,
$$-[2] = \{y : -y - r \notin [2]\} = \{y : -y - r \geq 2\} = \{y : y \leq -2 - r\}$$
$$= \{y : y < -2\} = [-2],$$

as expected. Now we show that this indeed defines the additive inverse. A similar argument shows that for all $r \in \mathbb{Q}$,
$$-[r] = [-r].$$

Lemma 1.4.6. *For each α in $\mathbb{R}$, $\alpha + (-\alpha) = [0]$.*

Proof. Let $s \in \alpha + (-\alpha)$. Then $s = p + q$ where $p \in \alpha$ and $q \in -\alpha$. Since $-q - r \notin \alpha$ for some $r > 0$, then $-q \notin \alpha$. This means that $p < -q$, or $p + q < 0$, so $s \in [0]$. The converse is left to the reader. □

We are not quite ready to define the product, as this is defined in parts, and needs the notion of a positive cut. Thus we first need to define an order.

Definition 1.4.7. Define the **positive reals** $\mathbb{R}^+$ to consist of all cuts α in $\mathbb{R}$ so that some element of α is a positive rational number. Write $\alpha > [0]$ if $\alpha \in \mathbb{R}^+$, and $\alpha > \beta$ if $\alpha - \beta > 0$. Also, we write $\alpha \geq \beta$ if $\alpha > \beta$ or $\alpha = \beta$, and $\alpha < \beta$ if $\beta > \alpha$.
◇

We need to show that $\mathbb{R}^+$ is a positive set in the field $\mathbb{R}$. We start with the trichotomy property. For the other property of a positive set we need the product of two elements in $\mathbb{R}$ in Definition 1.4.10.

1.4. Construction of the Real Numbers

Lemma 1.4.8. *Let α be a cut. Then exactly one of the following holds: $\alpha = [0]$, or $\alpha > [0]$, or $-\alpha > [0]$.*

Proof. Let α be a cut. If $\alpha = [0]$, then $\alpha \notin \mathbb{R}^+$ since it does not contain any positive elements, and $-\alpha = -[0] = \alpha$ is similarly not in $\mathbb{R}^+$. Suppose now that $\alpha \neq [0]$. If α contains a positive rational number, then $\alpha \in \mathbb{R}^+$. If not, every element of α is negative, and as α is not $[0]$, there exists $q < 0$ such that $q \notin \alpha$. We claim that $-q \in -\alpha$. In fact $r = -q > 0$, and $q - r = 2q \notin \alpha$ since $q < 2q$ and q is not in α. Thus $-q$ is in $-\alpha$. Finally, suppose $\alpha \in \mathbb{R}^+$. If $-\alpha$ is also in $\mathbb{R}^+$, then there exists $p \in -\alpha$ such that $p > 0$. But for all $r > 0$, we have that $-p - r < 0$, so $-p - r \in \alpha$, a contradiction to the choice of p. $\square$

Question 1.4.9. Let α be a cut. Show that if $\alpha > [0]$, then $-\alpha < [0]$.

We now define the product of two cuts.

Definition 1.4.10. Let α and β be two cuts. First assume that α and β are in $\mathbb{R}^+$. Then define
$$\alpha \cdot \beta = \{r \in \mathbb{Q} : r \leq pq \text{ for some } p \in \alpha, p > 0, \text{ and } q \in \beta, q > 0\}.$$
When $-\alpha$ and $-\beta$ are both in $\mathbb{R}^+$, we define $\alpha \cdot \beta = (-\alpha) \cdot (-\beta)$. Other cases are defined similarly: when α is positive and $-\beta$ is positive, we define $\alpha \cdot \beta = -(\alpha \cdot (-\beta))$, and when $-\alpha$ is positive and β is positive, we define $\alpha \cdot \beta = -((-\alpha) \cdot \beta)$. Finally, for any α and β, $\alpha \cdot [0]$ and $[0] \cdot \beta$ are defined to be $[0]$. $\diamond$

One has to check of course that this product is a cut.

Lemma 1.4.11. *For each α in $\mathbb{R}$, $\alpha \cdot [1] = \alpha$.*

Proof. Let $p \in \alpha \cdot [1]$. Then $p \leq rq$ for $r \in \alpha$ and $0 < q < 1$. Hence $p \leq r$. As r is in α, then $p \in \alpha$. Conversely, let $p \in \alpha$. Then there is $p' \in \alpha$ with $p < p'$, and we may assume $p' \neq 0$. Hence $p = p' \cdot \frac{p}{p'}$. As $p/p' \in [1]$, then $p \in \alpha \cdot [1]$. $\square$

Now the reader is invited to complete the proof of the following proposition.

Proposition 1.4.12. *The set of Dedekind cuts $\mathbb{R}$ with the operations $+$ and $\cdot$ that have been defined form a field.*

Lemma 1.4.13. *The set $\mathbb{R}^+$ is a positive set and makes $\mathbb{R}$ into an ordered field.*

Proof. We show that $\mathbb{R}^+$ satisfies the properties of Definition 1.1.8. Let α and β be in $\mathbb{R}^+$. Then there exists $q_1 \in \alpha, q_2 \in \beta$ such that $q_1 > 0$ and $q_2 > 0$. Thus $q_1 + q_2 > 0$, and since $q_1 + q_2 \in \alpha + \beta$, it follows that $\alpha + \beta$ is in $\mathbb{R}^+$. For the product, we similarly have that $pq > 0$ and $pq \in \alpha \cdot \beta$, so $\alpha \cdot \beta \in \mathbb{R}^+$.

The trichotomy property is in Lemma 1.4.8. $\square$

The following properties give a useful alternative characterization of order.

Question 1.4.14. Let α and β be two cuts in $\mathbb{R}$.

(a) Show that for all $r \in \mathbb{Q}$, $r \in \alpha$ if and only if $[r] \leq \alpha$.
(b) Show that $\beta \leq \alpha$ if and only if $\beta \subseteq \alpha$.

We now prove that $\mathbb{R}$ is order complete.

Proposition 1.4.15. *The set $\mathbb{R}$ with the order $<$ satisfies the supremum property.*

Proof. Let A be a nonempty subset of $\mathbb{R}$ that is bounded above with respect to the order $<$. Hence A is a set of cuts. To construct its supremum in $\mathbb{R}$, let α^* be union of all the cuts in A, i.e., $x \in \alpha^*$ if and only if $x \in \alpha$ for some $\alpha \in A$. We first verify α^* is a cut. It is nonempty as A must contain some cut. Let γ be an upper bound for A. As every cut in A is a subset of γ, it follows that $\alpha^* \subseteq \gamma$, so $(\alpha^*)^c$ is nonempty since it contains γ^c.

Now let $x \in \alpha^*$ and suppose $y < x$. As x must be in some cut $\alpha \in A$, then y is in α, so it is in α^*. For the last cut property let $x \in \alpha^*$. Then again x must be in some cut $\alpha \in A$. Hence, there is $z \in \alpha$ such that $x < z$, but this z must be in α^*. This completes the proof that α^* is a cut, so it is an element of $\mathbb{R}$.

It remains to show that α^* is the supremum of A. It is clearly an upper bound since if α is in A, then it must be a subset of α^*, so $\alpha < \alpha^*$. Now suppose γ is any other upper bound for A. Then every cut in A must be a subset of γ, so α^* is a subset of γ, or $\alpha^* \leq \gamma$. $\square$

Exercises: Construction of the Real Numbers

1.4.1 Prove that for all $\alpha, \beta \in \mathbb{R}$, $\alpha + \beta$ as defined is a cut and that $\alpha + \beta = \beta + \alpha$ and $\alpha + [0] = \alpha$.

1.4.2 Prove that for all $\alpha, \beta \in \mathbb{R}$, $\alpha \cdot \beta$ as defined is a cut and that $\alpha \cdot \beta = \beta \cdot \alpha$.

1.4.3 Complete the proof of Lemma 1.4.6.

1.4.4 Complete the proof of Proposition 1.4.12 that $\mathbb{R}$ is a field.

* 1.4.5 Let $\mathbb{R}$ denote the set of real numbers defined in this section. Let F be any complete ordered field. Prove that there exists a bijection $\phi : F \to \mathbb{R}$ such that for all $x, y \in F$, $\phi(x + y) = \phi(x) + \phi(y)$, $\phi(xy) = \phi(x)\phi(y)$ and if $x <_F y$, then $\phi(x) < \phi(y)$ (where $<_F$ is the order in F).

1.4.6 (Dedekind cut property) Let F be an ordered field. Prove that F is order complete if and only if whenever A and B are two nonempty subsets of F such that $A \cup B = F$, and for all $a \in A$ and $b \in B$ one has $a < b$, then there exists an element $c \in F$ such that $a \in A$ is equivalent to $a \leq c$.

1.5. The Complex Numbers

The complex numbers form an important and interesting field that is not ordered.

Definition 1.5.1. The set of **complex numbers** $\mathbb{C}$ consists of all ordered pairs of real numbers (a, b), where $a, b \in \mathbb{R}$. A real number a is identified with $(a, 0) \in \mathbb{C}$. $\Diamond$

In this way the real numbers can be seen as a subset of the complex numbers. As a set $\mathbb{C}$ is the same as $\mathbb{R} \times \mathbb{R}$ (or $\mathbb{R}^2$) but the notation we now introduce emphasizes

1.5. The Complex Numbers

the fact that we endow $\mathbb{C}$ with two operations that make it a field. Rather than writing a complex number $z \in \mathbb{C}$ as $z = (a, b)$, we write
$$z = a + bi.$$

Definition 1.5.2. We call $i = 0 + 1 \cdot i$ the **imaginary unit** and it corresponds to $(0, 1)$. The **real part** of a complex number $z = a + bi$ is $\mathfrak{Re}(z) = a$, and its **imaginary part** is $\mathfrak{Im}(z) = b$. ◇

Both the real and imaginary parts of a complex number are real numbers. Hence, a complex number z is real if and only if $\mathfrak{Im}(z) = 0$.

Definition 1.5.3. We define two operations on $\mathbb{C}$ called **complex addition** and **complex multiplication**. They are defined by
$$(a + bi) + (c + di) = (a + c) + (b + d)i,$$
$$(a + bi) \cdot (c + di) = (ac - bd) + (ad + bc)i.$$
◇

For example, $i \cdot i = -1$, or $i^2 = -1$.

Theorem 1.5.4. *The set of complex numbers $\mathbb{C}$, with complex addition, additive inverse 0, and complex multiplication, with unit 1, forms a field.*

Proof. The fact that the complex addition is commutative follows immediately from the commutativity of addition of real numbers. Associativity of addition is similar and is left to the reader to verify. It is also clear that $(a + bi) + 0 = a + bi$.

We show that every nonzero element $z = a + bi$ of $\mathbb{C}$ has a multiplicative inverse. Let
$$w = \frac{a - bi}{a^2 + b^2}.$$
As $z \neq 0$, $a^2 + b^2 \neq 0$, so w is defined. We calculate
$$z \cdot w = (a + bi) \cdot \frac{a - bi}{a^2 + b^2} = \frac{a^2 - abi + abi - bi^2}{a^2 + b^2} = 1.$$
The remaining properties are left as an exercise. □

Definition 1.5.5. Given a number $z = a + bi \in \mathbb{C}$, define its **complex conjugate** by
$$\bar{z} = a - bi$$
and its **modulus** by
$$|z| = \sqrt{a^2 + b^2}.$$
It follows that
$$z \cdot \bar{z} = a^2 + b^2 = |z|^2.$$
◇

We have the following properties.

Lemma 1.5.6. *For complex numbers z and w, the following hold.*

(1) $\overline{z + w} = \bar{z} + \bar{w}$
(2) $\overline{zw} = \bar{z}\bar{w}$
(3) $\bar{\bar{z}} = z$
(4) $|z + w| \leq |z| + |w|$
(5) $|zw| = |z||w|$

Proof. Let $z = a + bi$ and $w = c + di$. For part (1), we compute

$$\overline{z+w} = \overline{a+bi+c+di} = \overline{a+c+(b+d)i}$$
$$= a+c-(b+d)i$$
$$= a-bi+c-di = \overline{z}+\overline{w}.$$

For part (4), we first calculate

$$|z+w|^2 = (z+w)\overline{(z+w)}$$
$$= (z+w)(\overline{z}+\overline{w})$$
$$= z\overline{z} + z\overline{w} + w\overline{z} + w\overline{w}$$
$$= |z|^2 + z\overline{w} + w\overline{z} + |w|^2.$$

Now note that $\overline{(z\overline{w})} = w\overline{z}$, so by Exercise 1.5.8, $z\overline{w} + w\overline{z} = 2\mathfrak{Re}(z\overline{w})$ (one could just do this by direct computation). We will also need that $\mathfrak{Re}(z\overline{w}) \leq |z\overline{w}|$. Hence,

$$|z+w|^2 \leq |z|^2 + 2\mathfrak{Re}(z\overline{w}) + |w|^2$$
$$\leq |z|^2 + 2|z||w| + |w|^2$$
$$= (|z|+|w|)^2.$$

The remaining parts are left as exercises. $\square$

Exercises: The Complex Numbers

1.5.1 Prove that there is a bijection from $\mathbb{R}$ to $\mathbb{C}$.

1.5.2 Prove that the set $R = \{(a,0) : a \in \mathbb{R}\}$, with the operations it inherits from $\mathbb{C}$, is a field. Prove that the map $\phi : R \to \mathbb{R}$ defined by $\phi((a,0)) = a$ is a bijection that satisfies $\phi(z+w) = \phi(z) + \phi(w)$ and $\phi(z \cdot w) = \phi(z) \cdot \phi(w)$ for all $z, w \in R$.

1.5.3 Is the set $B = \{(0,b) : b \in \mathbb{R}\}$, with the operations it inherits from $\mathbb{C}$, a field?

1.5.4 Complete the proof of Theorem 1.5.4.

1.5.5 Prove that there is no order that can be defined on the field of complex numbers $\mathbb{C}$.

1.5.6 Complete the proof of Lemma 1.5.6.

1.5.7 Prove that for all complex numbers z and w,
$$\overline{\left(\frac{z}{w}\right)} = \frac{\bar{z}}{\bar{w}}.$$

1.5.8 Prove that for all complex numbers z and w,
$$\mathfrak{Re}(z) = \frac{z + \bar{z}}{2} \quad \text{and} \quad \mathfrak{Im}(z) = \frac{z - \bar{z}}{2}.$$

1.5.9 Prove that for all complex numbers z, $z = \bar{z}$ if and only if z is real.

1.5.10 Let $p(z) = a_n z^n + \cdots + a_0$ be a polynomial with real coefficients (i.e., $a_i \in \mathbb{R}$). Prove that if w is a root of p (i.e., $p(w) = 0$), then $\bar{w}$ is also a root of p. What if the coefficients are complex but not real?

Chapter 2

Sequences

We have already seen in Section 1.3.1 that a sequence with values in a set A is a function $a : \mathbb{N} \to A$. We write $a_1, a_2, \ldots$ instead of $a(1), a(2), \ldots$. We recall that a_n is called a term of the sequence, and we write the sequence as (a_n) or $(a_n)_{n \in \mathbb{N}}$. In this chapter we are interested in sequences with values in $\mathbb{R}$.

2.1. Limits of Sequences

2.1.1. Limit of a Sequence. Given a sequence (a_n) we shall mainly be interested in its eventual behavior, i.e., the behavior of a_n for n large. Informally, we can think of asking what we see if we plot the values of a sequence. For example, if $a_n = (-1)^n$, when we plot the values of this sequence we see the numbers 1 and -1. But consider the following sequence (b_n) defined by

$$b_n = \begin{cases} 3^n & \text{if } 1 \leq n \leq 100, \\ (-1)^n & \text{if } n > 100. \end{cases}$$

In the case of (b_n), if we were to plot the first 100 terms, we would see the numbers $3, 3^2, \ldots, 3^{100}$, but on reflection this is misleading as far as the eventual behavior of the sequence is concerned; "eventually" we see 1 and -1. We need to formalize the notion of "eventual behavior".

First we formalize what we mean by a term a_n being "close to or around" some number L.

Definition 2.1.1. Let L be a number in $\mathbb{R}$, and let $\varepsilon > 0$. (We think of ε as a "small" number.) We say that a_n is ε**-close to L** (see Figure 2.1) if

$$L - \varepsilon < a_n < L + \varepsilon, \text{ or equivalently}$$

$$|a_n - L| < \varepsilon.$$

We shall think of the sequence (a_n) as converging to a number L if for each prespecified degree of closeness ε, a_n is ε-close to L eventually. We are ready for the definition.

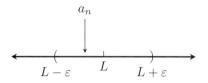

Figure 2.1. a_n is ε-close to L.

Definition 2.1.2. We define a sequence (a_n) to **converge** to a number $L \in \mathbb{R}$ if
for every $\varepsilon > 0$ there exists $N \in \mathbb{N}$ such that
$$n \geq N \text{ implies } |a_n - L| < \varepsilon.$$

This definition is important enough that we discuss an equivalent way to write it. First we note that the definition uses two *quantifiers*: *for all* and *there exists*.

We recall that the symbol $\forall$ denotes "for all" and $\exists$ stands for "there exists."

Then we can say that (a_n) converges to L if and only if

(2.1) $\qquad \forall\, \varepsilon > 0 \ \exists N \in \mathbb{N} \text{ such that } n \geq N \implies |a_n - L| < \varepsilon.$

We read this as "for all positive ε there exists a natural number N such that if $n \geq N$, then $|a_n - L| < \varepsilon$". We also think of this as "for all positive ε, a_n is ε-close to L for all large enough n." The notion of eventual behavior is captured by the existence of the number N (note that the required condition holds for all $n \geq N$).

Proposition 2.1.4 below shows that if (a_n) converges to number in $\mathbb{R}$, then this number is unique, so we can use the notation
$$\lim_{n \to \infty} a_n = L, \text{ or sometimes } a_n \to L,$$
when the sequence (a_n) converges to L.

The definition is flexible as the following shows.

Question 2.1.3. Show that $\lim_{n \to \infty} a_n = L$ if and only if

(2.2) $\qquad \forall\, \varepsilon > 0 \ \exists N \in \mathbb{N} \text{ such that } n \geq N \implies |a_n - L| \leq \varepsilon,$

and if and only if

(2.3) $\qquad \forall\, \varepsilon > 0 \ \exists N \in \mathbb{N} \text{ such that } n > N \implies |a_n - L| \leq \varepsilon,$

and if and only if

(2.4) $\qquad \forall\, \varepsilon > 0 \ \exists N \in \mathbb{N} \text{ such that } n > N \implies |a_n - L| < \varepsilon.$

There is an equivalent way to write (2.1), which is as

(2.5) $\qquad \forall\, \varepsilon > 0 \ \exists\, N \in \mathbb{N} \text{ such that } |a_n - L| < \varepsilon \text{ whenever } n \geq N.$

Proposition 2.1.4. *Limits are unique: if (a_n) converges to L_1 and (a_n) converges to L_2, then $L_1 = L_2$.*

Proof. We start by thinking of arbitrary approximations and do this by considering a number $\varepsilon > 0$. As the sequence converges to L_1, if we take a positive number, say $\varepsilon/2$, we know there exists $N_1 \in \mathbb{N}$ so that if $n \geq N_1$, then $|a_n - L_1| < \varepsilon/2$. Similarly, using that the sequence converges to L_2, we know there is a number $N_2 \in \mathbb{N}$ so that if $n \geq N_2$, then $|a_n - L_2| < \varepsilon/2$. To get both inequalities to hold, we choose an integer N such that $N = \max\{N_1, N_2\}$. Now we use a technique that will be applied often and that could be called "addition of zero". To compare L_1 with L_2 we add and subtract a_n so that we can compare L_1 with a_n and a_n with L_2. Then,

$$n \geq N \implies |L_1 - L_2| = |L_1 - a_n + a_n - L_2|$$
$$\leq |L_1 - a_n| + |a_n - L_2|$$
$$< \frac{\varepsilon}{2} + \frac{\varepsilon}{2} = \varepsilon.$$

As ε is arbitrary, $|L_1 - L_2| = 0$ (see Exercise 1.1.6), so $L_1 = L_2$. □

We note that since $|a_n - L| = |-1||a_n - L| = |(-a_n) - (-L)|$, it follows from the definition that $\lim_{n \to \infty} a_n = L$ implies $\lim_{n \to \infty} -a_n = -L$.

To further understand the definition of limit we explore its negation:

(a_n) does not converge to L if and only if

(2.6) $\qquad \exists \, \varepsilon > 0 \ \forall \, N \in \mathbb{N}, \text{ for some } n \geq N, |a_n - L| \geq \varepsilon.$

As we shall see, it is typically easier to show that a sequence does not converge to L than to show that it converges to L. This is because to show that the sequence does not converge, we need to find *one* $\varepsilon > 0$ so that for any given N, there is one $n \geq N$ for which the condition fails (or equivalently, one $\varepsilon > 0$ so that for infinitely many n the condition fails). On the other hand, to show convergence, we need to verify the condition *for each* $\varepsilon > 0$ and *for each* $n \geq N$.

For example, to show that $a_n = (-1)^n$ does not converge to 1, we can let $\varepsilon = \frac{1}{2}$ (in fact, any ε with $0 < \varepsilon < 2$ would work) and observe that for each $N \in \mathbb{N}$ we can choose $n > N$ odd so that $|a_n - 1| = |-1 - 1| = 2 > \varepsilon$. Similarly, to show it does not converge to -1, we choose $n > N$ even. Now, if we wanted to show it does not converge at all, we need to consider that there could be another number $L \in \mathbb{R}$ to which the sequence could converge, so to prove that it does not converge to any number, we still have to consider the cases when $L \neq 1, -1$. These are the easier cases: take any $\varepsilon > 0$ such that

$$0 < \varepsilon < \min\{|L - 1|, |L - (-1)|\}.$$

Then, by choice, ε is smaller than both the distance from L to 1 and the distance from L to -1, so $|a_n - L| > \varepsilon$ for all $n \in \mathbb{N}$. (It is possible to write a more elegant proof that would take care of all three cases at the same time.)

Question 2.1.5. Suppose that $\lim_{n \to \infty} a_n = L$ and $M \neq L$. Show that (a_n) cannot converge to M. Give two proofs, one using Proposition 2.1.4 and another using the definition.

Now we write what it means for a sequence not to converge at all.

A sequence (a_n) does not converge (we also say it **diverges**) if and only if

(2.7) $\qquad \forall \, L \in \mathbb{R} \ \exists \, \varepsilon > 0 \ \forall \, N \in \mathbb{N}, \text{ for some } n \geq N, |a_n - L| \geq \varepsilon.$

Here ε may depend on L.

We make a remark that will become clearer after reading Subsection 2.1.3. It may happen that the limit of a sequence is ∞; in this case we will say it diverges to infinity. This is a special case of what we define here as diverging. We do not think of this as "converging to infinity". When the limit does not exist as a real number nor as an infinite number, we can say that it diverges and does not diverge to ∞ or to $-\infty$.

2.1.2. Examples and Basic Properties. The following examples show the existence of limits for some well-known sequences.

Example 2.1.6. We show that $\lim_{n\to\infty} \frac{1}{\sqrt{n}} = 0$. We start the proof by letting $\varepsilon > 0$. We need to construct $N \in \mathbb{N}$ to satisfy the conditions of convergence; in general, N will depend on ε.

As this is the first example in this kind of proof, we discuss a method we use to find N. We call this the *development*; it is not part of the proof when it is finally written. For this we start at the end:
$$\left| \frac{1}{\sqrt{n}} - 0 \right| < \varepsilon.$$
Then we simplify it and find it is equivalent to
$$\frac{1}{\sqrt{n}} < \varepsilon \quad \text{or} \quad n > \frac{1}{\varepsilon^2}.$$
So we see that we should take N to satisfy $N > \frac{1}{\varepsilon^2}$. This is the end of the development, and we are ready to start the proof.

Let $\varepsilon > 0$. Choose N to be any natural number such that
$$N > \frac{1}{\varepsilon^2}.$$
(This can be done by the Archimedean property.) Hence, if $n \geq N$, $n > 1/\varepsilon^2$, or $\sqrt{n} > 1/\varepsilon$, then
$$\left| \frac{1}{\sqrt{n}} - 0 \right| = \frac{1}{\sqrt{n}} < \varepsilon,$$
completing the proof.

Example 2.1.7. We show that for $|a| < 1$, $\lim_{n\to\infty} a^n = 0$. We may assume that $a \neq 0$. Let $\varepsilon > 0$.

For the development we again start at the end:
$$|a^n| < \varepsilon.$$
It is not clear how to proceed here, as we cannot use logarithms since they will not be developed until Chapter 6. We wish to look for a simpler upper bound for $|a^n|$, i.e., an expression, typically depending on n that we temporarily call x_n such that $|a^n| < x_n$. If we find conditions so that $x_n < \varepsilon$, it will follow that $|a^n| < \varepsilon$.

Now we think of the inequalities we have proved, and remember Bernoulli's inequality (see Exercise 1.1.11), which gives that
$$(1+\alpha)^n \geq 1 + n\alpha \text{ for } \alpha > -1.$$

2.1. Limits of Sequences

As $|a| < 1$, this suggests we write
$$|a| = \frac{1}{1+\alpha} \text{ for some } \alpha > 0.$$
Then
$$|a|^n = \frac{1}{(1+\alpha)^n} \leq \frac{1}{n\alpha + 1} < \frac{1}{n\alpha}.$$
So $\frac{1}{n\alpha}$ is the simpler expression that bounds $|a^n|$ and that we had called x_n. From this we see that it suffices to choose N so that
$$\frac{1}{N\alpha} < \varepsilon.$$
We are ready to start the proof.

Let $\varepsilon > 0$. Choose N to be any integer such that
$$N > \frac{1}{\varepsilon\alpha}.$$
Hence, if $n \geq N$, from the calculations we have completed and that one would include in writing the proof, we see that $1/(n\alpha) < \varepsilon$. So $|a^n - 0| < \varepsilon$, completing the proof.

Example 2.1.8. This shows a property about the supremum that will be useful later. Let $A \subseteq \mathbb{R}$ be a bounded nonempty set, and let
$$\beta = \sup A.$$
We show that there exists a sequence (a_n) in A such that
$$\lim_{n \to \infty} a_n = \beta.$$
(A similar property holds for the infimum.) By Lemma 1.2.10, for each $n \in \mathbb{N}$, there exists an element in A that we may denote by a_n such that $a_n > \beta - 1/n$. Clearly, $a_n \leq \beta$, so $0 \leq \beta - a_n < 1/n$. We show this means that (a_n) converges to β. Let $\varepsilon > 0$. Choose $N \in \mathbb{N}$ such that $1/N < \varepsilon$. If $n \geq N$, then
$$|\beta - a_n| = \beta - a_n < 1/n \leq 1/N < \varepsilon.$$

Example 2.1.9. Let (a_n) be a sequence, and let $x \in \mathbb{R}$. Suppose that $a_n < x$ for all $n \in \mathbb{N}$ and some $x \in \mathbb{R}$. We show that if $\lim_{n\to\infty} a_n = L$, then $L \leq x$. In fact, if $L > x$, set $\varepsilon = L - x > 0$. Then there exists N such that if $n \geq N$, then $|a_n - L| < L - x$. So $-(L-x) < a_n - L$, or $x < a_n$, a contradiction. We note that we cannot hope to have a strict inequality. For example, we know $-\frac{1}{n} < 0$ for all $n \in \mathbb{N}$, but $\lim_{n\to\infty} -\frac{1}{n} = 0$.

Proposition 2.1.10. *A convergent sequence is bounded.*

Proof. Let (a_n) be a sequence and suppose it converges to L. We will show that there is a number $A \in \mathbb{R}$, $A > 0$, such that
$$|a_n| \leq A \text{ for all } n \in \mathbb{N}.$$
The idea is that eventually the sequence is going to be near L, say 1-close to L, and there are only finitely many terms that are not 1-close to L. Let $\varepsilon = 1$. There exists $N \in \mathbb{N}$ so that $|a_n - L| < 1$ for all $n \geq N$. Then
$$L - 1 < a_n < L + 1 \text{ for } n \geq N.$$

Using that $-|L| \leq L \leq |L|$, we have
$$-|L| - 1 < a_n < |L| + 1,$$
$$-(|L| + 1) < a_n < (|L| + 1).$$
So
$$|a_n| < (|L| + 1) \text{ for } n \geq N.$$
There are only finitely many terms that may not satisfy this inequality, namely the terms $a_1, a_2, \ldots, a_{N-1}$. So we let
$$A = \max\{|L| + 1, |a_1|, \ldots, |a_{N-1}|\}.$$
Then $|a_n| \leq A$ for all $n \in \mathbb{N}$. $\square$

Proposition 2.1.11 lists the basic properties of limits of a sequence.

Proposition 2.1.11. *Let (a_n) and (b_n) be sequences whose limits exist in $\mathbb{R}$, and write*
$$\lim_{n \to \infty} a_n = L_1 \text{ and } \lim_{n \to \infty} b_n = L_2, \text{ where } L_1, L_2 \in \mathbb{R}.$$
Then the following hold:

(1) *For each $c \in \mathbb{R}$, the sequence $(c \cdot a_n)$ converges and*
$$\lim_{n \to \infty} c \cdot a_n = c \cdot \lim_{n \to \infty} a_n = cL_1.$$

(2) *The sequence $(a_n + b_n)$ converges and*
$$\lim_{n \to \infty} (a_n + b_n) = \lim_{n \to \infty} a_n + \lim_{n \to \infty} b_n = L_1 + L_2.$$

(3) *The sequence $(a_n \cdot b_n)$ converges and*
$$\lim_{n \to \infty} a_n \cdot b_n = \lim_{n \to \infty} a_n \cdot \lim_{n \to \infty} b_n = L_1 L_2.$$

(4) *If $b_n \neq 0$ for all $n \in \mathbb{N}$ and $L_2 \neq 0$, then the sequence $\left(\frac{a_n}{b_n}\right)$ converges and*
$$\lim_{n \to \infty} \frac{a_n}{b_n} = \frac{\lim_{n \to \infty} a_n}{\lim_{n \to \infty} b_n} = \frac{L_1}{L_2}.$$

(5) *Squeeze property: Let (c_n) be a sequence in $\mathbb{R}$ such that*
$$a_n \leq c_n \leq b_n \text{ for all } n \in \mathbb{N}.$$
If $L_1 = L_2$, then (c_n) also converges to $L_1 = L_2$.

Proof.

(1) The conclusion is clear when $c = 0$, so we may assume that $c \neq 0$. Write $L = \lim_{n \to \infty} a_n$. As $\varepsilon / |c| > 0$, there exists $N \in \mathbb{N}$ such that
$$(2.8) \qquad n \geq N \implies |a_n - L| < \frac{\varepsilon}{|c|}.$$
Then $|ca_n - cL| < \varepsilon$, showing that (ca_n) converges to cL.

Remark. When building the proof, we start with the condition that we need to have satisfied: $|ca_n - cL| < \varepsilon$ or $|a_n - L| < \varepsilon / |c|$. From this we know how to choose N in (2.8).

2.1. Limits of Sequences

(2) Let $\varepsilon > 0$. Then there exists $N_1 \in \mathbb{N}$ so that
$$n \geq N_1 \implies |a_n - L_1| < \frac{\varepsilon}{2},$$
and there exists $N_2 \in \mathbb{N}$ so that
$$n \geq N_2 \implies |b_n - L_2| < \frac{\varepsilon}{2}.$$
Let $N = \max\{N_1, N_2\}$. Hence, if $n \geq N$,
$$|(a_n + b_n) - (L_1 + L_2)| \leq |a_n - L_1| + |b_n - L_2| < \frac{\varepsilon}{2} + \frac{\varepsilon}{2} = \varepsilon.$$
This means that $\lim_{n \to \infty} a_n + b_n = L_1 + L_2$.

(3) Before we start the proof, we give the idea for how to find the upper bounds. This process is what we have called the development and proceeds backward. We need to show that $|a_n b_n - L_1 L_2| < \varepsilon$ for n sufficiently large. We first use the technique of adding zero so we can compare a_n with L_1 and b_n with L_2. Thus we note that
$$|a_n b_n - L_1 L_2| = |a_n b_n - a_n L_2 + a_n L_2 - L_1 L_2|.$$
Then we have
$$|a_n b_n - L_1 L_2| \leq |a_n||b_n - L_2| + |a_n - L_1||L_2|.$$
As we have two terms in the sum, we want to bound each term above by $\frac{\varepsilon}{2}$. For the first term, we observe that $|a_n|$ can be assumed to be bounded by some constant A (which we may assume positive), so we need to choose n sufficiently large so that $|b_n - L_2| < \varepsilon/2A$. For the second term we want to choose n sufficiently large so that $|a_n - L_1||L_2| < \varepsilon/2$. Now we observe that $|L_2|$ could be 0. We could divide the proof into two cases (one when $|L_2| = 0$ and the other when $|L_2| > 0$), or keep it as one case by considering the stronger condition that $|a_n - L_1|(|L_2| + 1) < \varepsilon/2$ (which works even when $|L_2| = 0$). We decide to not break the proof into two cases, so we need to choose n large enough such that
$$|a_n - L_1| < \frac{\varepsilon}{2(|L_2| + 1)}.$$

Now having finished the development, we are ready to start the proof. By Proposition 2.1.10, there exists a constant $A > 0$ such that $|a_n| \leq A$ for all $n \in \mathbb{N}$. For $\varepsilon > 0$ there exists $N_1 \in \mathbb{N}$ such that
$$|b_n - L_2| < \frac{\varepsilon}{2A}$$
for all $n \geq N_1$. Similarly, for $\frac{\varepsilon}{2(|L_2|+1)} > 0$ there exists $N_2 \in \mathbb{N}$ such that
$$|a_n - L_1| < \frac{\varepsilon}{2(|L_2| + 1)}$$
for all $n \geq N_2$.

Finally, choose $N = \max\{N_1, N_2\}$. If $n \geq N$, then
$$|a_n b_n - L_1 L_2| = |(a_n b_n - a_n L_2) + (a_n L_2 - L_1 L_2)|$$
$$\leq |a_n||b_n - L_2| + |a_n - L_1||L_2|$$
$$< A \frac{\varepsilon}{2A} + \frac{\varepsilon}{2(|L_2| + 1)} |L_2| < \varepsilon,$$
completing the proof.

(4) First we show that $(1/b_n)$ converges and
$$\lim_{n\to\infty} \frac{1}{b_n} = \frac{1}{L_2}.$$

A similar argument to the one used to show that a convergent sequence is bounded can be used to show that as $L_2 \neq 0$, the sequence $(|b_n|)$ is bounded below by a positive constant, i.e., there exists B in $\mathbb{R}$, with $B > 0$, such that $|b_n| > B$ for all $n \in \mathbb{N}$.

Let $\varepsilon > 0$ and choose N so that $|b_n - L_2| < \varepsilon B|L_2|$ for $n \geq N$. Then
$$\left|\frac{1}{b_n} - \frac{1}{L_2}\right| = \frac{|L_2 - b_n|}{|b_n||L_2|}$$
$$\leq \frac{|L_2 - b_n|}{B|L_2|}$$
$$< \varepsilon.$$

The proof of part (4) is completed by applying part (3) to the sequences (a_n) and $(1/b_n)$.

(5) Write $L = L_1 = L_2$. Let $\varepsilon > 0$. Then there exists $N_1, N_2 \in \mathbb{N}$ so that $|a_n - L| < \varepsilon$ for $n \geq N_1$ and $|b_n - L| < \varepsilon$ for $n \geq N_2$. Let $N = \max\{N_1, N_2\}$. Hence, if $n \geq N$,
$$L - \varepsilon < a_n < L + \varepsilon \quad \text{and} \quad L - \varepsilon < b_n < L + \varepsilon.$$

Then $L - \varepsilon < a_n \leq c_n \leq b_n < L + \varepsilon$. Therefore $|c_n - L| < \varepsilon$ for all $n \geq N$, completing the proof. $\square$

Question 2.1.12. Let (a_n), (b_n), and (c_n) be sequences such that $a_n + b_n = c_n$ for all $n \in \mathbb{N}$. Give examples where (c_n) converges but (a_n) and (b_n) do not. Show that if (b_n) and (c_n) converge, then (a_n) converges and its limit is $\lim_{n\to\infty} c_n - \lim_{n\to\infty} b_n$.

Question 2.1.13. Let (a_n) and (b_n) be sequences such that $\lim_{n\to\infty} a_n = 0$ and (b_n) is bounded. Prove that $\lim_{n\to\infty} a_n b_n = 0$. Give examples to show that if (b_n) is not bounded, the limit may not exist.

Question 2.1.14. Let (a_n) and (b_n) be two sequences such that $\lim_{n\to\infty} a_n = L \in \mathbb{R}$ and $\lim_{n\to\infty} b_n = L$. Suppose sequences (c_n) and (d_n) satisfy $c_n \geq 0, d_n \geq 0$, and $c_n + d_n = 1$ for all $n \in \mathbb{N}$. Show that $\lim_{n\to\infty} a_n c_n + b_n d_n = L$.

Example 2.1.15.

(a) Prove that the following limit exists, and find its value:
$$\lim_{n\to\infty} \frac{5n^3 - 3n^2 + 7}{2n^3 + 5n + 3}.$$

After dividing top and bottom by n^3, we see
$$\frac{5n^3 - 3n^2 + 7}{2n^3 + 5n + 3} = \frac{5 - 3/n + 7/n^3}{2 + 5/n^2 + 3/n^3}.$$

2.1. Limits of Sequences

We have seen that $\lim_{n\to\infty} 1/n = 0$, so by Proposition 2.1.11,
$$\lim_{n\to\infty} 2 + 5/n^2 + 3/n^3 = 2 \neq 0.$$
Similarly,
$$\lim_{n\to\infty} 5 - 3/n + 7/n^3 = 5.$$
Therefore,
$$\lim_{n\to\infty} \frac{5 - 3/n + 7/n^3}{2 + 5/n^2 + 3/n^3} = \frac{5}{2}.$$

(b) Find $\lim_{n\to\infty} \sqrt{n} - \sqrt{n-1}$, if it exists.

We note that
$$\sqrt{n} - \sqrt{n-1} = \sqrt{n} - \sqrt{n-1} \cdot \frac{\sqrt{n} + \sqrt{n-1}}{\sqrt{n} + \sqrt{n-1}}$$
$$= \frac{n - (n-1)}{\sqrt{n} + \sqrt{n-1}} = \frac{1}{\sqrt{n} + \sqrt{n-1}}.$$

For $n \in \mathbb{N}$,
$$0 < \frac{1}{\sqrt{n} + \sqrt{n-1}} < \frac{1}{\sqrt{n}}.$$
So, by the squeeze property,
$$\lim_{n\to\infty} \frac{1}{\sqrt{n} + \sqrt{n-1}} = 0.$$
It follows that
$$\lim_{n\to\infty} \sqrt{n} - \sqrt{n-1} = 0.$$

(c) Let $a > 1$. We prove that
$$\lim_{n\to\infty} \sqrt[n]{a} = 1.$$
Let $x_n = a^{1/n}$ for $n \in \mathbb{N}$. (Note again that we cannot assume knowledge of logarithms, which would simplify the argument.) Write $y_n = x_n - 1$. By Bernoulli's inequality,
$$a = x_n^n = (y_n + 1)^n \geq ny_n + 1.$$
Then $0 < y_n \leq (a-1)/n$. So, by the squeeze property, $\lim_{n\to\infty} y_n = 0$ and $\lim_{n\to\infty} x_n = 1$.

2.1.3. Infinite Limits. We consider the case when the value of the limit is $\pm\infty$. It was implicit in our definition of a convergent sequence that the limit of the sequence is a real number (thus finite). There are two reasons a sequence may fail to converge: it may oscillate (such as the sequence $((-1)^n)$ or the sequence $((-1)^n n)$), or it may diverge to ∞ or $-\infty$ (such as the sequences (n) or $(-n)$, respectively). One might be tempted to say in this case that the sequence converges to ∞, but we prefer to say it diverges to ∞. Now we define what it means for the sequence to diverge to ∞.

Definition 2.1.16. A sequence (a_n) is said to **diverge to** ∞, and we write
$$\lim_{n \to \infty} a_n = \infty,$$
if for each $\alpha \in \mathbb{R}$ there is an integer $N \in \mathbb{N}$ so that $a_n > \alpha$ for all $n \geq N$. (We think of α as large, and there is no loss of generality in assuming that α is positive.)

Equivalently, $\lim_{n \to \infty} a_n = \infty$ if and only if

(2.9) $\qquad \forall\, \alpha > 0 \; \exists\, N \in \mathbb{N} \text{ such that } n \geq N \implies a_n > \alpha.$

Note that since a_n is approaching ∞, we cannot compare the difference between a_n and ∞, as in the case when a_n approaches a real number L, so instead we say that $a_n > \alpha$ for any given α and for all sufficiently large n.

Question 2.1.17. The constant α can be required to be in $\mathbb{N}$. Show that
$$\lim_{n \to \infty} a_n = \infty$$
if and only if for each $\alpha \in \mathbb{N}$ there is an integer $N \in \mathbb{N}$ so that $a_n > \alpha$ for all $n \geq N$.

Question 2.1.18. Let (a_n) be a sequence. Show that if
$$\lim_{n \to \infty} a_n = \infty,$$
then (a_n) does not converge, and that the converse does not hold in general.

Example 2.1.19. Let $a_n = n^2$. We prove that $\lim_{n \to \infty} n^2 = \infty$. Let $\alpha > 0$. Choose $N > \sqrt{\alpha}$. If $n \geq N$, then $n > \sqrt{\alpha}$ or $n^2 > \alpha$. Therefore, n^2 diverges to ∞ as $n \to \infty$.

Definition 2.1.20. We define a sequence (a_n) to **diverge to** $-\infty$ and write
$$\lim_{n \to \infty} a_n = -\infty,$$
if for each $\beta \in \mathbb{R}$ there is an integer $N \in \mathbb{N}$ so that $a_n < \beta$ for all $n \geq N$. (We think of β as negatively large, and there is no loss of generality in assuming that β is negative.)

Equivalently, $\lim_{n \to \infty} a_n = -\infty$ if and only if

(2.10) $\qquad \forall\, \beta < 0 \; \exists\, N \in \mathbb{N} \text{ such that } n \geq N \implies a_n < \beta.$

We conclude the section with some properties of these limits.

Example 2.1.21. Let (a_n) and (b_n) be sequences. Suppose that
$$\lim_{n \to \infty} a_n = L \in \mathbb{R} \quad \text{and} \quad \lim_{n \to \infty} b_n = \infty.$$

(a) Assume that $L > 0$. We show that $\lim_{n \to \infty} a_n b_n = \infty$. Let $\alpha > 0$. For $\varepsilon = L/2$ there exists $N_1 \in \mathbb{N}$ so that
$$|a_n - L| < \frac{L}{2} \text{ for all } n \geq N_1.$$
Hence,
$$\frac{L}{2} < a_n < \frac{3L}{2} \text{ for all } n \geq N_1.$$

2.1. Limits of Sequences

As (b_n) diverges to ∞, there exists $N_2 \in \mathbb{N}$ so that
$$b_n > \frac{2\alpha}{L} \text{ for all } n \geq N_2.$$
Therefore, if $n \geq N = \max\{N_1, N_2\}$,
$$a_n b_n > \frac{L}{2} \frac{2\alpha}{L} = \alpha.$$
It follows that $\lim_{n \to \infty} a_n b_n = \infty$.

(b) Assume that $L \neq 0$. We show that $\lim_{n \to \infty} \frac{a_n}{b_n} = 0$. First suppose that $L > 0$. Let $\varepsilon > 0$. As before, for $\varepsilon_1 = L/2$ there exists $N_1 \in \mathbb{N}$ so that
$$0 < \frac{L}{2} < a_n < \frac{3L}{2} \text{ for all } n \geq N_1.$$
Also there exists $N_2 \in \mathbb{N}$ so that
$$b_n > \frac{3L}{2\varepsilon} > 0 \text{ for all } n \geq N_2.$$
Equivalently,
$$0 < \frac{1}{b_n} < \frac{2\varepsilon}{3L} \text{ for all } n \geq N_2.$$
Therefore if $n \geq N = \max\{N_1, N_2\}$,
$$0 < \frac{a_n}{b_n} < \frac{3L}{2} \cdot \frac{2\varepsilon}{3L} = \varepsilon.$$
It follows that $\lim_{n \to \infty} a_n/b_n = 0$. When $L < 0$, it suffices to apply the result we just proved to $(-a_n)$ and (b_n), as $\lim_{n \to \infty} -a_n = -L > 0$.

Definition 2.1.22. A sequence (x_n) is said to be **increasing** if $x_n \leq x_{n+1}$ for all $n \in \mathbb{N}$; sometimes we may call it **monotone increasing**. We say (x_n) is **strictly increasing** if $x_n < x_{n+1}$ for all $n \in \mathbb{N}$. Similarly, a **decreasing** sequence (or **monotone decreasing**) (x_n) satisfies $x_n \geq x_{n+1}$ for all $n \in \mathbb{N}$. The sequence (x_n) is **strictly decreasing** if $x_n > x_{n+1}$ for all $n \in \mathbb{N}$. A sequence is **monotonic** if it is either monotone decreasing or monotone increasing.

For example, $x_n = 2$ and $x_n = 1 - 1/n$ are both increasing sequences. (Some authors say that a sequence is *nondecreasing* when we say it is increasing, and similarly for *nonincreasing*.)

Example 2.1.23. Let (a_n) be a monotone increasing sequence. We prove that
$$\lim_{n \to \infty} a_n = \sup\{a_n : n \in \mathbb{N}\},$$
showing a connection between the limit and the supremum in the case of monotone increasing sequences. First we consider the case when the sequence is not bounded above. Then from the definition of supremum, we have that $\sup\{a_n : n \in \mathbb{N}\} = \infty$. Now let $\alpha \in \mathbb{R}$; since the sequence does not have an upper bound, there exits $N \in \mathbb{N}$ such that $a_N > \alpha$. As the sequence is monotone increasing, for all $n \geq N$, $a_n \geq a_N$, so $a_n > \alpha$, which implies $\lim_{n \to \infty} a_n = \infty$.

Now suppose that the sequence is bounded above. We can let
$$L = \sup\{a_n : n \in \mathbb{N}\}.$$

If $\varepsilon > 0$, by the property of the supremum, there exists $N \in \mathbb{N}$ such that $a_N > L - \varepsilon$. As the sequence is monotone increasing, for all $n \geq N$, $a_n \geq a_N > L - \varepsilon$. Using that L is the supremum of the sequence, we get that $a_n \leq L < L + \varepsilon$ for all n. Thus, for all $n \geq N$, $L - \varepsilon < a_n < L + \varepsilon$, or $|a_n - L| < \varepsilon$. Therefore, $\lim_{n \to \infty} a_n = L$.

2.1.4. Subsequences

Definition 2.1.24. A **subsequence** of a sequence (a_n) is a sequence of the form (a_{n_k}), where (n_k) is a *strictly increasing* sequence of elements of $\mathbb{N}$.

For example, if $a_n = 2n, n \in \mathbb{N}$, and $n_k = 2k, k \in \mathbb{N}$, then a_{n_k} is the sequence of multiples of 4. But, for example, the sequence (b_n) defined by $b_n = 2$ for $n \in \mathbb{N}$ is not a subsequence of (a_n) even though every term of (b_n) is a term of (a_n). A property that is useful to keep in mind is that if (a_{n_k}) is a subsequence, then $n_k \geq k$ for all $k \in \mathbb{N}$. We note that a subsequence (a_{n_k}) is itself a sequence indexed by k. So to show that (a_{n_k}) converges to a number $L \in \mathbb{R}$, we need to show that for all $\varepsilon > 0$ there exists $N \in \mathbb{N}$ so that if $k > N$, then $|a_{n_k} - L| < \varepsilon$.

The following is a basic property of subsequences.

Lemma 2.1.25. *Let (a_n) be a sequence converging to $L \in \mathbb{R}$. Then every subsequence of (a_n) converges to L.*

Proof. Let (a_{n_k}) be a subsequence of (a_n). Let $\varepsilon > 0$. Since (a_n) converges to L, we know there exists $N \in \mathbb{N}$ such that $|a_n - L| < \varepsilon$ for all $n \geq N$. As (n_k) is strictly increasing, if $k \geq N$, then $n_k \geq N$. It follows that

$$|a_{n_k} - L| < \varepsilon \text{ for } k \geq N.$$

Therefore, (a_{n_k}) converges to L. $\square$

Example 2.1.26. We give another proof that for $|a| < 1$, $\lim_{n \to \infty} a^n = 0$. We may assume $a \neq 0$. As $|a| < 1$,

$$|a|^{n+1} < |a|^n$$

for all $n \in \mathbb{N}$. (This may be verified by the reader by induction.) It follows that the sequence $(|a|^n)$ is decreasing. Since it is bounded below by 0, by Theorem 2.2.1, it converges to a limit that we may denote by L. As the subsequence $(|a|^{n+1})$ converges to the same limit L,

$$L = \lim_{n \to \infty} |a|^{n+1} = \lim_{n \to \infty} |a||a|^n = |a| \lim_{n \to \infty} |a|^n = |a|L.$$

The fact that $0 < |a| < 1$ implies $L = 0$.

Exercises: Limits of Sequences

2.1.1 Prove that $\lim_{n \to \infty} |x_n| = 0$ if and only if $\lim_{n \to \infty} x_n = 0$.

2.1.2 Use the definition to show that $\lim_{n \to \infty} \frac{n}{n+1} = 1$.

2.1. Limits of Sequences

2.1.3 Let $r \in (0, \infty)$. Use the definition to show that $\lim_{n\to\infty} n^{-r} = 0$. (*Hint*: First consider r rational, then discuss informally the case when r is an arbitrary real number.)

2.1.4 Let (x_n), $x_n \geq 0$, be a sequence and suppose it converges to L. Show that $\lim_{n\to\infty} \sqrt{x_n} = \sqrt{L}$.

2.1.5 Show that $\lim_{n\to\infty} \sqrt[n]{n} = 1$.

2.1.6 Determine whether or not the following sequences are convergent and, if so, find their limit.
   (a) $\lim_{n\to\infty} (-1)^n \frac{5}{n}$
   (b) $\lim_{n\to\infty} \frac{5n^2 - 3n + 2}{n^2 - n}$
   (c) $\lim_{n\to\infty} \frac{n^2 + 5n + 2}{5n^3 + n}$
   (d) $\lim_{n\to\infty} (-1)^n n$
   (e) $\lim_{n\to\infty} \frac{1}{(-1)^n n + 2n}$
   (f) $\lim_{n\to\infty} \sqrt[3]{n+1} - \sqrt[3]{n}$

2.1.7 Let (b_n) be a monotone decreasing sequence. Prove that $\lim_{n\to\infty} b_n = \inf\{b_n : n \in \mathbb{N}\}$.

2.1.8 Show that a set $D \subseteq \mathbb{R}$ is dense if and only if for each $x \in \mathbb{R}$ there exists a sequence $(a_n), a_n \in D$, such that $a_n \to x$.

2.1.9 Let (x_n) be a sequence.
   (a) Prove that if $\lim_{n\to\infty} x_n = L \in \mathbb{R}$, then
   $$\lim_{n\to\infty} \frac{1}{n} \sum_{i=1}^n |x_i - L| = 0.$$
   (b) Is the converse true?

2.1.10 Let (x_n) be a bounded sequence.
   (a) Prove that if
   $$\lim_{n\to\infty} \frac{1}{n} \sum_{i=1}^n |x_i - L| = 0,$$
   then
   $$\lim_{n\to\infty} \frac{1}{n} \sum_{i=1}^n x_i = L.$$
   (b) Is the converse true?

2.1.11 Let $a > 1$. Prove that
$$\lim_{n\to\infty} a^n = \infty.$$
What if $a < 1$?

2.1.12 Let $0 < a < 1$ and $x_n = a^{1/n}$ for $n \in \mathbb{N}$. Prove that $\lim_{n\to\infty} x_n$ exists, and find its value.

2.1.13 Let (a_n) and (b_n) be sequences. Suppose that
$$\lim_{n\to\infty} a_n = L \quad \text{and} \quad \lim_{n\to\infty} b_n = \infty.$$
Assume that $L < 0$ and show that $\lim_{n\to\infty} a_n b_n = -\infty$.

2.1.14 Let $a > 1$ and $k \in \mathbb{N}$. Prove that $\lim_{n \to \infty} \frac{n^k}{a^n} = 0$. (*Hint*: Write $a = b + 1$ and use the binomial theorem to obtain an inequality for $(b+1)^n$.)

2.1.15 Let (a_n) be a sequence with $a_n > 0$. Suppose that $\lim_{n \to \infty} \frac{a_{n+1}}{a_n} = L$. Prove that if $L < 1$, then $\lim_{n \to \infty} a_n = 0$. Use this to give another proof of the limit in Exercise 2.1.14.

2.1.16 Discuss what happens if in Exercise 2.1.15, $L > 1$ and $L = 1$.

2.1.17 Let $k \in \mathbb{N}$. Prove that the following limit exists and find its value:
$$\lim_{n \to \infty} n^{\frac{1}{n^k}}.$$

2.1.18 Let $(a_n)_{n \in \mathbb{N}}$ be a sequence, and let $L \in \mathbb{R}$. Prove that if every subsequence $(a_{n_k})_{k \in \mathbb{N}}$ has a subsequence $(a_{n_{k_\ell}})_{\ell \in \mathbb{N}}$ that converges to L, then the sequence $(a_n)_{n \in \mathbb{N}}$ converges to L.

2.1.19 A sequence of complex numbers (z_n) is said to converge to $z \in \mathbb{C}$ if for all $\varepsilon > 0$ there exists $N \in \mathbb{N}$ such that $n \geq N$ implies $|z - z_n| < \varepsilon$. Prove that a sequence (z_n), where $z_n = a_n + b_n i$, with $a_n, b_n \in \mathbb{R}$, converges to $z = a + bi$ if and only if (a_n) converges to a and (b_n) converges to b.

2.2. Three Consequences of Order Completeness

This section introduces three theorems that have many applications and that are consequences of the completeness property of the real numbers. We study them in this chapter as they use the notion of a sequence in a fundamental way. The three theorems are the *monotone sequence theorem*, the *nested intervals theorem*, and the *Bolzano–Weierstrass theorem*. (In fact, we should count four since we also have the *strong nested intervals theorem*.)

We studied the order completeness property in Section 1.2. We saw that the field of rational numbers was not large enough for analysis, and that in addition to the algebraic and order properties of $\mathbb{Q}$ we need an additional property, called order completeness. We note that there are several other properties that are equivalent to order completeness, and, as we see below, one of them is the monotone sequence property.

It is possible to start this book with this chapter, and in that case the reader can use the monotone sequence property, instead of the supremum property, as a characterization of order completeness. The statement of this property is in Subsection 2.2.1.

For the reader who has covered Chapter 1, the next subsection uses the supremum property, to prove the monotone sequence theorem, and then we also show how the monotone sequence theorem can be used to prove the supremum property.

2.2.1. The Monotone Sequence Theorem.
Recall that a sequence (x_n) is monotone increasing if $x_n \leq x_{n+1}$ for all $n \in \mathbb{N}$ and monotone decreasing if $x_n \geq x_{n+1}$ for all $n \in \mathbb{N}$.

Theorem 2.2.1 (Monotone sequence theorem). *Every bounded monotone sequence in $\mathbb{R}$ converges.*

2.2. Three Consequences of Order Completeness

Proof. Let (a_n) be a bounded monotone sequence. First, suppose that (a_n) is monotone increasing. Then let $b = \sup\{a_n : n \in \mathbb{N}\}$. We claim that $\lim_{n \to \infty} a_n = b$. Let $\varepsilon > 0$. By Lemma 1.2.10, there exists $N \in \mathbb{N}$ with $b - \varepsilon < a_N$. But $a_n \geq a_N$ for all $n \geq N$. So, for all $n \geq N$, $|b - a_n| < \varepsilon$. The case when (a_n) is monotone decreasing is left as an exercise. $\square$

Example 2.2.2. Let $a_1 = \sqrt{2}$, and for $n \in \mathbb{N} \setminus \{1\}$, $a_n = \sqrt{2 + a_{n-1}}$. Prove that the limit $\lim_{n \to \infty} a_n$ exists, and find its value. First we show by induction that $a_n \leq 2$ for all $n \in \mathbb{N}$. It is clearly true for $n = 1$ and suppose it holds for n. Then $a_{n+1} \leq \sqrt{2 + 2} = 2$. It follows that the sequence is monotone increasing since $a_n^2 = 2 + a_{n-1} \geq 2a_{n-1} \geq a_{n-1}^2$. Now the monotone sequence theorem implies that the sequence converges. Let L be the limit. Since $\lim a_{n-1} = L$ and $a_n^2 = 2 + a_{n-1}$, we obtain that $L^2 = 2 + L$, so $L = 2$.

Remark 2.2.3. The *monotone sequence property* for an ordered field F states that every bounded monotone sequence in F converges. The monotone sequence theorem says that $\mathbb{R}$ satisfies the monotone sequence property. Now we give an argument to show that if an ordered field F satisfies the monotone sequence property, then it satisfies the supremum property. This justifies that fact that one can take the monotone sequence property as a version of order completeness of the real line, which is what readers starting with this chapter may do.

Let A be a bounded nonempty set in an ordered field F that satisfies the monotone sequence property. Then we can choose $a_1 \in A$ and b_1 an upper bound of A. Let $m_1 = (a_1 + b_1)/2$ be the midpoint. If m_1 is an upper bound of A, set $a_2 = a_1$ and $b_2 = m_1$. Otherwise, there must exist an element of A that is greater than m_1, which we denote by a_2, and we set $b_2 = b_1$. Continuing in this way by induction, we obtain two bounded monotonic sequences (a_n) and (b_n) such that $a_n \in A$, b_n is an upper bound of A, and $|a_{n+1} - b_{n+1}| \leq (b_1 - a_1)/2^n$. The sequence (a_n) is monotone increasing and bounded, so it must have a limit and we can set $p = \lim_{n \to \infty} a_n$. Note also that $\lim_{n \to \infty} b_n = p$. Now we show that p is the supremum of A. If p were not an upper bound of A there would exist $a \in A$ with $p < a$. But $a \leq b_n$ for all n, and this contradicts that the b_n converge to p. Now suppose there is an upper bound q of A with $q < p$. Then it must be that for all n, $a_n \leq q$. This implies $p \leq q$, a contradiction. It follows that the field F satisfies the supremum property.

2.2.2. The Nested Intervals Theorem

Definition 2.2.4. A sequence of intervals (I_n) in $\mathbb{R}$ is said to be **nested** if
$$I_{n+1} \subseteq I_n \text{ for all } n \in \mathbb{N}.$$
We may also write this as
$$I_1 \supseteq I_2 \supseteq I_3 \supseteq \cdots.$$

For example, the sequences of intervals $I_n = (0, 1/n)$ and $J_n = [1 - 1/n, 1 + 1/n]$ are nested, while the intervals $K_n = [n, n+1]$ are not nested.

Theorem 2.2.5 (Nested intervals theorem). *Let $I_n = [a_n, b_n]$ be a nested sequence of closed bounded intervals in $\mathbb{R}$. Then they have at least one point $p \in \mathbb{R}$ in*

common:
$$p \in \bigcap_{n=1}^{\infty} I_n.$$

Proof. As the intervals are nested, we have
$$a_n \leq a_{n+1} \leq b_{n+1} \leq b_n \text{ for all } n \in \mathbb{N}.$$

In particular, the sequence (a_n) is bounded above. In fact, we note that it is bounded by any b_k. For certainly $a_1 \leq a_2 \leq \cdots \leq a_k \leq b_k$, and if $n > k$, $a_n \leq b_n \leq b_k$.

The sequence (a_n) is monotone and bounded, so we can set
$$p = \lim_{n \to \infty} a_n.$$

We have, $p \geq a_n$ for all $n \in \mathbb{N}$. We claim that $p \leq b_n$ for all $n \in \mathbb{N}$. This follows from the fact that each b_n is an upper bound of $\{a_n : n \in \mathbb{N}\}$. It follows that $p \in [a_n, b_n]$ for all $n \in \mathbb{N}$. $\square$

The following theorem adds a hypotheses in the nested intervals theorem. If in addition to the usual hypotheses we assume that the length of the intervals decreases to 0, then there is only one point in the intersection. Sometimes this additional condition is stated as part of the nested intervals theorem.

Theorem 2.2.6 (Strong nested intervals theorem). *Let $I_n = [a_n, b_n]$ be a nested sequence of closed bounded intervals in $\mathbb{R}$ such that $\lim_{n \to \infty} |I_n| = 0$. Then*
$$\bigcap_{n=1}^{\infty} I_n = \{p\}.$$

Proof. By Theorem 2.2.5 we know there is a point p in $\bigcap_{n=1}^{\infty} I_n$. Suppose there exists $q \in \bigcap_{n=1}^{\infty} I_n$ with $q \neq p$. Suppose $q > p$. (In fact, this must be the case by the definition of p in the proof of Theorem 2.2.5, but a similar contradiction can be obtained if $p > q$.) It follows that $a_n \leq p < q \leq b_n$ for all $n \in \mathbb{N}$. Then the interval $[p, q]$ is a subset of I_n for all $n \in \mathbb{N}$. But as the lengths of the intervals I_n converge to 0, there exists $k \in \mathbb{N}$ such that $|I_k| < q - p$, contradicting that $[p, q] \subseteq I_k$. Therefore, $q = p$, and the intersection of the intervals contains only one element. $\square$

In the hypothesis of Theorem 2.2.5, both properties of being closed and bounded are necessary. For example, the intervals in the sequence $I_n = (0, 1/n)$ are nested and bounded but not closed and $\bigcap_{n=1}^{\infty} I_n = \varnothing$. Also, $J_n = [n, \infty)$ is a nested sequence but the intervals are not bounded and $\bigcap_{n=1}^{\infty} I_n = \varnothing$. In addition, this theorem does not hold for the rational numbers. For example, let (a_n) be an increasing sequence of rational numbers converging to $\sqrt{2}$, and let (b_n) be a decreasing sequence of rational numbers converging to $\sqrt{2}$. Then the intervals $[a_n, b_n]$ intersect to $\sqrt{2}$, i.e., $\bigcap_{n \in \mathbb{N}} [a_n, b_n] = \{\sqrt{2}\}$, which is not in $\mathbb{Q}$.

For the following example we will need the notation of dyadic intervals.

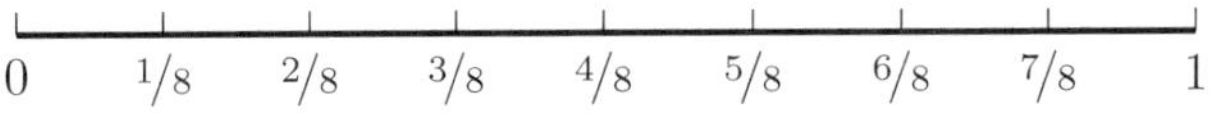

2.2. Three Consequences of Order Completeness

```
|----|----|----|----|----|----|----|----|
0   1/8  2/8  3/8  4/8  5/8  6/8  7/8   1
```

Figure 2.2. Endpoints of the dyadic intervals of rank 3 in $[0,1]$.

Definition 2.2.7. A **dyadic rational** is a number of the form
$$\frac{k}{2^n} \text{ where } k \in \mathbb{Z}, \text{ and } n \in \mathbb{N}.$$

We shall be mainly interested in the dyadic rationals that are in the unit interval $[0,1]$. The dyadic rationals in $[0,1]$ are obtained when we subdivide $[0,1]$ in half and continue subdividing each of the new subintervals in half. So we obtain $1/2, 1/4, 3/4, \ldots$, etc.

Definition 2.2.8. The (closed) **dyadic intervals** (in $[0,1]$) of **rank** n are the intervals of the form
$$\left[\frac{k}{2^n}, \frac{k+1}{2^n}\right] \text{ for } k = 0, \ldots, 2^n - 1, n \in \mathbb{N}.$$
We denote this dyadic interval by $D(n, k)$.

Each dyadic interval of rank n has two dyadic subintervals of rank $n+1$, namely $D(n+1, k)$ and $D(n+1, k+1)$. Figure 2.2 shows the endpoints of the dyadic intervals of rank 3 in $[0,1]$.

Example 2.2.9 (Binary fractional representation). Let α be an infinite sequence (string) of 0's and 1's (i.e., $\alpha \in \{0,1\}^{\mathbb{N}}$). We associate with α a unique point $x = \rho(\alpha) \in [0,1]$, thereby defining a function
$$\rho : \{0,1\}^{\mathbb{N}} \to [0,1].$$
To define ρ, we first construct a nested sequence of closed dyadic intervals (I_n). The idea is to progressively subdivide $[0,1]$ in halves, and choose the left half when we see a 0 and the right half when we have a 1. To start, if $\alpha_1 = 0$, let $I_1 = [0, 1/2]$, and if $\alpha_1 = 1$, let $I_1 = [1/2, 1]$. If $\alpha_2 = 0$, let I_2 be the closed dyadic interval that is to the left of the midpoint of I_1, and if $\alpha_2 = 1$, let I_2 be the closed dyadic interval that is to the right of the midpoint of I_1.

For the inductive step assume I_n has been defined. If $\alpha_{n+1} = 0$, let I_{n+1} be the closed dyadic interval that is to the left of the midpoint of I_n, and if $\alpha_{n+1} = 1$, let I_{n+1} be the closed dyadic interval that to the right of the midpoint of I_n. It is clear that (I_n) is a nested sequence of closed bounded intervals with $\lim_{n\to\infty} |I_n| = \lim_{n\to\infty} 1/2^n = 0$. Then by the strong nested intervals theorem there is a unique point x in $\bigcap_{n=1}^{\infty} I_n$. Then set $\rho(\alpha) = x$. For example, if $\alpha = 01\bar{1}$ (where $\bar{1}$ denotes that 1 is repeated infinitely often), then the point associated with this α is $1/2$. We note that ρ is surjective but not injective (Exercise 2.2.7).

We now describe the reverse process: given a point $x \in [0,1]$, we show how to associate a sequence of 0's and 1's. We will define a function
$$d : [0,1] \to \{0,1\}^{\mathbb{N}}.$$
For each $x \in [0,1], d(x)$ is a sequence of 0's and 1's; to simplify notation we will denote the ith digit of this sequence by $d_i(x)$. First for $x = 1 \in [0,1]$, set

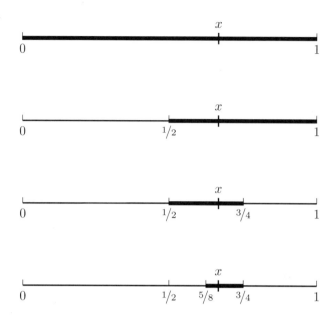

Figure 2.3. Sequence of dyadic intervals of increasing rank associated with $x = 2/3$.

$d(1) = 1\bar{1}\cdots$. Consider now subintervals of $[0,1)$ that are left-closed, right-open dyadic subintervals: $[k/2^n, (k+1)/2^n)$. We consider subintervals that are open at the right so that there is only one choice in the case when x is a binary rational. To find $d_1(x)$, we examine if x is in the left half (then $d_1(x) = 0$) or right half (then $d_1(x) = 1$) of $[0,1)$. (When $x = 1/2$, for example, since we are using subintervals that are open at the right endpoint, we choose the right subinterval, so $d_1(1/2) = 1$.) In general

$$(2.11) \qquad d_n(x) = \begin{cases} 0 & \text{if } x \in [k/2^n, (k+1)/2^n) \text{ and } k \text{ is even,} \\ 1 & \text{if } x \in [k/2^n, (k+1)/2^n) \text{ and } k \text{ is odd.} \end{cases}$$

Figure 2.3 shows the sequence of dyadic intervals in the unit interval of increasing rank associated with the point $x = 2/3$. For example, for the point $x = 2/3$, since $2/3$ is in $[1/2, 1)$ the first digit of $d(x)$ is 1. Next we find it is in $[1/2, 3/4)$ so the second digit is 0. Thus we find $d(x) = 10\overline{1001}\cdots$. We have chosen left-closed, right-open dyadic subintervals so that there is no ambiguity when defining $d(x)$ when x is a dyadic rational, so, for example, $d(1/2) = 1\bar{0}$. This map is injective but not surjective (Exercise 2.2.8).

Corollary 2.2.10. *The unit interval $[0, 1]$ is uncountable; hence, $\mathbb{R}$ is uncountable.*

Proof. We give two proofs. The first one is based on Cantor's diagonalization argument and uses the fact that we can identify points in $[0, 1]$ with infinite sequences of 0's and 1's. It can be shown that the sets $[0, 1]$ and $\{0, 1\}^{\mathbb{N}}$ are in a bijective correspondence, but this does not follow immediately from the existence of the function ρ in Example 2.2.9, since there are some numbers that have two binary representations. The numbers that have two representations are precisely the dyadic rationals. So let D be the set of dyadic rational numbers that are contained in $[0,1]$, and let E be the subset of $\{0,1\}^{\mathbb{N}}$ consisting of the sequences that end in either all 0's or in all 1's. Exercise 2.2.9 shows that there is a bijection from $[0,1] \setminus D$ to $\{0,1\}^{\mathbb{N}} \setminus E$. As the set E is countable and by Cantor's theorem (Theorem 1.3.16) the set $\{0,1\}^{\mathbb{N}}$ is uncountable, it follows that $\{0,1\}^{\mathbb{N}} \setminus E$ is uncountable. Then Proposition 1.3.4 implies that $[0,1]$ is uncountable. Therefore, $\mathbb{R}$ is uncountable.

The second proof is also due to Cantor. If $[0,1]$ were countable, there would exist a sequence (a_n) such that for each $x \in [0,1]$, $x = a_n$ for some $n \in \mathbb{N}$. Now let (a_n) be an arbitrary sequence in $[0,1]$. We will construct a point $p \in [0,1]$ such that $p \neq a_n$ for all $n \in \mathbb{N}$. This will show that $[0,1]$ cannot be countable. To construct this point, first subdivide $[0,1]$ into three equal-length closed subintervals. Then there exists at least one of these closed subintervals that does not contain a_1; call this subinterval I_1. (If we were to subdivide $[0,1]$ into only two subintervals and a_1 were $1/2$, then we would not be guaranteed that we could choose a subinterval not containing a_1.) Subdivide I_1 into three equal-length subintervals, and let I_2 be one of the subintervals that does not contain a_2. For the inductive step assume I_n has been defined. Subdivide I_n into three equal-length subintervals, and let I_{n+1} be one of the subintervals that does not contain a_{n+1}. This process generates a nested sequence of closed subintervals (I_n). By the nested intervals theorem there exists a point p in their intersection. By construction, p must be different from a_n for all $n \in \mathbb{N}$. $\square$

2.2.3. The Bolzano–Weierstrass Theorem. Now we state the third theorem of this section; it uses the notion of a subsequence.

Theorem 2.2.11 (Bolzano and Weierstrass). *Every bounded sequence of real numbers has a convergent subsequence.*

Proof. Let (x_n) be a bounded sequence of real numbers. Then there is a positive real number a such that $x_n \in [-a,a]$ for all $n \in \mathbb{N}$. We start with an outline of the proof. First we construct a nested sequence of closed intervals I_k in $[-a,a]$ with lengths decreasing to 0. By the strong nested intervals theorem there is a unique point p in their intersection. For each $k \in \mathbb{N}$, we choose $n_k \in \mathbb{N}$ so that n_k is the smallest integer such that x_{n_k} is in I_k. Then we show that the subsequence $(x_{n_k})_k$ must converge to p.

To make sure we can choose x_{n_k} in I_k for all $k \geq 1$, we must carefully choose the intervals I_k. For this we note that one of the following (or both) must be true: either there exist infinitely many integers i so that $x_i \in [-a,0]$ or there exist infinitely many integers i so that $x_i \in [0,a]$. Let I_1 be the subinterval containing infinitely many terms of the sequence, and let x_{n_1} be the first such term in I_1. (If there are infinitely many terms in both subintervals, just choose the left subinterval.) We know that the length $|I_1|$ is a.

For the inductive step suppose that a subinterval I_k, of length $a/2^{k-1}$, has been constructed such that it contains infinitely many terms of the sequence (x_n), and term x_{n_k} has been chosen in I_k. Then either there are infinitely many i so that x_i is in the closed subinterval that is the first half of I_k or there are infinitely many i so that x_i is in the closed subinterval that is the second half of I_k (or both). Let I_{k+1} be the closed subinterval with infinitely many terms (it has length $a/2^k$), and let n_{k+1} be the smallest integer greater than n_k such that $x_{n_{k+1}}$ is in I_{k+1}. This process generates a nested sequence of closed bounded intervals I_k, with $\lim_{n\to\infty} |I_k| = 0$, and a subsequence (x_{n_k}) with $x_{n_k} \in I_k$. The strong nested intervals theorem implies that $\bigcap_{k=1}^{\infty} I_k = \{p\}$ for some $p \in \mathbb{R}$.

We conclude by showing that $\lim_{k\to\infty} x_{n_k} = p$. Let $\varepsilon > 0$, and choose $K \in \mathbb{N}$ so that $|I_K| < \varepsilon$. Then for all $\ell \geq K$, p and x_{n_ℓ} are in the same interval I_K, which means $|p - x_{n_\ell}| < \varepsilon$. □

The following proposition is of independent interest, and it also can be used to give another proof of Theorem 2.2.11 (Exercise 2.2.12).

Proposition 2.2.12. *Every sequence of real numbers contains a subsequence that is either monotone decreasing or monotone increasing.*

Proof. We start with a special case. Suppose that x_n converges to a point $p \in \mathbb{R}$. Then either there are infinitely many n's such that x_n is in $(p-1, p)$, or infinitely many n's such that x_n is in $(p, p+1)$. Suppose there are infinitely many in $(p, p+1)$. Then one can expect to choose a monotone decreasing subsequence in $(p, p+1)$ converging to p. Now we describe how to choose such a subsequence but will do so for the general case of an arbitrary sequence.

Given a sequence (x_n), say that a term x_k is a *decreasing peak* if $x_k \geq x_m$ for all $m > k$. It is a *peak* in the sense that any term whose index is greater than k (we will say it comes after x_k or after k) is not greater than x_k; also, later peaks have the same value or decrease. Thus we can think of the peaks decreasing to some number or to $-\infty$. It follows that any sequence of decreasing peaks must be monotone decreasing. If there are infinitely many decreasing peaks, then they give a monotone decreasing subsequence, and we are done.

Now suppose that there are finitely many decreasing peaks, and let k be the largest integer so that x_k is a decreasing peak. We claim that then there must exist a strictly increasing subsequence. We know that x_k must be greater than or equal to everything that follows, but as it is the last peak there must exist a later term in the sequence that is not a peak, so there exist $n_1 > k$ and $n_2 > n_1$ such that $x_{n_1} < x_{n_2}$. But if $x_{n_2} \geq x_m$ for all $m > n_2$, then x_{n_2} would be a peak. As this cannot happen, there must exist $n_3 > n_2$ such that $x_{n_2} < x_{n_3}$. Continuing this by induction generates a strictly increasing subsequence. □

2.2.4. Properties Equivalent to Order Completeness. We have discussed four very useful consequences of the order completeness property. One may ask if each of these properties is equivalent to order completeness. We have already seen that if we have an ordered field such that every monotone bounded sequence converges, then it follows that every nonempty bounded set has a supremum. Therefore, we can say that the monotone sequence property is equivalent to order completeness (the supremum property).

We will also consider the other properties and start by showing that if an ordered field satisfies the strong nested intervals property, then it satisfies the supremum property. We outline the proof and ask the reader to complete the details in Exercise 2.2.18. Let A be a nonempty set that is bounded above. Choose $a_1 \in A$,

2.2. Three Consequences of Order Completeness

and let b_1 be an upper bound for A. Let m_1 be the midpoint of $[a,b]$. If m_1 is an upper bound of A, let $a_2 = a_1$ and $b_2 = m_1$, else let $a_2 = m_1$ and $b_2 = b_1$. This generates a nested decreasing sequence of closed bounded intervals whose intersection is the supremum of A, showing that the supremum of A exists.

Next we show that if an ordered field satisfies the monotone sequence property, then it satisfies the strong nested intervals property. Again we give an outline and the details are left for Exercise 2.2.19. Let F be an ordered field, and let $[a_n, b_n]$ be a sequence of nested intervals whose length converges to 0. The sequence (a_n) is monotone increasing, so it converges to a point a, and the sequence (b_n) is monotone decreasing so it converges to a point b. One can show that $a = b$, and therefore there is a unique point in their intersection.

Finally, we show that the Bolzano–Weierstrass property implies the monotone sequence property. Let (a_n) be a monotone sequence. By the Bolzano–Weierstrass property, the sequence has a subsequence (a_{n_k}) that converges to a point p. Exercise 2.2.21 implies that the monotone sequence converges to p.

We have shown that in an ordered field the following properties are equivalent to the supremum property: strong nested intervals property, monotone sequence property, and the Bolzano–Weierstrass property. The nested intervals property by itself does not imply the supremum property, though this is harder to see; the interested reader may refer to the Bibliographical Notes.

2.2.5. The Geometric Series. We will cover infinite series in Chapter 7 where we also discuss convergence tests for series, but here we briefly introduce a special series called the geometric series, since we will need it in Section 3.5.

Definition 2.2.13. If $(a_i)_{i=1}^{\infty}$ is a sequence of real numbers, we can form the corresponding **sequence of partial sums** corresponding to (a_i), defined by

$$S_n = \sum_{i=1}^{n} a_i = a_1 + a_2 + \cdots + a_n.$$

If the sequence (S_n) converges to a number S, we say that the series $\sum_{i=1}^{\infty} a_i$ **converges** to S; otherwise, we say it **diverges**. The **geometric series** is the case where

$$a_i = r^{i-1} \text{ for some fixed real number } r.$$

As we see presently, the sequence of partial sums for the geometric series converges when $|r| < 1$.

We saw in Exercise 1.1.12 that

$$\sum_{i=1}^{n} r^{i-1} = \frac{1 - r^n}{1 - r}.$$

When $|r| < 1$, we have that $\lim_{n \to \infty} r^n = 0$. Therefore, we can calculate the limit of the partial sums

(2.12) $$\lim_{n \to \infty} \sum_{i=1}^{n} r^{i-1} = \lim_{n \to \infty} \frac{1 - r^n}{1 - r} = \frac{1}{1 - r}.$$

We will mainly work with the limit of the sequence of partial sums, and will not need to use the full infinite series notation until Chapter 7. The reader who is already familiar with series will note that we may also write $\sum_{i=0}^{\infty} r^i = \sum_{i=1}^{\infty} r^{i-1} = \frac{1}{1-r}$ for $-1 < r < 1$.

Example 2.2.14. We find $\lim_{n\to\infty} \sum_{i=2}^{n} \frac{1}{2^i}$. We note that

$$\sum_{i=2}^{n} \frac{1}{2^i} = \frac{1}{2^2} \sum_{i=1}^{n-1} \frac{1}{2^{i-1}} = \frac{1}{4}\left(\frac{1-(1/2)^{n-1}}{1-1/2}\right).$$

Hence,

$$\lim_{n\to\infty} \sum_{i=2}^{n} \frac{1}{2^i} = \lim_{n\to\infty} \frac{1}{4}\left(\frac{1-(1/2)^{n-1}}{1-1/2}\right) = \frac{1}{2}.$$

2.2.6. Limit Superior and Limit Inferior. This subsection uses the notion of supremum and infimum and generalizes the limit of a sequence. We have seen that if a sequence (a_n) converges to a point $p \in \mathbb{R}$, then every subsequence of (a_n) converges to p. When the limit does not exist, there will be subsequences converging to various numbers. Of these, two will be important: the "largest" one and the "smallest" one, in a sense to be made precise below.

Definition 2.2.15. Let $LP(a_n)$ consist of all points q, where q could be in $\mathbb{R}$ or be $+\infty$ or $-\infty$, such that there is a subsequence (a_{n_k}) of (a_n) that converges to q.

We think of $LP(a_n)$ as the set of points that are limits of subsequences of (a_n); such points are sometimes called *subsequential limit points*.

Definition 2.2.16. We define the **lim inf** and **lim sup** of a sequence (a_n) by

$$\liminf_{n\to\infty} a_n = \inf LP(a_n),$$
$$\limsup_{n\to\infty} a_n = \sup LP(a_n).$$

We first note that $LP(a_n)$ is always nonempty. If the sequence (a_n) is bounded above and below, then by the Bolzano–Weierstrass theorem (Theorem 2.2.11) it has a subsequence converging to a point in $\mathbb{R}$. If it is not bounded above, then it has a subsequence converging to ∞, and if not bounded below, it has a subsequence converging to $-\infty$. Next, note that if $LP(a_n)$ is bounded below, its infimum exists as a real number, and if it is not bounded below, its infimum is $-\infty$, so the lim inf of a sequence is always defined. A similar statement can be made about the lim sup.

It follows then from the definition that

$$\liminf_{n\to\infty} a_n \leq \limsup_{n\to\infty} a_n \quad \text{and}$$

$$\liminf_{n\to\infty} a_n = \limsup_{n\to\infty} a_n, \text{ if and only if } \lim_{n\to\infty} a_n \text{ exists in } \mathbb{R} \cup \{\pm\infty\},$$

where by existing in $\mathbb{R} \cup \{\pm\infty\}$ we mean that we allow the limit to possibly be infinite. For example, $\liminf_{n\to\infty} n^2 = \limsup_{n\to\infty} n^2 = \infty = \lim_{n\to\infty} n^2$. We may sometimes write $\liminf a_n$ and $\limsup a_n$ instead of $\liminf_{n\to\infty} a_n$ and $\limsup_{n\to\infty} a_n$, respectively.

2.2. Three Consequences of Order Completeness

While the limit of a sequence may not exist, the lim inf and the lim sup are always defined (with the understanding that we allow infinite values). For example, while the sequence $a_n = (-1)^n$ does not converge, we have that $\liminf (-1)^n = -1$ and $\limsup (-1)^n = 1$. Also, $\liminf (-1)^n n = -\infty$ and $\limsup (-1)^n n = \infty$.

We observe that there is always a subsequence of (a_n) converging to lim inf and another converging to lim sup (see Exercise 2.2.31).

The following proposition offers a useful characterization for the limit inferior and limit superior.

Proposition 2.2.17. *Let (a_n) be a sequence of real numbers. Then*

$$\limsup a_n = \inf\{b_k : k \in \mathbb{N}\}, \text{ where } b_k = \sup\{a_n : n \geq k\}$$

and

$$\liminf a_n = \sup\{c_k : k \in \mathbb{N}\}, \text{ where } c_k = \inf\{a_n : n \geq k\}.$$

Proof. We prove the statement for the case of lim sup; a similar proof works for lim inf and is left as an exercise. Set

$$\beta = \limsup a_n \quad \text{and} \quad \gamma = \inf\{b_k : k \in \mathbb{N}\}.$$

We first prove that $\beta \geq \gamma$. Observe first that from its definition the sequence (b_k) is decreasing:

$$b_1 \geq b_2 \geq b_3 \geq \cdots.$$

Therefore, we also have that $\gamma = \lim_{k \to \infty} b_k$. For each $\ell \in \mathbb{N}$ there exists $k_\ell \in \mathbb{N}$ so that $|b_{k_\ell} - \gamma| < \frac{1}{2\ell}$. Then for this b_{k_ℓ}, from the definition of supremum, we have an element that we may denote by a_{n_ℓ} such that $0 \leq b_{k_\ell} - a_{n_\ell} < \frac{1}{2\ell}$. By the triangle inequality,

$$|\gamma - a_{n_\ell}| \leq |\gamma - b_{k_\ell}| + |b_{k_\ell} - a_{n_\ell}|$$
$$< \frac{1}{2\ell} + \frac{1}{2\ell} = \frac{1}{\ell}.$$

This implies that the subsequence (a_{n_ℓ}) converges to γ. The definition of β implies that $\beta \geq \gamma$.

To show the other inequality, note that there is a subsequence (a_{n_i}) such that $\beta = \lim_{i \to \infty} a_{n_i}$. Also, $\gamma = \lim_{i \to \infty} b_{n_i}$ (since $\gamma = \lim b_k$). As $a_{n_i} \leq b_{n_i}$, we have that $\beta \leq \gamma$. $\square$

Exercises: Three Consequences of Order Completeness

2.2.1 Prove Theorem 2.2.1 in the case of monotone decreasing sequences.

2.2.2 Prove that there is a bijection from $[0,1]$ to $(0,1)$. Provide a proof without using the Cantor–Schröder–Bernstein theorem (see Theorem 0.5.18) and a second proof using this theorem.

2.2.3 Prove that there is a bijection from $[0,1]$ to $[0,1)$. Provide one proof without using the Cantor–Schröder–Bernstein theorem and another using this theorem.

2.2.4 Prove that there is a bijection from $[0,1]$ to $[0,1] \times [0,1]$. Provide one proof without using the Cantor–Schröder–Bernstein theorem and another using this theorem.

2.2.5 Prove that there is a bijection from $(0,1)$ to $\mathbb{R}$. Provide one proof without using the Cantor–Schröder–Bernstein theorem and another using this theorem.

2.2.6 Prove that for all real numbers x and y such that $x < y$ the set $\{z \in \mathbb{R} : x < z < y$ and z is irrational$\}$ is uncountable.

2.2.7 Prove that the map $\rho : \{0,1\}^{\mathbb{N}} \to [0,1]$ is surjective but not injective.

2.2.8 Prove that the map $d : [0,1] \to \{0,1\}^{\mathbb{N}}$ is injective but not surjective.

2.2.9 Let D be the set of dyadic rationals in $[0,1]$, and let E be the subset of $\{0,1\}^{\mathbb{N}}$ consisting of sequences that end in all 0's or all 1's. Prove that the map $d : [0,1] \setminus D \to \{0,1\}^{\mathbb{N}} \setminus E$ is a bijection.

2.2.10 Prove that there is a bijection from $[0,1]$ to $\{0,1\}^{\mathbb{N}}$. (*Hint*: Use Exercise 2.2.9.)

2.2.11 Generalize the definitions of the functions d and ρ from Example 2.2.9 to base 10.

2.2.12 Use Proposition 2.2.12 to give another proof of Theorem 2.2.11. (*Hint*: If there is an increasing subsequence, show that it must converge to its supremum.)

2.2.13 Prove that for each $x \in \mathbb{R}$ there exists a sequence of dyadic rationals (q_n) with $\lim_{n \to \infty} q_n = x$.

2.2.14 Let (x_n) be a sequence. Suppose the subsequence (x_{2n}) converges to $L \in \mathbb{R}$ and the subsequence (x_{2n+1}) converges to L. Prove that the sequence (x_n) converges to L.

2.2.15 *Newton's method*: Let $a > 0$, and define a sequence (x_n) by
$$x_{n+1} = \frac{1}{2}\left(x_n + \frac{a}{x_n}\right),$$
where x_1 is chosen to be any number greater than $\sqrt{a}$. Prove that
$$\lim_{n \to \infty} x_n = \sqrt{a}.$$
(*Hint*: Prove that the sequence (x_n) is monotone and bounded.)

2.2. Three Consequences of Order Completeness

* 2.2.16 Prove that $\lim_{n\to\infty} \left(1 + \frac{1}{n}\right)^n$ exists. (*Hint*: Show the sequence is monotone and bounded. Use the binomial theorem. This limit converges to Euler's constant e.)

* 2.2.17 Let a be a nonnegative real number. Let $x \in \mathbb{R}$, and let (r_n) be a sequence of rational numbers converging to x.
(a) Prove that $\lim_{n\to\infty} a^{r_n}$ exists and, furthermore, if (s_n) is any other sequence of rational numbers converging to x, then $\lim_{n\to\infty} a^{s_n}$ is equal to $\lim_{n\to\infty} a^{r_n}$. Argue that this defines a^x for every real number x. Show that this definition agrees with the definition in Exercise 1.2.12.
(b) Prove that $a^{-x} = 1/a^x$ and $a^{x+y} = a^x a^y$.

2.2.18 Prove that the strong nested intervals property implies the completeness property; i.e., show that if F is an ordered field where Theorem 2.2.6 holds, then every nonempty set that is bounded above has a supremum. (It is a harder problem and is not assigned here to find an ordered field that satisfies the nested intervals property but not the strong nested intervals property.)

2.2.19 Prove that the monotone sequence property implies the strong nested intervals property (see Subsection 2.2.4).

2.2.20 *The cut property*: Prove that the following property, called the cut property, is equivalent to the completeness property. An ordered field F is said to satisfy the cut property if whenever there exist disjoint sets A and B whose union is F and such that any element of A is less than any element of B (under the order of F), then there exists an element c in F such that if $x < c$, then $x \in A$, and if $x > c$, then $x \in B$.

2.2.21 Prove that if a monotone sequence has a subsequence that converges, then the monotone sequence converges. Argue why this proof holds not only in $\mathbb{R}$ by in any ordered field.

2.2.22 For $n \in \mathbb{N}$, let $a_n = \sum_{i=1}^{n} \frac{1}{i(i+1)}$. Prove that the limit $\lim_{n\to\infty} a_n$ exists, and find its value.

2.2.23 Let $|r| < 1$, and fix $k \in \mathbb{N}$. For $n \in \mathbb{N}, n \geq k$, let $s_n = \sum_{i=k}^{n} r^i$. Prove that the limit $\lim_{n\to\infty} s_n$ exists, and find its value.

2.2.24 Find $\lim_{n\to\infty} \sum_{i=3}^{n} \frac{2^i}{3^{i+1}}$.

2.2.25 Let (a_n) be a sequence. Prove that $\inf a_n \leq \liminf a_n \leq \limsup a_n \leq \sup a_n$.

* 2.2.26 Let (a_n) be a sequence. It is said to be *subadditive* if $a_{n+m} \leq a_n + a_m$ for all $m, n \in \mathbb{N}$. Prove that if (a_n) is a subadditive sequence, then $\lim \frac{a_n}{n}$ exists and equals $\inf a_n$.

2.2.27 A sequence of complex numbers (z_n) is bounded if there exists a constant $M \in \mathbb{R}$ such that $|z_n| \leq M$ for all $n \in \mathbb{N}$. Prove that if (z_n) is a sequence of complex numbers that is bounded, then it has a subsequence that converges to a complex number.

2.2.28 Make sense of following expression as a limit, and find its value.
$$1 + \cfrac{1}{2 + \cfrac{1}{2 + \cfrac{1}{2 + \cdots}}}$$

2.2.29 Make sense of following expression as a limit and find its value.
$$\sqrt{1 + \sqrt{1 + \sqrt{1 + \cdots}}}$$

2.2.30 Prove the statement for lim inf in Proposition 2.2.17.

2.2.31 Let (a_n) be a sequence. Prove that
$$\limsup a_n = \lim_{k \to \infty} a_{n_k},$$
$$\liminf a_n = \lim_{\ell \to \infty} a_{n_\ell},$$
for some subsequences (a_{n_k}) and (a_{n_ℓ}) of (a_n).

2.2.32 Let $(a_n), (b_n)$ be sequences. Prove that
$$\limsup(a_n + b_n) \leq \limsup a_n + \limsup b_n,$$
$$\liminf(a_n + b_n) \geq \liminf a_n + \liminf b_n.$$
Also show that strict inequalities are possible.

2.2.33 Let $(a_n), (b_n)$ be sequences. Prove that
$$\limsup(a_n \cdot b_n) \leq \limsup a_n \cdot \limsup b_n, \text{ provided } a_n \geq 0, b_n \geq 0.$$

2.2.34 Let $(a_n), (b_n)$ be sequences and suppose $\lim_{n \to \infty} b_n = b$. Prove that
$$\limsup(a_n + b_n) = \limsup a_n + b,$$
$$\limsup(a_n \cdot b_n) = \limsup a_n \cdot b, \text{ provided } a_n \geq 0, b_n \geq 0.$$

2.3. The Cauchy Property for Sequences

This section may be postponed to be read after Section 7.1. Augustin Louis Cauchy (1789–1857) introduced a method that is extremely useful for showing that a sequence converges when we do not know its limit. We know that a monotone sequence that is bounded converges, but when a sequence is not monotone and we do not have a candidate for a limit, we can still verify convergence of the sequence using the Cauchy property. If a sequence converges, we know that its terms must get and stay close to each other. Cauchy formulated the correct converse of this.

Definition 2.3.1. Let (a_n) be a sequence in $\mathbb{R}$. We say that (a_n) is a **Cauchy sequence** or that it has the **Cauchy property** if

for every $\varepsilon > 0$ there exists $N \in \mathbb{N}$ such that
$$|x_n - x_m| < \varepsilon \text{ whenever } n, m \geq N.$$

We first show the following lemma.

2.3. The Cauchy Property for Sequences

Lemma 2.3.2. *Let (a_n) be a sequence in $\mathbb{R}$. If (a_n) converges, then (a_n) is a Cauchy sequence.*

Proof. Let $L = \lim_{n \to \infty} a_n$. For $\varepsilon > 0$ there exists $N \in \mathbb{N}$ such that if $n \geq N$, then
$$|a_n - L| < \frac{\varepsilon}{2}.$$
Then for any $n, m \geq N$,
$$|a_n - a_m| = |a_n - L + L - a_m| \leq |a_n - L| + |L - a_m| < \frac{\varepsilon}{2} + \frac{\varepsilon}{2} = \varepsilon.$$
Therefore (a_n) is a Cauchy sequence. $\square$

The first property we show of Cauchy sequences is that they are bounded.

Lemma 2.3.3. *Let (a_n) be a Cauchy sequence in $\mathbb{R}$. Then (a_n) is bounded.*

Proof. Since (a_n) is a Cauchy sequence, for $\varepsilon = 1$ there exists a positive integer N so that after N, all terms are at most 1-apart. In particular, for $n > N$, a_n is at most 1-apart from a_N, implying that (a_n) is bounded. $\square$

The next theorem depends in a fundamental way on the nature of $\mathbb{R}$: we show that Cauchy sequences in $\mathbb{R}$ converge. It is interesting to note that a Cauchy sequence may not converge in $\mathbb{Q}$. The fact that Cauchy sequences converge in $\mathbb{R}$ makes the Cauchy property very important as we now have a property that can be used to show convergence of a sequence without knowledge of the limit of the sequence. We start with a technical lemma that uses the Cauchy property in a fundamental way.

Lemma 2.3.4. *Let (a_n) be a Cauchy sequence in $\mathbb{R}$. If (a_n) has a subsequence that converges to a point p in $\mathbb{R}$, then the sequence (a_n) converges to p.*

Proof. Let (a_{n_k}) be a subsequence converging to $p \in \mathbb{R}$. Let $\varepsilon > 0$. There exists $N_1 \in \mathbb{N}$ so that
$$|a_{n_k} - p| < \frac{\varepsilon}{2} \text{ for all } k \geq N_1.$$
Also, there exists $N_2 \in \mathbb{N}$ so that
$$|a_n - a_m| < \frac{\varepsilon}{2} \text{ for all } n, m \geq N_2.$$
Let $N = \max\{N_1, N_2\}$. If $n \geq N$, fixing $k > N$ (which implies $n_k > N$),
$$|a_n - p| = |a_n - a_{n_k} + a_{n_k} - p| \leq |a_n - a_{n_k}| + |a_{n_k} - p| < \frac{\varepsilon}{2} + \frac{\varepsilon}{2} = \varepsilon.$$
Therefore (a_n) converges to p. $\square$

Theorem 2.3.5. *Let (a_n) be a sequence in $\mathbb{R}$. Then (a_n) is a Cauchy sequence if and only if it converges (in $\mathbb{R}$).*

Proof. We have already shown that a convergent sequence is Cauchy. Assume now that (a_n) is a Cauchy sequence. By Lemma 2.3.3, (a_n) is bounded. Therefore, by the Bolzano–Weierstrass theorem, (a_n) has a subsequence (a_{n_k}) converging to a point p in $\mathbb{R}$. By Lemma 2.3.4, (a_n) must converge to p. $\square$

The following proposition for Cauchy sequences in $\mathbb{R}$ follows immediately from Theorem 2.3.5, but it can be shown in general.

Proposition 2.3.6. *Let (a_n) and (b_n) be Cauchy sequences. Then $(\alpha a_n + \beta b_n)$ is a Cauchy sequence for all $\alpha, \beta \in \mathbb{R}$. Also $(a_n b_n)$ is a Cauchy sequence.*

The following remark gives a useful technique for showing that a sequence is Cauchy.

Remark 2.3.7. Let (a_n) be a sequence in $\mathbb{R}$, and suppose there is a $k \in \mathbb{N}$ such that
$$|a_{n+1} - a_n| < \frac{1}{2^n}$$
for all $n \geq k$. Then (a_n) is a Cauchy sequence.

Let $\varepsilon > 0$, and choose $N \in \mathbb{N}$ such that $2^{-N+1} < \varepsilon$ and $N \geq k$. Then, for $n \geq N$ and $\ell > 1$,
$$|a_{n+\ell} - a_n| \leq |a_{n+\ell} - a_{n+\ell-1}| + |a_{n+\ell-1} - a_{n+\ell-2}| + \cdots + |a_{n+1} - a_n|$$
$$< \frac{1}{2^{n+\ell-1}} + \frac{1}{2^{n+\ell-2}} + \cdots + \frac{1}{2^n}$$
$$< \frac{1}{2^{n-1}} < \varepsilon.$$

Exercises: The Cauchy Property for Sequences

2.3.1 Give an example of a Cauchy sequence of elements of $\mathbb{Q}$ that does not converge in $\mathbb{Q}$.

2.3.2 Prove, without using Theorem 2.3.5, that every subsequence of a Cauchy sequence is Cauchy.

2.3.3 Find an example of a sequence (a_n) such that $\lim_{n\to\infty} |a_{n+1} - a_n| = 0$ but (a_n) is not Cauchy.

2.3.4 Let $a_n = \sum_{i=0}^{n} \frac{1}{i!}$ for $n \in \mathbb{N}$. Prove that $\lim_{n\to\infty} a_n$ exists. Give one proof by showing it is a Cauchy sequence and another by showing it is monotone and bounded. The limit defines **Euler's constant** e:
$$e = \lim_{n\to\infty} \sum_{i=0}^{n} \frac{1}{i!}.$$

2.3.5 Prove that
$$\lim_{n\to\infty} \sum_{i=0}^{n} \frac{1}{i!} = \lim_{n\to\infty} \left(1 + \frac{1}{n}\right)^n.$$
(Compare with Exercise 2.2.16.)

2.3.6 Let e be the limit of the sequence in Exercise 2.3.4. Prove that e is irrational. (*Hint*: If $e = p/q$ for $p, q \in \mathbb{N}$, write $q!e$ as the sum of an integer plus a positive number less than 1.)

2.3.7 Let (x_n) be a bounded and increasing sequence. Prove that it is Cauchy from the definition. Give another proof using the monotone sequence theorem.

2.3.8 Define what it means for a sequence of complex numbers (z_n) to be a Cauchy sequence. Prove that if (z_n) is a Cauchy sequence of complex numbers, then it converges to a complex number.

Chapter 3

Topology of the Real Numbers and Metric Spaces

The notions of open and closed sets are now ubiquitous in mathematics and will play a crucial role in our study of the real numbers. It is possible to place the study of open and closed sets in a more general context, for us in the context of metric spaces. One advantage of doing so is that one can also study these sets for the plane and other Euclidean spaces, in addition to the real line. At the same time it is important to first develop one's intuition in a concrete setting, in our case that of the real line. Open and closed sets in $\mathbb{R}$ are introduced in Section 3.2, and one can start with that section without reading Section 3.1. The sections covering metric spaces are 3.1, 3.3, and 3.7, and they can be skipped by the reader not studying metric spaces, or they can be left as optional enrichment sections.

3.1. Metrics

A basic notion in the study of convergence of sequences is the notion of the distance between two points. For points x and y in $\mathbb{R}$ we have used the absolute value function to achieve a notion of distance; namely, the distance between x and y is given by $|x - y|$. It turns out that the notion of distance can be generalized considerably. Our main interest will be in the higher-dimensional Euclidean spaces $\mathbb{R}^2$ and $\mathbb{R}^n$, but we will also briefly consider other spaces, such as symbolic spaces.

The principal properties of a *distance* are characterized by the following definition of a *metric*.

Let X be a set. If we think of $d(x, y)$ as the distance between x and y, d needs to be a function from the Cartesian product $X \times X$ to the interval $[0, \infty)$ (in this way the distance is defined between any two points of X). The first property that is required is that if $d(x, y) = 0$, then x and y should be the same point, and that the distance from a point to itself should be 0 (i.e., $d(x,x) = 0$). The next property is that the notion of distance should be symmetric, i.e., $d(x,y) = d(y,x)$. The last property is usually the hardest one to establish but is crucial to a notion of distance. It is called the *triangle inequality*: the distance between any two points $d(x,y)$ should be bounded above by the sum of the distance between the first point and an arbitrary point and the distance between the arbitrary point and the second point.

We state the definition more formally.

Definition 3.1.1. Let X be a set. A function $d : X \times X \to [0, \infty)$ is a **metric** on X if the following three properties hold for all points x, y, z in X:

(1) $d(x,y) = 0$ if and only if $x = y$;

(2) $d(x,y) = d(y,x)$;

(3) *Triangle inequality*: $d(x,y) \leq d(x,z) + d(z,y)$. ◇

Example 3.1.2 (Euclidean metric for $\mathbb{R}$). Define d on $\mathbb{R} \times \mathbb{R}$ by
$$d(x,y) = |x-y|.$$
We verify that d is a metric. First note that $d(x,x) = |x-x| = 0$, and if $d(x,y) = 0$, then $|x-y| = 0$, so $x = y$. For the second property, note that $d(x,y) = |x-y| = |y-x| = d(y,x)$. For the triangle inequality, we already know the triangle inequality for the absolute value is
$$|a+b| \leq |a| + |b| \text{ for all } a, b \in \mathbb{R}.$$
Now we calculate
$$d(x,y) = |x-y| = |x-z+z-y| \leq |x-z| + |z-y| = d(x,z) + d(z,y).$$
Therefore, d satisfies the triangle inequality and is a metric.

The next set we are interested in is $\mathbb{R}^n$, where n is a positive integer defined as
$$\mathbb{R}^n = \mathbb{R} \times \cdots \times \mathbb{R} \ (n \text{ times}).$$
An element $x \in \mathbb{R}^n$ can be written as $x = (x_1, x_2, \ldots, x_n)$ where $x_i \in \mathbb{R}$ for $i = 1, \ldots, n$. Addition of elements of $\mathbb{R}^n$ is defined by
$$(x_1, x_2, \ldots, x_n) + (y_1, y_2, \ldots, y_n) = (x_1 + y_1, \ldots, x_n + y_n).$$
Multiplication of an element of $\mathbb{R}^n$ by an element α of $\mathbb{R}$ (a *scalar*) is defined by
$$\alpha(x_1, \ldots, x_n) = (\alpha x_1, \ldots, \alpha x_n).$$

The notion of absolute value in $\mathbb{R}$ is generalized to the notion of norm in $\mathbb{R}^n$. The **norm** of $x \in \mathbb{R}^n$, denoted $\|x\|$, is defined by
$$\|x\| = \sqrt{x_1^2 + x_2^2 + \cdots + x_n^2}.$$
(Here x_1^2 denotes $(x_1)^2$, etc.) See Figure 3.1. For example, the norm of $x = (1, -2) \in \mathbb{R}^2$ is $\|(1, -2)\| = \sqrt{5}$.

To prove properties about the norm in $\mathbb{R}^n$, it is useful to introduce the dot product of two elements $x, y \in \mathbb{R}^n$.

3.1. Metrics

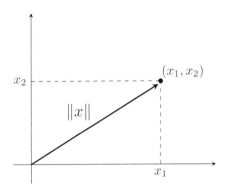

Figure 3.1. The norm of a point in $\mathbb{R}^2$.

Definition 3.1.3. The **dot product** of x and y in R^n, denoted $x \cdot y$, is defined by
$$x \cdot y = x_1 y_1 + x_2 y_2 + \cdots + x_n y_n.$$
◇

The following proposition states some basic properties of the dot product.

Proposition 3.1.4. *Let $x, y, z \in \mathbb{R}^n$. Then*

(1) $x \cdot x = \|x\|^2$.
(2) $x \cdot y = y \cdot x$.
(3) $x \cdot (y + z) = x \cdot y + x \cdot z$.

Proof. Parts (1) and (2) are clear from the definition. For part (3),
$$\begin{aligned} x \cdot (y + z) &= (x_1, \ldots, x_n) \cdot ((y_1, \ldots, y_n) + (z_1, \ldots, z_n)) \\ &= (x_1, \ldots, x_n) \cdot (y_1 + z_1, \ldots, y_n + z_n) \\ &= (x_1(y_1 + z_1), \ldots, x_n(y_n + z_n)) \\ &= (x_1 y_1 + x_1 z_1, \ldots, x_n y_n + x_n z_n) \\ &= (x_1 y_1, \ldots, x_n y_n) + (x_1 z_1, \ldots, x_n z_n) = x \cdot y + x \cdot z. \end{aligned}$$
□

Example 3.1.5. (Euclidean metric for $\mathbb{R}^n$.) We are ready to define the Euclidean distance or metric in the case of $\mathbb{R}^n$. For $x, y \in \mathbb{R}^n$, we define
$$d(x, y) = \|x - y\| = \sqrt{(x_1 - y_1)^2 + \cdots + (x_n - y_n)^2}.$$

The first two properties of a metric are shown in a similar way to the case of the Euclidean metric in $\mathbb{R}$.

We establish the third property, the triangle inequality for d, which is equivalent to

(3.1) $$\|x - y\| \le \|x - z\| + \|z - y\|.$$

Inequality (3.1) will follow if we show that

(3.2) $$\|a + b\| \le \|a\| + \|b\| \text{ for all } a, b \in \mathbb{R}^n.$$

(Just let $a = x - z$ and $b = z - y$ in (3.2)). Now to show (3.2), we square both sides and simplify the left-hand side,

$$(3.3) \qquad \|a+b\|^2 = (a+b) \cdot (a+b) = a \cdot a + a \cdot b + b \cdot a + b \cdot b$$
$$= \|a\|^2 + 2a \cdot b + \|b\|^2.$$

For the right-hand side of (3.2) we have

$$(3.4) \qquad (\|a\| + \|b\|)^2 = \|a\|^2 + 2\|a\|\|b\| + \|b\|^2.$$

Therefore, (3.2) will follow if we show that

$$(3.5) \qquad a \cdot b \leq \|a\|\|b\|.$$

The proof of inequality (3.5) (a special case of the Cauchy–Schwartz inequality, see Exercise 3.1.2) will follow from the observation that the norm of any element is nonnegative, so by considering the norm of $\|b\|a - \|a\|b$, we obtain

$$(\|b\|a - \|a\|b) \cdot (\|b\|a - \|a\|b) = \|\|b\|a - \|a\|b\|^2 \geq 0.$$

So

$$\|b\|^2 a \cdot a - 2\|a\|\|b\|a \cdot b + \|a\|^2 b \cdot b \geq 0,$$

which implies that

$$2\|a\|\|b\|\ a \cdot b \leq 2\|a\|^2\|b\|^2,$$

and, as we may assume $a, b \neq 0$, after the corresponding cancellations we obtain (3.5), which in turn implies that d satisfies the triangle inequality. Therefore, d is a metric.

Our next example is a metric on $\{0,1\}^\mathbb{N}$, the set of all sequences of 0's and 1's; recall that each element x of $\{0,1\}^\mathbb{N}$ has the form $x = x_1 x_2 \cdots$ where $x_i \in \{0,1\}$ for $i \geq 1$.

Example 3.1.6. We define a metric on the symbolic space $\{0,1\}^\mathbb{N}$. The idea for defining a distance on $\{0,1\}^\mathbb{N}$ is to regard two infinite sequences as "close" when they agree at the beginning at "several" places. For example, $000\cdots$ is closer to $001\cdots$ than it is to $010\cdots$, as it agrees with $001\cdots$ in two places while it agrees with $010\cdots$ in one place (we only count from the beginning until the first time they disagree). To do this, we start by defining the first place $I(x,y)$ where the two sequences x and y, $x \neq y$, differ by

$$(3.6) \qquad I(x,y) = \min\{i \in \mathbb{N} : x_i \neq y_i\}.$$

So the larger $I(x,y)$ is, the more x and y agree. For example,

$$I(000\cdots, 001\cdots) = 3 \text{ while } I(000\cdots, 010\cdots) = 2.$$

We are ready to define the metric. For $x, y \in \{0,1\}^\mathbb{N}$, $x \neq y$, define

$$(3.7) \qquad \delta(x,y) = 2^{-I(x,y)},$$

and set $\delta(x,x) = 0$. (We can also interpret $I(x,x)$ as $I(x,x) = \infty$.)

3.1. Metrics

So, for the examples that we have,

$$\delta(000\cdots, 001\cdots) = 2^{-3},$$
$$\delta(000\cdots, 010\cdots) = 2^{-2}, \text{ and}$$
$$\delta(001\cdots, 010\cdots) = 2^{-2}.$$

To show that δ is a metric, we first note that clearly $\delta(x, y) = 0$ if and only if $x = y$, and $\delta(x, y) = \delta(y, x)$. To show the triangle inequality, we instead verify the stronger inequality,

(3.8) $$\delta(x, y) \leq \max\{\delta(x, z), \delta(z, y)\}.$$

We first observe that this inequality indeed implies the triangle inequality. In fact, if $\delta(x, y) \leq \max\{\delta(x, z), \delta(z, y)\}$, since $\max\{\delta(x, z), \delta(z, y)\} \leq \delta(x, z) + \delta(z, y)$, it follows that $\delta(x, y) \leq \delta(x, z) + \delta(z, y)$. (A metric that satisfies the stronger inequality (3.8) is called an **ultrametric**.)

To show (3.8), it suffices to show

(3.9) $$I(x, y) \geq \min\{I(x, z), I(y, z)\}.$$

If $I(x, y) \geq I(x, z)$, we are done. If not, $I(x, z) > I(x, y)$, so x agrees with z more than it agrees with y, thus z and y cannot differ before $I(x, y)$, implying that $I(x, y) \geq I(z, y)$, which means that (3.9) holds. It follows that δ is a metric.

Example 3.1.7. There is a simple metric that can be defined on any nonempty set X by setting $d(x, y) = 1$ when $x \neq y$ and $d(x, x) = 0$ for all $x \in X$.

Definition 3.1.8. A **metric space** (X, d) consists of a set X with a metric d defined on X. ◇

Example 3.1.9. The first examples we have seen of metric spaces are $\mathbb{R}^n$ with Euclidean distance d and the symbolic space $\{0, 1\}^{\mathbb{N}}$ with the distance δ. Also, if (X, d) is a metric space and A is a subset of X, then (A, d) is also a metric space as we can consider the metric d restricted to $A \times A$.

Exercises: Metrics

3.1.1 Prove that for every $a, b \in \mathbb{R}^n$,
$$\big|\|a\| - \|b\|\big| \leq \|a + b\|.$$

3.1.2 *Cauchy–Schwartz inequality*: Prove that for every $a, b \in \mathbb{R}^n$,
$$|a \cdot b| \leq \|a\| \, \|b\|.$$

3.1.3 Let $a, b \in \{0, 1\}^{\mathbb{N}}$, and define $\rho : \{0, 1\}^{\mathbb{N}} \times \{0, 1\}^{\mathbb{N}} \to [0, \infty)$ by
$$\rho(a, b) = \sum_{i=1}^{\infty} \frac{|a_i - b_i|}{2^i}.$$

Prove that ρ is a metric.

3.1.4 *Taxicab metric*: Let $a, b \in \mathbb{R}^n$, and define $d_t : \mathbb{R}^n \times \mathbb{R}^n \to [0, \infty)$ by

$$d_t(a,b) = \sum_{i=1}^{n} |a_i - b_i|.$$

Prove that d_t is a metric.

3.1.5 Prove that $I(x, y)$ as defined in (3.6) is not a metric.

3.1.6 *Sup metric*: Let $a, b \in \mathbb{R}^n$, and define $d_s : \mathbb{R}^n \times \mathbb{R}^n \to [0, \infty)$ by

$$d_s(a,b) = \sup\{|a_i - b_i| : i = 1, \ldots, n\}.$$

Prove that d_s is a metric. (Note that as this is a finite set, it is the same as a max.)

3.1.7 Let $x \in \mathbb{R}^2$ and $\varepsilon > 0$. Draw the open balls in $\mathbb{R}^2$ of radius ε centered at x for the Euclidean, taxicab, and sup metrics.

3.1.8 Let d be a metric on a set X. Define $d_1 : X \times X \to [0, \infty)$ by

$$d_1(x, y) = \min\{1, d(x, y)\}.$$

Prove that d_1 is also a metric on X.

3.1.9 Let d be a metric on a set X. Define $d_2 : X \times X \to [0, \infty)$ by

$$d_2(x, y) = \frac{d(x,y)}{1 + d(x,y)}.$$

Prove that d_2 is also a metric on X.

3.1.10 *The p-adic metric on $\mathbb{Q}$*: Let p be a fixed prime. Write each $x \in \mathbb{Q}$, $x \neq 0$ in the form

$$x = p^i \frac{r}{s},$$

where r and s are integers that do not have p as a factor. Set

$$v_p(x) = p^{-i} \quad \text{and} \quad v_p(0) = 0.$$

Define $d_p : \mathbb{Q} \times \mathbb{Q} \to [0, \infty)$ by

$$d_p(x, y) = v_p(x - y).$$

Prove that d_p is a metric on $\mathbb{Q}$.

3.1.11 Let (X_1, d_1) and (X_2, d_2) be metric spaces. Define d_3 on $X_1 \times X_2$ by

$$d_3((x_1, x_2), (y_1, y_2)) = \sqrt{d_1(x_1, y_1)^2 + d_2(x_2, y_2)^2}.$$

Prove that d_3 is a metric. Generalize this statement to the case of the product of n metric spaces, and prove your claim.

3.1.12 State and prove a version of the nested intervals theorem in $\mathbb{R}^n$.

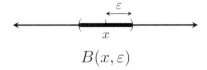

Figure 3.2. Open ball of radius ε in $\mathbb{R}$.

3.2. Open and Closed Sets in $\mathbb{R}$

This section introduces open and closed sets, and some important related notions such as accumulation points, closure, and interior, and can be read independently of Section 3.1. We study them in the context of the real numbers $\mathbb{R}$ with the Euclidean distance, or metric. The same notions, in the more general context of arbitrary metric spaces, are studied in Section 3.3, for which Section 3.1 will be needed.

3.2.1. Open Sets.
We start with the definition of an open ball in $\mathbb{R}$.

Definition 3.2.1. Given $x \in \mathbb{R}$ and $\varepsilon > 0$, the **open ball** centered at x with radius ε is
$$B(x, \varepsilon) = \{y \in \mathbb{R} : |x - y| < \varepsilon\}.$$ ◊

One can verify that
$$B(x, \varepsilon) = (x - \varepsilon, x + \varepsilon),$$
which is the open interval centered at x of radius ε; see Figure 3.2. Also, if $I = (a, b)$ is a nonempty open interval in $\mathbb{R}$, then $p = (a + b)/2$ is its midpoint and I is the open ball $B(p, b - p)$. So in $\mathbb{R}$, open balls are nothing but nonempty open intervals.

Definition 3.2.2. A subset G of $\mathbb{R}$ is defined to be an **open set** if for every $x \in G$ there is an open ball centered at x that is contained in G; namely, for each $x \in G$ there is a number $\varepsilon = \varepsilon(x) > 0$ such that $B(x, \varepsilon) \subseteq G$. (We write $\varepsilon = \varepsilon(x)$ to clarify that ε depends on x.) ◊

Proposition 3.2.3.

(1) *The empty set $\varnothing$ and the set $\mathbb{R}$ are open.*

(2) *Every open interval in $\mathbb{R}$ is an open set.*

Proof.

(1) $\mathbb{R}$ is clearly open as for every $x \in \mathbb{R}$, for any $\varepsilon > 0$, the open ball $B(x, \varepsilon)$ is in $\mathbb{R}$. We see that the empty set $\varnothing$ is open by contradiction: if it were not open, then there would exist an $x \in \varnothing$ for which there is no such ε, but this is a contradiction since there is no x in $\varnothing$; therefore, $\varnothing$ must be open.

(2) Let I be a nonempty open interval, and let $x \in I$. Then I can be a bounded or an unbounded interval. If it is bounded, then it is of the form $I = (a, b)$, for some $a, b \in \mathbb{R}$. In this case we can choose ε to be any positive number less than the distance from x to the endpoints, i.e., $\varepsilon < \min\{x - a, b - x\}$. Then the ball centered

at x of radius ε must be in I. If $I = (a, \infty)$, let $\varepsilon = x - a$, and if $I = (-\infty, b)$, let $\varepsilon = b - x$. In both cases we have that $B(x, \varepsilon) \subseteq I$. Therefore I is open. □

Question 3.2.4. Let $A \subseteq \mathbb{R}$. Prove that if A is countable and nonempty, then it is not open.

The next proposition states basic properties of open sets.

Proposition 3.2.5. *The open sets in $\mathbb{R}$ satisfy the following properties.*

(1) *The union of any collection of open sets is open.*

(2) *The intersection of any finite collection of open sets is open.*

Proof.

(1) Let $\{G_\alpha\}_{\alpha \in \Gamma}$ be an arbitrary collection of open sets indexed by a set Γ. To show that $\bigcup_{\alpha \in \Gamma} G_\alpha$ is open, we take a point x in this union. So $x \in G_\alpha$ for some $\alpha \in \Gamma$. Since G_α is an open set, there exists $\varepsilon > 0$ such that $B(x, \varepsilon) \subseteq G_\alpha$. Then

$$B(x, \varepsilon) \subseteq \bigcup_{\alpha \in \Gamma} G_\alpha,$$

showing that the union is open.

(2) Let $G_1, \ldots, G_n$ be a finite collection of open sets, and let $G = \bigcap_{i=1}^n G_i$. If $G = \varnothing$, then it is open, so assume that $G \neq \varnothing$, and let $x \in G$. For each $i = 1, \ldots, n$ there exists $\varepsilon_i > 0$ such that $B(x, \varepsilon_i) \subseteq G_i$. Set

$$\varepsilon = \min\{\varepsilon_1, \ldots, \varepsilon_n\}.$$

Then $\varepsilon > 0$ and $B(x, \varepsilon) \subseteq G_i$ for each $i = 1, \ldots, n$, so $B(x, \varepsilon) \subseteq G$. Therefore, G is open. □

Remark 3.2.6. The intersection of an infinite collection of open sets need not be open. For example, if $G_n = (-\frac{1}{n}, \frac{1}{n})$, for $n \in \mathbb{N}$, we have $\bigcap_{n=1}^\infty G_n = \{0\}$, a set that is not open.

3.2.2. Closed Sets

Definition 3.2.7. A set F in $\mathbb{R}$ is said to be **closed** if its complement in $\mathbb{R}$, F^c, is open. ◇

The following proposition gives us examples of closed sets.

Proposition 3.2.8.

(1) *The empty set $\varnothing$ and the set $\mathbb{R}$ are closed (and open).*

(2) *The intervals $[a, b], [a, \infty), (-\infty, b]$ (where $a, b \in \mathbb{R}$) are closed sets.*

(3) *The union of any finite collection of closed sets is closed.*

(4) *The intersection of any collection of closed sets is closed.*

Proof.

(1) The sets $\varnothing$ and $\mathbb{R}$ are closed since they are complements of each other.

(2) We note that $[a, b]^c = (-\infty, a) \cup (b, \infty)$, an open set by Proposition 3.2.5; so $[a, b]$ is closed. Similarly one can see that $[a, \infty)$ and $(-\infty, b]$ are closed.

(3) Let $F_1, \ldots, F_n$ be a finite collection of closed sets. Then, by De Morgan's law,
$$\left(\bigcup_{i=1}^{n} F_i\right)^c = \bigcap_{i=1}^{n} F_i^c.$$
Since each F_i^c is open, Proposition 3.2.5 implies that their intersection is open, so $\bigcup_{i=1}^{n} F_i$ is closed.

(4) This proof is similar to the one in part (3) in the sense that it also uses De Morgan's laws. In this case we consider an arbitrary family of closed sets $\{F_\alpha\}$ for α in some set Γ. Then
$$\left(\bigcap_{\alpha \in \Gamma} F_\alpha\right)^c = \bigcup_{\alpha \in \Gamma} F_\alpha^c.$$
As the sets F_α^c are open their union is open, so it follows that $\bigcap_{\alpha \in \Gamma} F_\alpha$ is closed. □

Remark 3.2.9. A set that is not open need not be closed. For example, $[a, b)$ is not open (the definition of open is not satisfied at the point a); however, its complement $(-\infty, a) \cup [b, \infty)$ is not open either, so $[a, b)$ is not closed. It can be shown that the only sets in $\mathbb{R}$ that are both open and closed are the empty set and $\mathbb{R}$.

3.2.3. Accumulation Points, Closure, Interior. Accumulation points play an important role in analysis and in many of the definitions we will study. They were called *Grenzpunkt* by Cantor, which is translated as "limit point", and sometimes accumulation points are called **limit points**, though we will use the term accumulation point.

Definition 3.2.10. Given a set $A \subseteq \mathbb{R}$, we say that a point $x \in \mathbb{R}$ is an **accumulation point** of A if there exists a sequence (a_n) of elements of A such that (a_n) converges to x and $a_n \neq x$ for all $n \in \mathbb{N}$. ◇

For example, if $A = (0, 1) \sqcup \{2\}$, then 0 is an accumulation point of A since the sequence $(1/n)$ converges to 0 and $1/n \in A, 1/n \neq 0$ for $n \in \mathbb{N}$. Of course, A has other accumulation points such as 1 and $1/2$. But 2, for example, is not an accumulation point of A.

Definition 3.2.11. A point x in A is an **isolated point** of A if it is not an accumulation point of A. ◇

In the previous example, 2 is an isolated point of A. (Note that while accumulation points need not belong to the set, isolated points are assumed to be in the set in question.)

Definition 3.2.12. The **derived set** of a set A, denoted A', consists of all accumulation points of A. ◇

For example $\mathbb{N}' = \emptyset$ and $\mathbb{Q}' = \mathbb{R}$.

A related notion is that of a closure point.

Definition 3.2.13. A point $x \in X$ is a **closure point** of a set $A \subseteq X$ if there is a sequence $a_n \in A$ converging to x, i.e., $\lim_{n \to \infty} a_n = x$. ◇

From the definition it follows directly that every accumulation point of A is a closure point of A. The difference is that in the case of closure points we do not require that the elements of the sequence be different from x. For example, if $A = (0,1) \sqcup \{2\}$, then 2 is a closure point of A but is not an accumulation point of A. On the other hand, if p is a closure point of A and p is not in A, then p is an accumulation point of A.

Definition 3.2.14. For $A \subseteq \mathbb{R}$, define the **closure** of A, denoted $\overline{A}$ (or sometimes also written $\mathrm{Cl}(A)$), by
$$\overline{A} = \{x \in \mathbb{R} : x \text{ is a closure point of } A\},$$
which is equivalent to
$$\overline{A} = A \cup \{x \in \mathbb{R} : x \text{ is an accumulation point of } A\} = A \cup A'. \qquad \diamond$$

It can be shown that $\overline{A}$ is the intersection of all closed sets containing A, so it is the smallest closed set containing A (in the sense that if F is closed and $A \subseteq F$, then $\overline{A} \subseteq F$; see Exercise 3.2.4). Also, A is closed if and only if $A = \overline{A}$; see Proposition 3.2.16.

Remark 3.2.15. In the case when a set A consists of the points in a sequence, then the closure of A consists of the points that are limits of subsequences of the sequence. A limit of convergent subsequences of a sequence is sometimes called a *subsequential limit*.

Proposition 3.2.16. *Let $F \subseteq \mathbb{R}$. F is closed if and only if every accumulation point of F is in F. Equivalently, F is closed if and only if F is equal to its set of closure points $\overline{F}$.*

Proof. Let F be a closed set, and let x be an accumulation point of F. Then there is a sequence (x_n) in F (i.e., $x_n \in F$ for $n \in \mathbb{N}$) converging to $x \in \mathbb{R}$. We need to show that $x \in F$. Suppose that $x \notin F$, then $x \in F^c$, and as F^c is open, there exists $\varepsilon > 0$ so that $B(x, \varepsilon) \subseteq F^c$. As $x_n \in F$ for all $n \in \mathbb{N}$, $x_n \notin B(x, \varepsilon)$. This means that $|x_n - x| \geq \varepsilon$ for all $n \in \mathbb{N}$. This implies that (x_n) cannot converge to x, a contradiction. Therefore, $x \in F$.

Now suppose that every accumulation point of F is in F. We need to show that F^c is open. Suppose that F^c is not open. Then there exists $x \in F^c$ so that for every $\varepsilon > 0$ there is a point, that we may denote by x_ε that is in $B(x, \varepsilon)$ and is not in F^c. Now we choose ε of the form $\varepsilon = 1/n$ for each $n \in \mathbb{N}$. Then we obtain a point, that now we denote by x_n, that is in $B(x, 1/n)$ and in F. We claim that the sequence (x_n) converges to x. In fact, if $\varepsilon > 0$, choose $N \in \mathbb{N}$ so that $N > 1/\varepsilon$. Then for every $n \geq N$,
$$|x_n - x| < 1/n \leq 1/N < \varepsilon,$$
showing that $\lim_{n \to \infty} x_n = x$. This shows that x is an accumulation point of F, thus $x \in F$, contradicting that $x \in F^c$. Therefore F^c is open and F is closed.

For the last part, we simply note that every point of a set F is a closure point of F. $\square$

Question 3.2.17. Prove that for every set A, its closure $\overline{A}$ is a closed set.

3.2. Open and Closed Sets in $\mathbb{R}$

We define another interesting set. For any set $A \subseteq \mathbb{R}$ define the **interior** of A, denoted A° (or sometimes also denoted $\text{Int}(A)$), by

$$A^\circ = \{x \in A : \text{ there exists } \varepsilon > 0 \text{ with } B(x, \varepsilon) \subseteq A\}.$$

Note that by definition, for any set $A \subseteq \mathbb{R}$,

$$A^\circ \subseteq A \subseteq \overline{A}.$$

It can be shown that A° is the union of all the open sets contained in A (see Exercise 3.2.2) and hence is itself an open set. We can think of A° as the largest open set contained in A (i.e., if G is open and $G \subseteq A$, then $G \subseteq A^\circ$).

Exercises: Open and Closed Sets in $\mathbb{R}$

3.2.1 Prove that a nonempty open set must be uncountable and a finite set must be closed. Give an example of a countably infinite set that is not closed.

3.2.2 Let $A \subseteq \mathbb{R}$, and let $U(A)$ be the union of all sets $G \subseteq A$ such that G is open, i.e.,

$$U(A) = \bigcup_{G \subseteq A, G \text{ open}} G.$$

Prove that A° is equal to $U(A)$.

3.2.3 Let $A \subseteq \mathbb{R}$. Prove that A is open if and only if $A = A^\circ$.

3.2.4 Let $A \subseteq \mathbb{R}$, and let $I(A)$ be the intersection of all closed sets containing A, i.e.,

$$I(A) = \bigcap_{F \supset A, F \text{ closed}} F.$$

Prove that $\overline{A}$ is equal to $I(A)$.

3.2.5 Let $A \subseteq \mathbb{R}$. Prove that A is closed if and only if $A = \overline{A}$.

3.2.6 Let $A \subseteq \mathbb{R}$. When is $A^\circ = \overline{A}$?

3.2.7 Is there a set whose interior is empty and whose closure is $\mathbb{R}$?

3.2.8 Find the closure and interior of each of the following sets.
  (a) $[0, 1) \cup (1, 2)$
  (b) $\mathbb{N}$
  (c) $\{1/n : n \in \mathbb{N}\}$
  (d) $\mathbb{Q} \cap (0, 1)$
  (e) $\mathbb{R} \setminus \mathbb{Q}$
  (f) $\{p/2^n : n \in \mathbb{N}, p = 0, \ldots, 2^n\}$

3.2.9 Prove or disprove (i.e., give a proof or a counterexample):

$$\overline{(A \cup B)} = \overline{A} \cup \overline{B}.$$

3.2.10 Prove or disprove (i.e., give a proof or a counterexample):

$$\overline{(A \cap B)} = \overline{A} \cap \overline{B}.$$

3.2.11 Prove or disprove (i.e., give a proof or a counterexample):
$$(A \cup B)^\circ = A^\circ \cup B^\circ.$$

3.2.12 Prove or disprove (i.e., give a proof or a counterexample):
$$(A \cap B)^\circ = A^\circ \cap B^\circ.$$

3.2.13 Let $A \subseteq \mathbb{R}$. Prove that the following are equivalent.
  (a) p is an accumulation point of A.
  (b) For each $\varepsilon > 0$ the intersection $B(p, \varepsilon) \cap A$ contains a point different from p.
  (c) For each $\varepsilon > 0$ the intersection $B(p, \varepsilon) \cap A$ contains infinitely many points.

3.2.14 Prove that every nonempty open subset G of $\mathbb{R}$ can be written as a countable union of disjoint open intervals. (*Hint*: For each x in G show that there exists a largest interval I_x with $x \in I_x \subseteq G$.) Conclude that if $A \subseteq \mathbb{R}$ is nonempty and both open and closed, then $A = \mathbb{R}$.

3.3. Open and Closed Sets in Metric Spaces

In this section we study open and closed sets in the context of metric spaces. As we shall see, the notion of convergence, as well as the notion of open balls and open sets, can be defined solely using a metric d. Our most important examples will be given by the Euclidean spaces $\mathbb{R}^n$. This section may be omitted on a first reading, but it depends on Section 3.1.

We first define open balls in the context of metric spaces.

Definition 3.3.1. Let (X, d) be a metric space. For each $x \in X$ and $\varepsilon > 0$, define the **open ball** around x of radius ε, denoted $B(x, \varepsilon)$, by
$$B(x, \varepsilon) = \{y \in X : d(x, y) < \varepsilon\}. \qquad \Diamond$$

Note that this generalizes the definition of an open ball in $\mathbb{R}$ with respect to the Euclidean distance. We define a closed ball similarly.

Definition 3.3.2. A **closed ball** around x of radius ε, denoted $B[x, \varepsilon]$, is defined by
$$B[x, \varepsilon] = \{y \in X : d(x, y) \leq \varepsilon\}. \qquad \Diamond$$

The notion of open balls are used to define open sets.

Definition 3.3.3. A set G in X is said to be **open** if for each $x \in G$ there is a real number $\varepsilon > 0$ (ε may depend on x) such that
$$B(x, \varepsilon) \subseteq G. \qquad \Diamond$$

We should say a few words about this definition. Formally, it looks the same as the definition of open sets in $\mathbb{R}$. However, there are new ideas. For example, let $X = [0, 1)$. We have already mentioned that the Euclidean metric d on $\mathbb{R}$ when restricted to any subset X of $\mathbb{R}$ is a metric on X, so $([0, 1), d)$ is a metric space. We observe that in this metric space, the set $G = [0, 1/2)$ is open. To see this, let x

be a point in G. When $x \neq 0$, proceed as before for the case of open intervals, and when $x = 0$, let $\varepsilon = 1/4$ (for example). Then $B(0, \varepsilon) \subseteq [0, 1)$. This is true because $B(x, \varepsilon)$ is defined as consisting of the points y in X such that $d(x, y) < \varepsilon$, so

$$B(0, 1/4) = \{y \in [0, 1) : d(0, y) < 1/4\} = [0, 1/4) \subseteq [0, 1/2).$$

Thus $[0, 1/2)$ is open in $([0, 1), d)$. However, we know that $[0, 1/2)$ is not open in $\mathbb{R}$, so we see that it is important to mention the metric space one is working on. The precise definition of the open ball depends on the metric space. To clarify these ideas, we have the following definition.

Definition 3.3.4. Let (X, d) be a metric space, and let Y be a subset of X. A set $G \subseteq Y$ is said to be **open relative to** Y if it is open in the metric space (Y, d). ◊

For example, $[0, 1/2)$ is open relative to $[0, 1)$, but it is not open relative to $(-1, 1)$ (in both cases with the Euclidean metric). If we know the universe we are working with is the space X, we only need to use the notion of open in X, and open relative to X is the same as open in X. However, if we are considering several subspaces of a metric space, the notion of open relative to a subspace is useful. It is convenient in this case to introduce the notation of an open ball where we explicitly mention the metric space we are working with.

Definition 3.3.5. Let (X, d) be a metric space, and let Y be a subset of X. Define

$$B_Y(x, \varepsilon) = \{y \in Y : d(x, y) < \varepsilon\}.$$ ◊

Using this notation we see that A is open relative to Y if and only if for all $x \in A$ there exists $\varepsilon > 0$ such that $B_Y(x, \varepsilon) \subseteq A$. Of course, if we only have one metric space, this notation is not needed, and $B_X(x, \varepsilon)$ is the same as $B(x, \varepsilon)$.

Example 3.3.6. As in the case of $\mathbb{R}$, we show that open balls are open sets. Let (X, d) be a metric space, and let $x \in X$ and $\varepsilon > 0$. Let $y \in B(x, \varepsilon)$. Let $\delta = \varepsilon - d(x, y) > 0$. We claim that the ball $B(y, \delta)$ is contained in $B(x, \varepsilon)$. Indeed, let $z \in B(y, \delta)$. Then

$$d(x, z) \leq d(x, y) + d(y, z) < d(x, y) + \varepsilon - d(x, y) = \varepsilon,$$

therefore $B(y, \delta) \subseteq B(x, \varepsilon)$ and $B(x, \varepsilon)$ is open in (X, d).

Example 3.3.7. In the case of $\mathbb{R}^2$ with the Euclidean distance, open balls $B(x, \varepsilon)$ consist of the interior of a circle centered at x and of radius ε. The set A of Figure 3.3 is not open as one can take x to be any corner point of A, and for all $\varepsilon > 0$, the open ball $B(x, \varepsilon)$ is not contained in A.

We state the following proposition, whose proof proceeds in the same way as the proof for the real case, and is left as an exercise.

Proposition 3.3.8. *Let (X, d) be a metric space. Then the following hold.*

(1) *The empty set $\varnothing$ and the set X are open sets.*

(2) *The union of an arbitrary collection of open sets is an open set.*

(3) *The intersection of a finite collection of open sets is an open set.*

Example 3.3.9. Let (X, d) be a metric space, and let Y be a subset of X. We show that a set $A \subseteq Y$ is open relative to Y if and only if A is of the form $A = Y \cap G$, where G is open in X. In fact, let A be open relative to Y. For each $x \in A$ there exists $\varepsilon_x > 0$ such that $B_Y(x, \varepsilon_x) \subseteq A$. From the definition we have that $B_Y(x, \varepsilon_x) = B_X(x, \varepsilon_x) \cap Y$, and we know that $B_X(x, \varepsilon_x)$ is open in X. So let $G = \bigcup_{x \in A} B_X(x, \varepsilon_x)$. By Proposition 3.3.8, G is open in X, and from the definition of G it follows that $G \cap Y = A$. For the converse suppose A is of the form $A = G \cap Y$, and let $x \in A$. As G is open in X, there exists $\varepsilon > 0$ such that $B_X(x, \varepsilon) \subseteq G$. Then $B_Y(x, \varepsilon) = B_X(x, \varepsilon) \cap Y \subseteq G \cap Y = A$. Therefore, A is open in Y.

We define closed sets in a similar way as before.

Definition 3.3.10. A set F in X is said to be **closed in** X if its complement F^c is open in X. If Y is a subset of X and A is in Y, we say that A is **closed relative to** Y if A is closed in the metric space (Y, d). Note that closed in X is the same as closed relative to X. ◇

We have the following basic properties. The proof of this proposition follows from Proposition 3.3.8 by the use of De Morgan's laws and is left to the reader.

Proposition 3.3.11. *Let (X, d) be a metric space. Then the following hold.*

(1) *The empty set $\varnothing$ and the set X are closed sets.*

(2) *The intersection of an arbitrary collection of closed sets is closed.*

(3) *The union of any finite collection of closed sets is closed.*

Now we investigate convergence in metric spaces.

Definition 3.3.12. Let (x_n) be a sequence of points in X. We say that (x_n) converges to a point $L \in X$ and write $\lim_{n \to \infty} x_n = L$, if for all $\varepsilon > 0$ there exists $N \in \mathbb{N}$ such that
$$d(x_n, L) < \varepsilon \text{ for all } n \geq N.$$
◇

This is the same definition of convergence that we studied in the case of $\mathbb{R}$; in that case we used $\mathbb{R}$ with the Euclidean metric defined by $d(x, y) = |x - y|$ for $x, y \in \mathbb{R}$. But now we have extended the notion to other spaces such as $(\mathbb{R}^n, d)$ and $(2^{\mathbb{N}}, \delta)$.

Example 3.3.13. Let $x_n = (\frac{2}{n}, 2 + \frac{1}{n}), n \in \mathbb{N}$, be a sequence in $\mathbb{R}^2$. We show that the sequence (x_n) in $\mathbb{R}^2$ converges to $(0, 2)$ (when the metric is not mentioned explicitly it is understood to be the Euclidean metric). Let $\varepsilon > 0$. Choose $N \in \mathbb{N}$ so that $N > \sqrt{5}/\varepsilon$. Hence, if $n > N$,

$$\|x_n - x\| = \left\|(\frac{2}{n}, 2 + \frac{1}{n}) - (0, 2)\right\|$$
$$= \left\|(\frac{2}{n}, \frac{1}{n})\right\|$$
$$= \sqrt{\frac{4}{n^2} + \frac{1}{n^2}} = \frac{\sqrt{5}}{n} < \varepsilon.$$

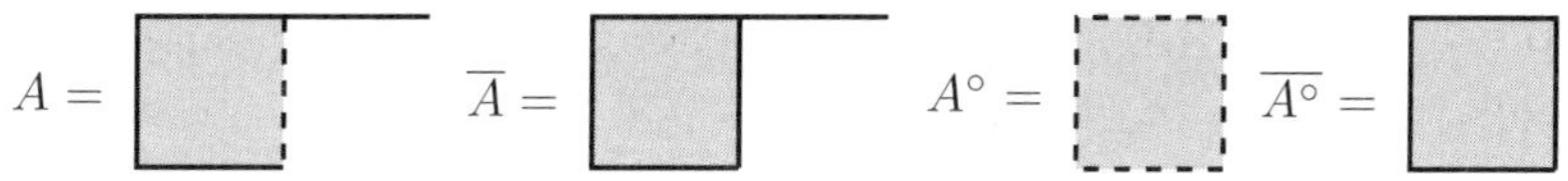

3.3. Open and Closed Sets in Metric Spaces

Figure 3.3. The closure and interior of a set in $\mathbb{R}^2$.

Example 3.3.14. In this example we work in the symbolic metric space defined in Example 3.1.6. Recall that $X = \{0,1\}^{\mathbb{N}}$, the space of infinite sequences of 0's and 1's, and the metric is given by δ in (3.7). Under this metric the distance between two infinite sequences gets smaller as the sequences agree more at the start. Recall that $\overline{0}$ means that the symbol 0 is repeated infinitely many times, and let 0^n denote 0 repeated n times. Let $x = 0\overline{0}$. Define a sequence of elements of X by
$$x_1 = 0\overline{1}, x_2 = 00\overline{1}, \ldots, x_n = 0^n\overline{1}, \ldots.$$
Then x and x_n agree in the first n places, so $\delta(x, x_n) = 2^{-n-1}$. Therefore, the sequence (x_n) converges to x in the δ metric.

We now show that, surprisingly, the open ball $B(0\overline{0}, 1/2)$ is also a closed set. Let y be a closure point of $B(0\overline{0}, 1/2)$. Then there exist $x_n \in B(0\overline{0}, 1/2)$ such that $\rho(x_n, y) \to 0$. Choose k so that $\rho(x_k, y) < 1/2$. This means that x_k and y agree in the first two symbols. Since x_k is in $B(0\overline{0}, 1/2)$, the first two symbols of x_k are 00. So the first two symbols of y are 00, which means that $\rho(y, 0\overline{0}) < 1/2$, thus y is in $B(0\overline{0}, 1/2)$. Therefore, the set is closed.

Definition 3.3.15. A point $x \in X$ is an **accumulation point** (sometimes called a **limit point**) of a set $A \subseteq X$ if there is a sequence $a_n \in A$ such that $\lim_{n\to\infty} a_n = x$, and $a_n \neq x$ for all $n \in \mathbb{N}$. $\diamond$

Definition 3.3.16. A point $x \in X$ is a **closure point** of a set $A \subseteq X$ if there is a sequence $a_n \in A$ converging to x, i.e., $\lim_{n\to\infty} a_n = x$. $\diamond$

We note that every accumulation point is a closure point, and a closure point of A that is not in A is an accumulation point of A.

The closure and interior of a set are defined similarly as in the case of subsets of $\mathbb{R}$.

Definition 3.3.17. The **closure** $\overline{A}$ of a set A in (X, d) consists of the set of closure points of A. The **interior** A° of A is the set of all points $x \in A$ such that $B(x, \varepsilon) \subseteq A$ for some $\varepsilon > 0$. $\diamond$

Figure 3.3 illustrates the closure and interior of a set in $\mathbb{R}^2$ with Euclidean metric.

Closed sets are characterized by sequences in the following lemma.

Lemma 3.3.18. *Let (X, d) be a metric space. A set F is closed if and only if whenever $x \in X$ is the limit of a sequence (x_n) of points in F, then $x \in F$ (i.e., F contains all its closure points).*

Proof. Suppose that F is closed, and let (x_n) be a sequence of points in F converging to $x \in X$. Suppose x is not in F, so $x \in F^c$. As F^c is open, there exists $\varepsilon > 0$ such that $B(x, \varepsilon) \subseteq F^c$. Since (x_n) converges to x, there exists $N \in \mathbb{N}$ so that $x_n \in B(x,\varepsilon)$ for all $n \geq N$. But this leads to a contradiction since $x_n \in F$ for all $n \in \mathbb{N}$, so x_n cannot be in $B(x,\varepsilon)$. Therefore, $x \in F$.

For the converse, suppose that F contains all its closure points. If F^c is not open, then there exists a point $x \in F^c$ so that every open ball centered at x cannot be wholly contained in F^c. This means that for each $n \in \mathbb{N}$, if we set $\varepsilon_n = \frac{1}{n}$, the ball $B(x,\varepsilon_n)$ has to contain a point that is not in F^c. So there is a point $x_n \in F$ with $d(x,x_n) < \varepsilon_n$. This implies that $\lim_{n\to\infty} x_n = x$. So x is a closure point of F, which implies that x is in F, a contradiction. Therefore, F^c is open and F is closed. $\square$

Exercises: Open and Closed Sets in Metric Spaces

3.3.1 Prove Proposition 3.3.8.

3.3.2 Complete the proof of Proposition 3.3.11.

3.3.3 Let (X,d) be a metric space, and let $A \subseteq X$. Prove that A° is the union of all sets $G \subseteq A$ such that G is open.

3.3.4 Let (X,d) be a metric space, and let $A \subseteq X$. Prove that A is open if and only if $A = A^\circ$.

3.3.5 Let (X,d) be a metric space, and let $A \subseteq X$. Prove that $\overline{A}$ is the intersection of all closed sets containing A.

3.3.6 Let (X,d) be a metric space, and let $A \subseteq X$. Prove that A is closed if and only if $A = \overline{A}$.

3.3.7 Find the closure and interior of each of the following sets as a subsets of $\mathbb{R}^2$.
   (a) $\mathbb{N} \times \mathbb{Q}$
   (b) $\mathbb{Z} \times \mathbb{Z}$
   (c) $\mathbb{Q} \times (0,1)$
   (d) $\{(x, \sin(\frac{1}{x})) : 0 < x < \pi\}$

3.3.8 Let (X,d) be a metric space, and let Y be a subset of X. Prove that a set $A \subseteq Y$ is closed relative to Y if and only if A is of the form $A = F \cap Y$, where F is a closed subset of X.

3.3.9 Let (X,d) be a metric space, and let $A \subseteq X$. Prove that p is an accumulation point of A if and only if for any $\varepsilon > 0$ the intersection $B(p,\varepsilon) \cap A$ contains a point different from p. Prove that p is an accumulation point of A if and only if for any $\varepsilon > 0$ the intersection $B(p,\varepsilon) \cap A$ contains infinitely many points.

3.3.10 Let (X,d) be a metric space. Let d_1 and d_2 be the metrics defined in Exercises 3.1.8 and 3.1.9. Let $A \subseteq X$. Prove that A is open in (X,d_1) if and only if it is open in (X,d_2).

3.3.11 Let (a_n) be a sequence in $\mathbb{R}^d$. Write $a_n = (a_{n,1}, a_{n,2}, \ldots, a_{n,d})$. Prove that (a_n) converges to $L = (L_1, \ldots, L_d)$ if and only if for each $i = 1, \ldots, d$, $(a_{n,i})_n$ converges to L_i.

3.3.12 Let A be a subset of $\mathbb{R}^n$. Prove that A is open and closed (with the Euclidean metric) if and only if $A = \varnothing$ or $A = \mathbb{R}^n$. Give an example of a set in $(\{0,1\}^{\mathbb{N}}, \delta)$ that is open and closed and not empty and not the whole space.

3.4. Compactness in $\mathbb{R}$

Compactness is one of the most important notions in analysis. It will have several applications. In $\mathbb{R}$ (and $\mathbb{R}^n$), the simplest characterization of a compact set is that a set is compact if and only if it is closed and bounded. Another characterization is in terms of sequences (this is called sequential compactness), and the most general one (which works in spaces more general than $\mathbb{R}^n$ and spaces more general than metric spaces) is in terms of open covers. Theorem 3.4.8 proves the equivalence of these three properties for the case of subsets of $\mathbb{R}$ (it is also true for subsets of $\mathbb{R}^n$). When we talk about topological properties of $\mathbb{R}$ (or $\mathbb{R}^n$) such as compactness, unless we explicitly say otherwise, we always mean with respect to the Euclidean metric.

3.4.1. Sequential Compactness. We start with a characterization of compactness in terms of sequences.

Definition 3.4.1. A set $A \subseteq \mathbb{R}$ is said to be **sequentially compact** if every sequence (a_n) of elements of A has a subsequence (a_{n_k}) converging to a point a in A. $\diamondsuit$

For example, $A = (0,1]$ is not sequentially compact since every subsequence of the sequence $(1/n)$ converges to 0, but $0 \notin A$. To show that a set is sequentially compact, one needs to verify the condition for every sequence of elements of the set. Therefore, the following simpler characterization is useful.

Theorem 3.4.2. *A set $A \subseteq \mathbb{R}$ is sequentially compact if and only if it is closed and bounded.*

Proof. Suppose that A is sequentially compact. To show that A is closed, let a be a closure point of A. Then there exists a sequence (a_n) of points of A converging to a. As every subsequence of (a_n) converges to a (by sequential compactness), it follows that $a \in A$, so A is closed. If A were not bounded, then it is either not bounded above or not bounded below. Suppose A is not bounded above. Then for each $k \in \mathbb{N}$ there exists an element $x_k \in A$ with $x_k > k$. This constructs a sequence (x_k). From the construction it follows that every subsequence of (x_k) is unbounded. Therefore, the sequence (x_k) does not contain a convergent subsequence. This would contradict that A is sequentially compact. Thus A is bounded above. The fact that A is bounded below is proved in a similar way and is left to the reader.

Now suppose that A is closed and bounded. Let (a_n) be a sequence of elements of A. By the Bolzano–Weierstrass theorem, (a_n) contains a convergent subsequence; let $a \in \mathbb{R}$ be the limit of this subsequence. As A is closed, $a \in A$. Therefore, A is sequentially compact. $\square$

The following is our first application of sequential compactness.

Lemma 3.4.3. *Let $A \subseteq \mathbb{R}$ be a nonempty sequentially compact set. Then A has a maximum and a minimum.*

Proof. We give two proofs. The first proof uses the supremum property. By Theorem 3.4.2, A is bounded. Let $\alpha = \inf A$ and $\beta = \sup A$. By Exercise 1.2.4 and Lemma 1.2.10, respectively, for each $n \in \mathbb{N}$ there exists $x_n, y_n \in A$ so that

$$\alpha \le x_n < \alpha + \frac{1}{n} \quad \text{and} \quad \beta - \frac{1}{n} < y_n \le \beta.$$

By the squeeze property (Proposition 2.1.11), $\lim_{n \to \infty} x_n = \alpha$ and $\lim_{n \to \infty} y_n = \beta$. Sequential compactness then implies that $\alpha, \beta \in A$. So A has a maximum β and a minimum α.

For the second proof we use the characterization of order completeness in terms of the monotone sequence property. The set A is nonempty and bounded. Let $a_1 \in A$, and let b_1 be an upper bound of A. Continue as in Remark 2.2.3 to obtain bounded monotonic sequences (a_n) and (b_n) such that $a_n \in A$, b_n is an upper bound of A, and $|a_{n+1} - b_{n+1}| \le (b_1 - a_1)/2^n$. Set $\beta = \lim_{n \to \infty} a_n$. Since A is closed, it follows that $\beta \in A$. Also, the same argument as in Remark 2.2.3 shows that β is an upper bound; therefore, it is the maximum of A. A similar argument works for the minimum. □

We note, for example, that the set $(0, 1)$ is bounded but has no maximum (i.e., there is no element of $(0, 1)$ that is greater than every element of the set) and similarly has no minimum.

3.4.2. Definition of Compactness and the Heine–Borel theorem.
Now we consider the general definition of compactness that, as mentioned earlier, is in terms of covers.

Definition 3.4.4. A collection of sets $\{G_\alpha\}_{\alpha \in \Gamma}$ (for some indexing set Γ) is said to be a **cover** of a set A if

$$A \subseteq \bigcup_{\alpha \in \Gamma} G_\alpha.$$ ◊

Definition 3.4.5. A collection of sets $\{G_\alpha\}_{\alpha \in \Gamma}$ is said to be an **open cover** of A if it is a cover and each G_α is open. It is a **countable open cover** if it is an open cover and Γ is a countable set. ◊

For example, if $G_n = (1/n, 1)$, then $\{G_n\}_{n \in \mathbb{N}}$ is a countable open cover of $(0, 1)$ but it is not a cover of $[0, 1)$; here Γ is $\mathbb{N}$. The indexing set Γ need not be countable. For example, for each $x \in (0, 1)$, we can consider the open set $G_x = (x - 1/x, x + 1/x)$, and if we set $\Gamma = (0, 1)$, then $\{G_x : x \in \Gamma\}$ is an open cover of $[0, 1]$.

Definition 3.4.6. A set $A \subseteq \mathbb{R}$ is said to be **compact** if for every open cover $\{G_\alpha\}_{\alpha \in \Gamma}$ of A there is a finite set $\Gamma^* \subseteq \Gamma$ so that $\{G_\alpha\}_{\alpha \in \Gamma^*}$ is still a cover of A, i.e.,

$$A \subseteq \bigcup_{\alpha \in \Gamma^*} G_\alpha.$$ ◊

3.4. Compactness in ℝ

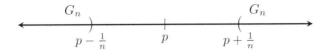

Figure 3.4. The open sets G_n.

Informally, we will say that A is compact if and only if "every open cover of A has a finite subcover".

As we have seen, an open cover for the interval $(0,1)$ can be given by the open sets $G_n = (1/n, 1)$ for $n \in \mathbb{N}$. However, this open cover has no finite subcover, since any finite subcover would have an element G_{n_0} where n_0 is the largest index in the subcover. But then an element such as $1/(2n_0)$ is in $(0,1)$ but not in the subcover.

The following is an important theorem that characterizes compact sets in $\mathbb{R}$.

Theorem 3.4.7 (Heine and Borel). *Let A be a set in $\mathbb{R}$. Then A is compact if and only if A is closed and bounded.*

Proof. Suppose that A is compact. If A is empty, it is clearly closed and bounded, so assume it is nonempty and fix a point x in A. The collection of open balls $(B(x,n))_{n\in\mathbb{N}}$ forms an open cover for A since each $B(x,n)$ is open, and for each y in A, $d(x,y) < n$ for some $n \in \mathbb{N}$ and so $y \in B(x,n)$. By compactness this cover has a finite subcover, which we may denote by $B(x, n_1), \ldots, B(x, n_k)$, for some $n_1 < n_2 < \cdots < n_k$. But from the definition of the open balls we have that $B(x, n_1) \subseteq B(x, n_2) \subseteq \cdots \subseteq B(x, n_k)$. So A is contained in $B(x, n_k)$, and thus it is a bounded set.

To see that A is closed, we take p an accumulation point of A. We want to show that p is in A and proceed by contradiction by assuming p is not in A. Then each point of A is at some positive distance from p. Define, then, for each $n \in \mathbb{N}$, the sets
$$G_n = \{x \in \mathbb{R} : |x - p| > \frac{1}{n}\}.$$
This is illustrated in Figure 3.4.

One can see that each G_n is an open set as its complement is closed. Also, since $1/(n+1) < 1/n$, we have
$$G_n \subseteq G_{n+1} \subseteq \cdots.$$
Using that p is not in A, we show that the collection $\{G_n\}_{n\in\mathbb{N}}$ is a cover of A. In fact, if $x \in A$, then $x \neq p$, so $|x - p| > 1/n$ for some $n \in \mathbb{N}$, which implies that $x \in G_n$. Therefore, (G_n) is an open cover of A, so compactness implies that there exists a finite subcover, $G_{n_1}, \ldots, G_{n_k}$, for some $n_1 < \cdots < n_k$. As the sets are increasing, G_{n_k} includes their union, so it covers A. This is a contradiction since, using that p is an accumulation point, we can find a point y in A such that $|y - p| < 1/n_k$, so by definition y is not in G_{n_k}, contradicting that G_{n_k} covers A. Therefore $p \in A$ and A is closed.

For the converse we give two proofs. Suppose A is closed and bounded.

For the first proof, we first reduce the problem to closed bounded intervals. That is, we claim that if we show that every closed bounded interval is compact, then it follows that every closed bounded set is compact. So suppose every closed bounded interval is compact, and let A be a closed bounded set. Let $\{G_\alpha\}_{\alpha \in \Gamma}$ be an open cover for A. As A is bounded, there is an interval $[a, b]$ so that $A \subseteq [a, b]$. Now we add one more open set to the cover of A to get a cover of $[a, b]$, namely add the set A^c, which is open. Now we have an open cover of $[a, b]$ consisting of A^c together with the open sets $\{G_\alpha\}_{\alpha \in \Gamma}$. Since we are assuming that every closed bounded interval is compact, then $[a, b]$ has a finite subcover that must consist of some $G_{\alpha_1}, \ldots, G_{\alpha_n}$ and A^c. Then the sets $G_{\alpha_1}, \ldots, G_{\alpha_n}$ must cover A, so there is a finite subcover of the collection $\{G_\alpha\}_{\alpha \in \Gamma}$, and therefore A is compact.

Now we show that every closed bounded interval $I = [a, b]$ is compact. Let $\{G_\alpha\}_{\alpha \in \Gamma}$ be an open cover for I. We proceed by contradiction and assume there is no finite subcover. We start by constructing a nested sequence of closed subintervals of I by induction. First subdivide I into two closed subintervals; at least one of these subintervals does not have a finite subcover (if both subintervals had a finite subcover, then there would be a finite subcover for I). Choose one of the subintervals (the leftmost if both have no finite subcover), and call it I_1. For the inductive step assume we have constructed closed subintervals

$$I_1 \supseteq I_2 \supseteq \cdots \supseteq I_n$$

so that I_n (and hence I_j for $j = 1, \ldots, n-1$) has no finite subcover. Subdivide I_n into two closed subintervals; at least one of these subintervals does not have a finite subcover. Choose one and call it I_{n+1}. Thus we have constructed an infinite sequence of nested closed subintervals I_n so that each one has no finite subcover. By the nested intervals theorem there is a point

$$p \in \bigcap_{n \geq 1} I_n.$$

There must exist $\alpha_0 \in \Gamma$ so that $p \in G_{\alpha_0}$. Since G_{α_0} is open, there is an open interval J centered at p with $J \subseteq G_{\alpha_0}$. Choose k large enough so that the length of I_k is less than $|J|/2$. As $p \in I_k$, it follows that $I_k \subseteq G_{\alpha_0}$, which would mean that I_k has a finite subcover (namely just the set G_{α_0}), a contradiction as each such subinterval has been constructed not to have a finite subcover. Therefore, since assuming that there is no finite subcover leads to a contradiction, there must be a finite subcover for A, showing that A is compact.

For the second proof we first use Exercise 3.4.7, which says that every open cover has a countable subcover. (This fact uses that $\mathbb{R}$ has a countable dense subset.) So given an open cover $\{G_\alpha\}_{\alpha \in \Gamma}$ of A, we can start by assuming that there is a countable set Γ' such that $\{G_\alpha\}_{\alpha \in \Gamma'}$ is an open cover of A. We write the countable subcover as $\{G_n\}_{n \in \mathbb{N}}$.

Next we proceed by contradiction by assuming that for each $k \in \mathbb{N}$ the collection $G_1, \ldots, G_k$ does not cover A. Then for each $k \in \mathbb{N}$, we can choose a point

$$x_k \in A \setminus \bigcup_{i=1}^{k} G_i.$$

This produces a sequence (x_k) of points in A. From the construction of the sequence it follows that for all $k \in \mathbb{N}$,

$$\text{if } m \geq k, \text{ then } x_m \notin G_k.$$

By Theorem 3.4.2, the sequence (x_k) must have a subsequence (x_{k_i}) that converges to a point $p \in A$. Then $p \in G_\ell$ for some $\ell \in \mathbb{N}$. Since G_ℓ is open, by convergence of the subsequence, the terms x_{k_i} must all be in G_ℓ for all i greater than some integer N. We can choose j such that $j > N$ and $k_j > \ell$. Then $x_{k_j} \in G_\ell$, a contradiction. Therefore, $\{G_n\}_{n \in \mathbb{N}}$ must have a finite subcover for A and A has to be compact. □

We state in the following theorem the equivalences of the main properties of compactness in $\mathbb{R}$ that we have shown.

Theorem 3.4.8. *Let A be a set in $\mathbb{R}$. The following are equivalent.*

(1) *A is closed and bounded.*

(2) *A is sequentially compact.*

(3) *A is compact.*

Proof. The proof follows from Theorems 3.4.2 and 3.4.7. □

Exercises: Compactness in $\mathbb{R}$

3.4.1 Find a countably infinite collection of closed intervals that cover $[0,1]$ such that no finite subcollection covers $[0,1]$.

3.4.2 Two intervals are said to be nonoverlapping if they intersect in at most one point. Prove that any nonempty open set in $\mathbb{R}$ can be written as a countable union of nonoverlapping closed bounded intervals.

3.4.3 Let E and F be two disjoint nonempty compact sets in $\mathbb{R}$. Prove that there exists a number $\delta > 0$ such that for every interval I, if $|I| < \delta$, then $I \cap E = \varnothing$ or $I \cap F = \varnothing$. Furthermore, find examples of disjoint closed sets E and F such that for all $\delta > 0$ there exists an interval I, such that $|I| < \delta$, $I \cap E \neq \varnothing$, and $I \cap F \neq \varnothing$.

3.4.4 Give another proof that a closed bounded interval $[a,b]$ is compact along the following lines. Let $\{G_\alpha\}_{\alpha \in \Gamma}$ be an open cover of $[a,b]$. Let $C = \{x \in [a,b] : [a,x] \text{ has a finite subcover}\}$. Note that C must have a supremum, set $c^* = \sup C$, and prove that $c^* \in C$ and $c^* = b$.

3.4.5 Prove that if A is compact, then every infinite subset of A has an accumulation point in A.

3.4.6 Let C_i be a nonempty compact set for each $i \in \mathbb{N}$ and suppose that $C_1 \supset C_2 \supset C_3 \cdots$. Prove that $\bigcap_i C_i \neq \varnothing$.

3.4.7 (Lindelöf) Prove that if $\{G_\alpha\}_{\alpha \in \Gamma}$ is an arbitrary collection of open sets in $\mathbb{R}$, then there exists a countable subset $\Gamma' \subseteq \Gamma$ such that $\bigcup_{\alpha \in \Gamma} G_\alpha = \bigcup_{\alpha \in \Gamma'} G_\alpha$. (*Hint*: Use open intervals with rational endpoints.)

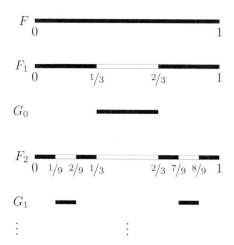

Figure 3.5. The first few steps in the construction of the Cantor set.

3.5. The Cantor Set

The Cantor set is one of the most remarkable sets in analysis. This set is also called the middle-thirds Cantor set since as we shall see there are other kinds of Cantor sets.

We define the (middle-thirds) Cantor set by an induction process. First let

$$F_0 = [0,1] \quad \text{and} \quad G_0 = (\frac{1}{3}, \frac{2}{3}).$$

The interval G_0 is called the *open middle-third* of F_0. For clarity of exposition we do the first few steps explicitly before passing to the inductive step.

Define
$$F_1 = F_0 \setminus G_0.$$
It will be convenient to denote F_0 by F.

The set F_1 consists of 2^1 closed intervals in $F = F_0$; we denoted these subintervals by $F[0]$ and $F[2]$ (from left to right), each of length $\frac{1}{3}$. We use this notation because we think of subdividing F into three closed subintervals of equal length; the first we call $F[0]$, the second we call $F[1]$, and the third we call $F[2]$, so $F_1 = F[0] \sqcup F[2]$.

Next, let
$$G_1 = (\frac{1}{9}, \frac{2}{9}) \sqcup (\frac{7}{9}, \frac{8}{9}),$$
the union of the middle-thirds of each of the subintervals of F_1. Then define
$$F_2 = F_1 \setminus G_1.$$

The set F_2 consists of 2^2 closed subintervals of F_1, each of length $\frac{1}{3^2}$ and denoted (in order from left to right) by $F[00]$, $F[02]$, $F[20]$, $F[22]$. So
$$F_2 = F[00] \sqcup F[02] \sqcup F[20] \sqcup F[22].$$

See Figure 3.5.

3.5. The Cantor Set

For the inductive step, suppose that the set F_{n-1} has been defined and consists of 2^{n-1} closed subintervals, each of length $\frac{1}{3^{n-1}}$, and denoted by (from left to right)
$$F[0 \cdots 0], F[0 \cdots 2], \ldots, F[2 \cdots 2].$$
Let G_{n-1} be the union of the open middle-thirds (each of length $\frac{1}{3^n}$) of each of the 2^{n-1} closed subintervals of F_{n-1}. Then let
$$F_n = F_{n-1} \setminus G_{n-1},$$
a union of 2^n closed subintervals, each of length $\frac{1}{3^n}$. Now the sequence of closed sets F_n has been defined.

Definition 3.5.1. Define the **middle-thirds Cantor set** K by

(3.10) $$K = \bigcap_{n=0}^{\infty} F_n. \qquad \diamond$$

Sometimes we may refer to K simply as the *Cantor set*. First note that K is closed as it is the intersection of closed sets. It is also bounded; therefore, K is compact. It follows from definition (3.10) that all endpoints of the subintervals in F_n belong to K. This is because $F_{n+1} \subseteq F_n$ and the endpoints of subintervals of F_n continue to be endpoints of subintervals of F_k for all $k > n$. Also note that for all $k \in \mathbb{N}$,
$$K = \bigcap_{n=k}^{\infty} F_n.$$
The next exercise provides an alternative way to think about K.

Question 3.5.2. Prove that the Cantor set is also given by
$$K = F \setminus \left(\bigcup_{n=0}^{\infty} G_n \right).$$
Thus, the Cantor set is the set of points in $[0, 1]$ that is obtained after removing all the open intervals comprising the sets G_n.

Definition 3.5.3. A set is **perfect** if it is equal to its set of accumulation points, i.e., every point of the set is an accumulation point and every accumulation point is in the set; equivalently, it is closed and every point is an accumulation point. $\diamond$

For example, the interval $[0, 1]$ is a perfect set; also, the empty set is perfect. A nonempty finite set cannot be perfect as its set of accumulation points is empty. The set $A = \{1/n : n \in \mathbb{N}\}$ is infinite but not perfect as it does not contain 0, which is an accumulation point of A. The set $A = \{1/n : n \in \mathbb{N}\} \cup \{0\}$ is not perfect either as 1 is not an accumulation point of this set. In fact, it can be shown that a nonempty perfect set must be uncountable (Exercise 3.5.8). As we shall see, one of the most interesting examples of a perfect set is the Cantor set.

It might seem that "most" points of $[0, 1]$ have been removed in the construction of K, but in fact, as we shall see in the theorem below, there are uncountably many points that are left in K.

Theorem 3.5.4. *The Cantor set K is an uncountable compact subset of $[0, 1]$. Furthermore, K contains no positive length intervals and is perfect.*

Proof. We have already seen that K is compact.

Suppose I is an interval contained in K. It follows that $I \subseteq F_n$ for all $n > 0$. Since F_n consists of 2^n intervals, each of length 3^{-n}, it follows that $|I| < 1/3^n$ for all $n > 0$. Taking the limit as $n \to \infty$ gives that $|I| = 0$. This means that K contains no intervals of positive length.

To show that K is uncountable, we define a function $\phi : \{0,2\}^{\mathbb{N}} \to K$, where $\{0,2\}^{\mathbb{N}}$ is the set of all sequences of 0's and 2's, and show it is injective. By identifying 2 with 1, we can define a bijection between $\{0,2\}^{\mathbb{N}}$ and $\{0,1\}^{\mathbb{N}}$, so $\{0,2\}^{\mathbb{N}}$ is uncountable.

For $\alpha \in \{0,2\}^{\mathbb{N}}$ write
$$\alpha = \alpha_1 \alpha_2 \cdots,$$
where $\alpha_i \in \{0,2\}$ for $i \in \mathbb{N}$. Define the map $\phi : \{0,2\}^{\mathbb{N}} \to K$ by

(3.11) $$\phi(\alpha) \in \bigcap_{n=1}^{\infty} F[\alpha_1 \alpha_2 \cdots \alpha_n].$$

To show that ϕ is a function, we need to verify that the intersection in (3.11) consists of a single point. We have already seen that the sets $F[x_1 x_2 \cdots x_n]$ are closed intervals of length $(1/3)^n$. So
$$\lim_{n \to \infty} |F_n| = 0.$$
Therefore, intersection $\bigcap_{n=1}^{\infty} F[\alpha_1 \alpha_2 \cdots \alpha_n]$ contains a unique point, and this point must be in K. So ϕ is a well-defined function from $\{0,2\}^{\mathbb{N}}$ into K. Now we show that ϕ is injective. Let $\alpha, \beta \in \{0,2\}^{\mathbb{N}}$. Write $\alpha = \alpha_1 \alpha_2 \cdots$ and $\beta = \beta_1 \beta_2 \cdots$. Suppose that $\alpha \neq \beta$. Then there exists some $k > 0$ such that $\alpha_k \neq \beta_k$. Suppose $\alpha_k = 0$ and $\beta_k = 2$. Now the sets $F[\alpha_1 \cdots \alpha_{k-1} 0]$ and $F[\beta_1 \cdots \beta_{k-1} 2]$ are disjoint, which implies that $\phi(\alpha) \neq \phi(\beta)$, so ϕ is injective. A similar argument holds if $\alpha_k = 2$ and $\beta_k = 0$. Since $\{0,2\}^{\mathbb{N}}$ is uncountable and ϕ is injective, it follows that K is uncountable.

Finally, we show that K is perfect. Let $x \in K$. We define a sequence (a_n) converging to x using the endpoints of the subintervals containing x. For each $n > 0$, let a_n be the left endpoint of the subinterval of F_n containing x if x is not a left endpoint, and the right endpoint of the subinterval of F_n containing x if x is a left endpoint. The sequence (a_n) has been constructed so that $|x - a_k| < 1/2^n$ for all $k \geq n$. This shows that (a_n) converges to x. Also, by construction $a_n \neq x$ and $a_n \in F_n$. This shows that K is perfect. $\square$

3.5.1. Sets of Measure Zero. Another remarkable property of K is that it is a set of **measure (or length) zero** (also called a *null set*, but this should not be confused with the empty set). In this subsection we define the notion of measure zero and will need basic properties of series. Series are covered in detail in Chapter 7; here we will only consider series of nonnegative terms and in the context of the limit of the sequence of partial sums as discussed in Subsection 2.2.5. We recall that given a sequence (a_n), when we write $\sum_{n=1}^{\infty} a_n$ we mean the limit $\lim_{n \to \infty} \sum_{i=1}^{n} a_i$ of its sequence of partial sums. As the sequence of partial sums is increasing (since we are assuming $a_i \geq 0$), the limit exists if and only if the

3.5. The Cantor Set

sequence is bounded. For calculations we only need the geometric series discussed in Subsection 2.2.5.

Definition 3.5.5. We say that $A \subseteq \mathbb{R}$ has **measure zero** if for all $\varepsilon > 0$ there exists a countable collection of open intervals I_j such that

$$A \subseteq \bigcup_{j=1}^{\infty} I_j \quad \text{and} \quad \lim_{n \to \infty} \sum_{j=1}^{n} |I_j| < \varepsilon. \qquad \diamond$$

The condition $\lim_{n \to \infty} \sum_{j=1}^{n} |I_j| < \varepsilon$ says that all finite sums of the intervals are bounded by ε. So the idea of a set of measure zero is one that, for every positive number ε, the set can be covered by intervals whose total length is less than the number ε.

As our first example we show that a set consisting of a single point has measure zero. Let $A = \{p\}$. Given $\varepsilon > 0$, set $I_1 = (p - \frac{\varepsilon}{4}, p + \frac{\varepsilon}{4})$ and $I_j = \varnothing$ for $j \geq 2$. Then $|I_1| = \frac{\varepsilon}{2}$ and $\sum_{j=1}^{n} |I_j| = \frac{\varepsilon}{2} < \varepsilon$ for all $n \in \mathbb{N}$. (If in the definition of measure zero the intervals are required to be nonempty, one can set $I_j = (p - \frac{\varepsilon}{2^{j+2}}, p + \frac{\varepsilon}{2^{j+2}})$; see Remark 3.5.6.) This idea can be used to show that any countable set has measure zero (Exercise 3.5.10).

Remark 3.5.6. The definition of measure zero is very robust, in the sense that it does not change if we require all the intervals to be nonempty or require them to be closed (Exercise 3.5.12), etc. To illustrate this idea, we give a proof that points have measure zero under a definition where all the intervals are required to be nonempty (and open). Again, let $A = \{p\}$. Given $\varepsilon > 0$, we choose I_1 to be an open interval around p of length $\varepsilon/4$, then I_2 an open interval around p of length $\varepsilon/8$, etc. So if we set $I_j = (p - \frac{\varepsilon}{2^{j+2}}, p + \frac{\varepsilon}{2^{j+2}})$, then $|I_j| = \frac{\varepsilon}{2^{j+1}}$, and, using (2.12), we have

$$\lim_{n \to \infty} \sum_{j=1}^{n} |I_j| = \lim_{n \to \infty} \sum_{j=1}^{n} \frac{\varepsilon}{2^{j+1}} = \frac{\varepsilon}{2^2} \lim_{n \to \infty} \sum_{j=1}^{n} \frac{1}{2^{j-1}}$$

$$= \frac{\varepsilon}{2^2} \lim_{n \to \infty} \frac{1 - 2^{-n}}{1 - \frac{1}{2}} = \frac{\varepsilon}{2} < \varepsilon.$$

To show that the Cantor set K has measure zero, we first recall that

$$K = \bigcap_{n=1}^{\infty} F_n,$$

where F_n is a union of 2^n closed intervals, each of length $1/3^n$. Given $\varepsilon > 0$, choose n so that $(2/3)^n < \varepsilon$. Then the sum of the lengths of the intervals comprising F_n is $(2/3)^n$, which is less than ε. Therefore, K is covered by a finite union of intervals whose total length is less than ε. As ε was arbitrary, K has measure zero.

A second computation is based on the fact that

$$G = \bigsqcup_{n=0}^{\infty} G_n.$$

Each G_n is a finite disjoint union of 2^n intervals, each of length $(1/3)^{n+1}$. So we can write that
$$|G_n| = \frac{2^n}{3^{n+1}}.$$
By the geometric sum formula (see Exercise 1.1.12), the sum of the lengths of the intervals in G_n when n ranges from $n=0$ to $n=k$ is
$$\sum_{n=0}^{k} |G_n| = \sum_{n=0}^{k} \frac{2^n}{3^{n+1}} = \frac{1}{3} \sum_{n=1}^{k+1} \left(\frac{2}{3}\right)^{n-1} = \frac{1}{3} \cdot \frac{1-(\frac{2}{3})^{k+1}}{1-\frac{2}{3}} = 1 - \left(\frac{2}{3}\right)^{k+1}.$$
It follows that the sum of the lengths of the intervals composing G is
$$\lim_{k \to \infty} \sum_{n=0}^{k} |G_n| = \lim_{k \to \infty} \left(1 - \left(\frac{2}{3}\right)^{k+1}\right) = 1.$$
Since $K = [0,1] \setminus G$, we can argue that we have removed a set of length 1 from $[0,1]$; it can be shown that this is equivalent to saying that K has measure zero (Exercise 3.5.9).

We point out, however, that knowing that K contains no intervals of positive length is not enough to prove that K has measure zero. In the exercises the reader will construct Cantor sets that contain no intervals but are not of measure zero. (The Cantor middle-thirds set is a set of measure zero but Cantor sets need not be of measure zero.)

Definition 3.5.7. A subset of $\mathbb{R}$ that contains no intervals (of positive length) and consists of at least two points is said to be **totally disconnected**. So the middle-thirds Cantor set is totally disconnected. ◊

Definition 3.5.8. A **Cantor set** in $\mathbb{R}$ is defined to be a nonempty subset of $\mathbb{R}$ that is compact, perfect, and is totally disconnected. ◊

We conclude with another representation of the middle-thirds Cantor set that is useful in exercises; for this we need infinite series. Let $x \in K$. Write x in base 3. If x is in $F[0]$, then its first digit in its base 3 representation is 0; if x is in $F[2]$, then its first digit in its base 3 representation is 2. In this way we obtain a sequence $x_i \in \{0, 2\}$, and we can write

(3.12) $$K = \left\{ x \in [0,1] : x = \lim_{n \to \infty} \sum_{i=1}^{n} \frac{x_i}{3^i}, \ x_i \in \{0,2\} \right\}.$$

Exercises: The Cantor Set

3.5.1 Define F_n as in the construction of the Cantor set K. Show that for all n, the endpoints of the closed subintervals in F_n belong to the Cantor set.

3.5.2 *Middle-fifths set—classic case*: Modify the construction of the Cantor middle-thirds set in the following way. At the first stage remove the central interval of length $\frac{1}{5}$, and at the nth stage, instead of removing the open middle-thirds of each of the 2^n remaining subintervals, remove the open

3.5. The Cantor Set

middle-fifth of each subinterval. So in this case $G_0 = (2/5, 3/5), F_1 = [0, 2/5] \cup [3/5, 1]$; then we remove each of the middle-fifths of the subintervals of F_1, so $G_1 = (4/25, 6/25) \cup (19/25, 21/25)$, etc. Prove that this set is a Cantor set (i.e., it is compact, totally disconnected, and perfect). Also show that it is of measure zero.

3.5.3 *Middle-fifths set—symmetric case*: Modify the construction of the Cantor middle-thirds set in the following way. At the first stage divide $[0, 1]$ into five subintervals of equal length and remove the central interval of length $\frac{1}{5}$. At the nth stage, for each of the 4^n remaining subintervals, divide them into five subintervals and remove the middle one. So in this case $G_0 = (2/5, 3/5), F_1 = [0, 1/5] \cup [1/5, 2/5] \cup [3/5, 4/5] \cup [4/5, 1]$; then we remove each of the middle-fifths of the four subintervals in F_1, so $G_1 = (2/25, 3/25) \cup (7/25, 8/25) \cup (17/25, 18/25) \cup (22/25, 23/25)$, etc. Prove that this set is a Cantor set (i.e., it is compact, totally disconnected, and perfect). Is it of measure zero?

3.5.4 Modify the construction of the middle-thirds Cantor set in the following way. At the first stage remove a central interval of length $\frac{1}{6}$ and at the nth stage continue breaking intervals into three subintervals but remove a subinterval in the middle of a length that is half of what you remove in the Cantor middle-thirds set. So in this case $G_0 = (5/12, 7/12)$, and in each of the subintervals $[0, 5/12]$ and $[7/12, 1]$, we remove an open subinterval of length $1/18$ (as in the Cantor middle-thirds case we would be removing a subinterval of length $1/9$). Prove that this set is a Cantor set. Also show that it is not of measure zero by showing that the sum of the lengths of the intervals that are removed is $1/2$.

3.5.5 Prove that the Cantor set K satisfies the base 3 representation in (3.12).

3.5.6 Prove that if K is the middle-thirds Cantor set, then $K + K = [0, 2]$, where $K + K = \{z : z = x + y \text{ for some } x, y \in K\}$.

3.5.7 Prove that the function ϕ in the proof of Theorem 3.5.4 is surjective.

* 3.5.8 Prove that a nonempty perfect set is uncountable.

3.5.9 Prove that $A \subseteq [0, 1]$ has measure zero if and only if $[0, 1] \setminus A$ can be written as a disjoint union $\bigsqcup_{j=1}^{\infty} I_j$ such that $\lim_{n \to \infty} \sum_{j=1}^{n} |I_j| = 1$.

3.5.10 Prove that every countable set in $\mathbb{R}$ has measure zero.

3.5.11 Let $A \subseteq \mathbb{R}$. Prove that in the definition that a set A has measure zero, one can allow some of the intervals I_j to be nonempty.

3.5.12 Let $A \subseteq \mathbb{R}$. Prove that A has measure zero if and only if for every ε there exist closed intervals K_j such that $A \subseteq \bigcup_{j=1}^{\infty} K_j$ and $\sum_{j=1}^{\infty} |K_j| < \varepsilon$. (Again here it does not make a difference if one requires the closed intervals to be of positive length or not.)

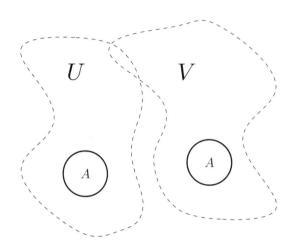

Figure 3.6. Open sets U and V separating A.

3.6. Connected Sets in $\mathbb{R}$

To define when a set is connected, it is easier to start with when it is not.

Definition 3.6.1. Let $A \subseteq \mathbb{R}$. We say that A is **disconnected** if there exist open sets U and V in $\mathbb{R}$ such that

$$A \cap U \neq \varnothing,$$
$$A \cap V \neq \varnothing,$$
$$A \cap U \cap V = \varnothing, \text{ and}$$
$$A \subseteq U \cup V. \qquad \diamond$$

We think of U and V as *separating* or *disconnecting* A; each of U and V contain a part of A. The condition that U and V both be open sets is important. If they were arbitrary sets, we could always disconnect any nonempty set A of more than one point; in that case we could simply let U consist of a point of A and let V be its complement, but here U would not be open in $\mathbb{R}$. Equivalently, though, one could require that both U and V be closed (simply consider their complements—see Exercise 3.7.12). Note that we only require that they be disjoint when intersected with A as illustrated in Figure 3.6.

An example of a disconnected set is given by $A = [0,1] \cup \{2\}$. In this case we can choose $U = (-1, 3/2)$ and $V = (5/3, 3)$. Then each of U and V intersect A, they are disjoint, and clearly $A \subseteq U \cup V$. Therefore, A is disconnected. Now, a point $\{p\}$ cannot be disconnected since if a set U intersects $\{p\}$, then, if a set V also intersects $\{p\}$, it cannot be disjoint from U.

Definition 3.6.2. We define a set $A \subseteq \mathbb{R}$ to be **connected** if it is not disconnected.
$\diamond$

We have already seen that a set consisting of a single point is connected, and that the empty set is trivially connected. The following proposition shows that the connected sets in $\mathbb{R}$ are precisely the intervals.

Theorem 3.6.3. *Let $A \subseteq \mathbb{R}$. A is connected if and only if A is an interval.*

Proof. Suppose first that A is an interval. We have already seen that if A consists of a single point or is empty, then it is connected. Assume now that there exist disjoint, when restricted to A, open sets U and V such that $A \cap U \neq \emptyset, A \cap V \neq \emptyset$, and $A \subseteq U \cup V$. Let $x \in A \cap U$ and $y \in A \cap V$. Without loss of generality we may assume $x < y$. We study the largest interval starting at x that is included in $A \cap U$ and does not contain y. For this define

$$\alpha = \sup\{z : x < z < y \text{ and } [x, z) \subseteq A \cap U\}.$$

The set above is bounded by y, and is nonempty since A is a nontrivial interval, so its supremum exists. Also, $x < \alpha \leq y$ and $\alpha \in A$. Now, if $\alpha \in U$, as U is open, there exists $\varepsilon_1 > 0$ such that $[\alpha, \alpha + \varepsilon_1) \subseteq U$, which implies $[x, \alpha + \varepsilon_1) \subseteq A \cap U$, contradicting that α is the supremum. So $\alpha \in V$. But as V is open, there exists $\varepsilon_2 > 0$ such that $(\alpha - \varepsilon_2, \alpha] \subseteq A \cap V$, contradicting that $U \cap V \cap A = \emptyset$. It follows that A is connected.

For the converse assume that A is a nonempty connected set. We will show that for all $x, y \in A$ with $x < y$, whenever $x < z < y$, we have that $z \in A$. This will prove that A is an interval (see Exercise 3.6.3). Now, if z were not in A, then the sets $U = (-\infty, z)$ and (z, ∞) would disconnect A, a contradiction. $\square$

Exercises: Connected Sets in $\mathbb{R}$

3.6.1 Prove, without using Theorem 3.6.3, that $\mathbb{Q}$ is disconnected.

3.6.2 Prove, without using Theorem 3.6.3, that if $A, B \subseteq \mathbb{R}$ are connected and $A \cap B \neq \emptyset$, then $A \cup B$ is connected.

3.6.3 Prove that a set $A \subseteq \mathbb{R}$ is an interval if and only if for all $x, y \in A$, if $x < z < y$, then $z \in A$. (Note that the condition is trivially satisfied when A is empty or consists of a single point.)

3.6.4 Let $A \subseteq \mathbb{R}$ consist of at least two points. Prove that A is totally disconnected if and only if the only nonempty connected sets in A consists of single points. Prove that the Cantor set satisfies this property.

3.6.5 Let $A \subseteq R$. Prove that if A is connected and $A \subseteq B \subseteq \bar{A}$, then B is connected.

3.6.6 Let $\{A_n\}_{n \in \mathbb{N}}$ be a collection of connected sets. Prove that if $\bigcap_{n \in \mathbb{N}} A_n \neq \emptyset$, then $\bigcup_{n \in \mathbb{N}} A_n$ is connected.

3.7. Compactness, Connectedness, and Completeness in Metric Spaces

In this section we generalize the notions of compactness, connectedness, and completeness to metric spaces.

3.7.1. Compactness

Definition 3.7.1. A metric space (X, d) is said to be **sequentially compact** if every sequence in X has a subsequence that converges to a point in X. ◇

This notion extends to any subset A of X as A can be considered a metric space with the metric d from X.

As in the case of $\mathbb{R}$ with the Euclidean metric, one can show that a sequentially compact set A in a metric space X is closed and bounded. However, the converse is not true in general; in Example 3.7.9 we show that the converse of Lemma 3.7.3 does not hold for some metric spaces.

Definition 3.7.2. A set A in X is **bounded** if A is a subset of $B(x, \varepsilon)$ for some $x \in X$ and $\varepsilon > 0$. ◇

The reader should verify that this agrees with our definition of bounded sets in $\mathbb{R}$.

Lemma 3.7.3. *Let (X, d) be a metric space. If A is a subset of X that is sequentially compact, then it is closed and bounded.*

Proof. If x is an accumulation point of A, there exists a sequence (a_n) in A converging to x. By sequential compactness this sequence has a subsequence that converges to a point of A. Since every subsequence of (a_n) must converge to x, it follows that x is in A. To show it is bounded, assume it is nonempty. Let $x \in A$. If A is not bounded, for each $n \in \mathbb{N}$ we can choose $x_n \in X \setminus B(x, n)$. So $d(x, x_\ell) \geq n$ for all $\ell \geq n$. By sequential compactness, there is a subsequence (x_{n_i}) converging to a point p in X. Then there exists an integer N so that $d(p, x_{n_i}) < 1$ for all $i \geq N$. Choose $K > d(x, p) + 1$. Then, for all $i \geq N$,

$$d(x, x_{n_i}) \leq d(x, p) + d(p, x_{n_i}) < d(x, p) + 1 < K.$$

Choosing i so that $n_i > K$ leads to a contradiction; therefore, the set A is bounded. □

By Theorem 3.4.8 we know that closed and bounded subsets of $\mathbb{R}$ with the Euclidean metric are sequentially compact; this is also true for closed and bounded subsets of $\mathbb{R}^n$ (by Exercise 3.7.4 and Theorem 3.7.6). However, in general the converse of Lemma 3.7.3 does not hold. In fact, by Lemma 3.7.8 there exists a metric in $\mathbb{R}^n$ that is topologically equivalent to the Euclidean metric, i.e., it has the same open and closed sets, but under this new metric every set is bounded (so there are bounded closed sets that are not compact).

A different example of a compact metric space is given below.

Example 3.7.4. We show that $\{0, 1\}^{\mathbb{N}}$, with its metric δ, is sequentially compact. Let (x_n) be a sequence of elements of $\{0, 1\}^{\mathbb{N}}$. Let $x_n(i)$ denote the ith entry of x_n ($x_n(i) \in \{0, 1\}$). We define, by induction, a subsequence of (x_n) that converges to an element of $\{0, 1\}^{\mathbb{N}}$. For the base case consider the sequence $\{x_n(1)\}_{n \geq 1}$. It must be the case that there are infinitely many integers n so that $x_n(1) = 0$ or there are infinitely many integers n so that $x_n(1) = 1$ (the "or" is inclusive of course). If $x_n(1) = 0$ for infinitely many n, then let n_1 be the first integer such that $x_{n_1}(1) = 0$.

If $x_n(1) = 1$ for infinitely many n, then n_1 is the first integer such that $x_{n_1}(1) = 1$. (If both possibilities occur, choose the first.) To illustrate the idea of the proof, we describe how to choose n_2 before passing to the inductive step. Suppose we have chosen $x_{n_1}(1) = 0$. There are infinitely many integers n such that

$$x_n(1) = 0 \quad \text{and} \quad x_n(2) = 0, \text{ or}$$
$$x_n(1) = 0 \quad \text{and} \quad x_n(2) = 1.$$

If the first case holds, choose n_2 the smallest integer greater than n_1 such that $x_{n_2}(2) = 0$; otherwise, choose it so that $x_{n_2}(2) = 1$.

Now, for the inductive step suppose $n_1, \ldots, n_i$ have been defined and we need to define n_{i+1}. Again, there are infinitely many integers n so that

$$x_n(1) = x_{n_1}(1), x_n(2) = x_{n_2}(2), \ldots, x_n(i) = x_{n_i}(1) \text{ and } x_n(i+1) = 0, \text{ or}$$
$$x_n(1) = x_{n_1}(1), x_n(2) = x_{n_2}(2), \ldots, x_n(i) = x_{n_i}(1) \text{ and } x_n(i+1) = 1.$$

If the first possibility occurs, let n_{i+1} be the first integer greater than n_i so that $x_{n_{i+1}}(i+1) = 0$, and if the second possibility occurs, set $x_{n_{i+1}}(i+1) = 1$. This generates a subsequence (x_{n_k}). Let x be the element of $\{0,1\}^{\mathbb{N}}$ such that $x(i) = x_{n_i}(i)$. From the definition it follows that (x_{n_i}) converges to x in $(\{0,1\}^{\mathbb{N}}, \delta)$. Therefore $\{0,1\}^{\mathbb{N}}$ is sequentially compact.

Definition 3.7.5. A metric space (X, d) is said to be **compact** if for every open cover $\{G_\alpha\}_{\alpha \in \Gamma}$ of X there exists a finite set $\Gamma' \subseteq \Gamma$ such that

$$X = \bigcup_{\alpha \in \Gamma'} G_\alpha. \qquad \diamond$$

Theorem 3.7.6. *A metric space is compact if and only if it is sequentially compact.*

Proof. Suppose the metric space (X, d) is compact, and let (x_n) be a sequence in X. If there is no convergent subsequence, then for each $p \in X$ there exists an open ball $B(p, \varepsilon_p)$, $\varepsilon_p > 0$, such that $B(p, \varepsilon_p)$ does not contain any elements of the sequence other than possibly p (when p appears in the sequence). (Since there is no subsequence converging to p, there exists $\varepsilon' > 0$ and $N \in \mathbb{N}$ such that $x_i \notin B(p, \varepsilon') \setminus \{p\}$ for all $i > N$; since there are at most finitely many points inside the ball one can choose a smaller radius ε_p that avoids all of them.) Then $\{B(p, \varepsilon_p) : p \in X\}$ forms an open cover of X. Compactness implies that there is a finite subcover $B(p_1, \varepsilon_1), \ldots, B(p_k, \varepsilon_k)$, for some $k \in \mathbb{N}$; this means that the sequence has only finitely many values, so one has to be attained infinitely often, meaning that there would be a converging subsequence, a contradiction.

For the converse we outline the ideas and ask the reader to complete the details. Let (X, d) be a sequentially compact metric space. By Exercise 3.7.5, (X, d) is separable. By an argument similar to Exercise 3.4.7, every open cover of X has a countable subcover. Then adapt to metric spaces the second proof in the last part of Theorem 3.4.7 to show that there is a finite subcover. $\square$

Definition 3.7.7. Let (X, d) be a metric space. A metric d' defined on X is said to be a **topologically equivalent metric** to d if for all $A \subseteq X$, the set A is open in (X, d) if and only if it is open in (X, d'). $\diamond$

We observe that if d and d' are topologically equivalent metrics, and a sequence (x_n) converges in d, then it converges in d' since the open sets are the same. Conversely, if a sequence converges with respect to d if and only if it converges for d', then the closed sets are the same, so are the open sets, and therefore the metrics are topologically equivalent.

Given a metric d on X, one can always define a bounded metric by

(3.13) $$\bar{d}(x,y) = \min\{d(x,y), 1\}.$$

Lemma 3.7.8. *Let (X, d) be a metric space, and let $\bar{d}$ be the corresponding bounded metric defined in (3.13). Then $\bar{d}$ is a metric on X, and it is topologically equivalent to d.*

Proof. To show that $\bar{d}$ is a metric it suffices to verify the triangle inequality:
$$\bar{d}(x,y) \le \bar{d}(x,z) + \bar{d}(z,y).$$

If $d(x,z) < 1$ and $d(z,y) < 1$, then
$$\bar{d}(x,y) \le d(x,y) \le d(x,z) + d(z,y) = \bar{d}(x,z) + \bar{d}(z,y).$$

If $d(x,z) \ge 1$ or $d(z,y) \ge 1$, the inequality trivially holds.

Now let $A \subseteq X$ be open with respect to d. If $x \in A$, then there exists $\varepsilon > 0$ such that $B_d(x,\varepsilon) \subseteq A$. Since $\bar{d}(x,y) \le d(x,y)$, for all $x,y \in X$, then for all $\varepsilon > 0$, $B_{\bar{d}}(x,\varepsilon) \subseteq B_d(x,\varepsilon)$, showing that A is open with respect to $\bar{d}$. Finally, let A be open with respect to $\bar{d}$. If $x \in A$, then there exists $\varepsilon > 0$ such that $B_{\bar{d}}(x,\varepsilon) \subseteq A$. Let $\varepsilon' = \min\{\varepsilon, 1\}$. Then $B_{\bar{d}}(x,\varepsilon') = B_d(x,\varepsilon)$, so A is open with respect to d. □

Example 3.7.9. Let d be Euclidean distance in $\mathbb{R}$, and let $\bar{d}$ be the corresponding bounded metric. Then any closed subset of $\mathbb{R}$ is bounded in $(\mathbb{R}, \bar{d})$, but it need not be compact (e.g., $[0, \infty)$ or even $\mathbb{R}$). Thus the Heine–Borel theorem does not hold in $(\mathbb{R}, \bar{d})$, showing that the equivalence of closed and bounded with compactness depends on the metric, so this theorem does not hold on an arbitrary metric space.

There is a notion stronger than topological equivalence, simply called equivalence.

Definition 3.7.10. Let (X, d) be a metric space. A metric d' defined on X is said to be an **equivalent metric** to d if there exist positive constants c_1 and c_2 such that

(3.14) $$c_1 d'(x,y) \le d(x,y) \le c_2 d'(x,y) \text{ for all } x,y \in X. \qquad \Diamond$$

The property is clearly reflexive as (3.14) implies that
$$(1/c_2) d(x,y) \le d'(x,y) \le (1/c_1) d(x,y).$$

Also, if d' is equivalent to d, then d' is topologically equivalent to d (Exercise 3.7.19).

3.7. Compactness, Connectedness, and Completeness in Metric Spaces

3.7.2. Connected Sets. Now we consider the notion of connectedness in the case of metric spaces.

Definition 3.7.11. Let (X, d) be a metric space. We say that (X, d) is **disconnected** if there exist nonempty open disjoint sets U and V such that $X = U \sqcup V$. A set $A \subseteq X$ is **disconnected** if (A, d) as a metric space is disconnected. ◇

Definition 3.7.12. We define A to be **connected** if A is not disconnected. ◇

The reader should verify that when A is a subset of $\mathbb{R}$ both definitions agree.

3.7.3. Completeness. The notion of Cauchy sequences of real numbers (defined in Section 2.3) can be extended in a natural way to metric spaces.

Definition 3.7.13. Let (X, d) be a metric space. A sequence (x_n) is said to be a **Cauchy sequence** if

for every $\varepsilon > 0$ there exists $N \in \mathbb{N}$ such that
$$d(x_n, x_m) < \varepsilon \text{ whenever } n, m \geq N.$$
◇

The following lemma has essentially the same proofs as Lemmas 2.3.2 and 2.3.3, and it is left as an exercise.

Lemma 3.7.14. *Let (x_n) be a sequence in a metric space (X, d).*

(a) *If (x_n) converges, then (x_n) is a Cauchy sequence.*

(b) *If (x_n) is a Cauchy sequence in (X, d), then (x_n) is bounded.*

We define a notion of completeness for metric spaces, different from order completeness, which was defined earlier and required the set to be a field with an order. To differentiate this property from order completeness, we may call it Cauchy completeness instead of simply calling it completeness. One can usually tell from the context, though; when we say completeness in a metric space we mean Cauchy completeness.

Definition 3.7.15. A metric space (X, d) is said to be **complete**, or **Cauchy complete**, if every Cauchy sequence in (X, d) converges to a point in X. ◇

By Theorem 2.3.5, the set of real numbers with the Euclidean metric forms a (Cauchy) complete metric space, but the rational numbers with the Euclidean metric is not (Cauchy) complete as one can choose a sequence of rational numbers converging to $\sqrt{2}$. However, the Euclidean space $\mathbb{R}^n$ is Cauchy complete.

Lemma 3.7.16. *Let (X, d) be a complete metric space. A set $A \subseteq X$ is closed if and only if (A, d) is a complete metric space.*

Proof. Suppose A is closed, and let (a_n) be a Cauchy sequence in (A, d). Then it is a Cauchy sequence in (X, d), and since X is complete, the sequence (a_n) converges to a point p in X. This means that p is a closure point of A, and as A is closed, p is in A, which means that (A, d) is complete.

Now suppose (A, d) is complete and p is a closure point of A. Then there is a sequence (a_n) in A that converges to p. As (a_n) is a convergent sequence in a metric space, it must be Cauchy in (A, d). Since (A, d) is complete, the sequence must converge to a point in A. This point must be p, so p is in A and therefore A is closed. □

Lemma 3.7.17. *Let (X, d) be a metric space. If it is compact, then it is complete.*

Proof. Let (x_n) be a Cauchy sequence in X. By compactness it has a convergent subsequence (x_{n_i}) to a point p in X. As (x_n) is a Cauchy sequence, it must converge to p (see Exercise 3.7.1). □

Given a metric space (X, d), it is possible to extend it to another metric space (X', d') that is complete; this is discussed in Subsection 4.5.1.

Exercises: Compactness and Completeness in Metric Spaces

3.7.1 Let (X, d) be a metric space. Prove that if a Cauchy sequence in (X, d) has a convergent subsequence, then the original Cauchy sequence converges. (*Hint*: Verify that this is essentially the same proof as in Lemma 2.3.4.)

3.7.2 A set D in a metric space (X, d) is said to be dense if for each $x \in X$ and each $\varepsilon > 0$ there exists $q \in D$ such that $d(x, q) < \varepsilon$. A metric space is **separable** if it has a countable dense subset. Prove that a compact metric space is separable.

3.7.3 Show that $(0, 1)$ is not complete with the Euclidean metric, and find a topologically equivalent metric so that it is complete.

3.7.4 *Heine–Borel for $\mathbb{R}^d$*: Prove that a subset of $\mathbb{R}^d$ with the Euclidean metric is compact if and only if it is closed and bounded.

3.7.5 A metric space (X, d) is **totally bounded** if for each $\varepsilon > 0$ there exists finitely many points $x_1, \ldots, x_n$ in X so that $X = B(x_1, \varepsilon) \cup \cdots \cup B(x_n, \varepsilon)$. Prove that if (X, d) is sequentially compact, then it is totally bounded. Use this to deduce that a sequentially compact metric space is separable. (It is easier to show that compactness implies totally bounded, but if you want to use this for Exercise 3.7.6, you should give a direct proof just using the sequential compactness property.)

3.7.6 Complete the details in the second part of the proof of Theorem 3.7.6.

3.7.7 Prove that open balls in $\mathbb{R}^n$ are connected, and find an example of an open set in $\mathbb{R}^n$ that is connected and is not an open ball.

3.7.8 Let $A = \{(x, \sin(\frac{1}{x})) : 0 < x < \pi\} \cup \{(0, y) : -1 \leq y \leq 1\}$. Prove that $A \subseteq \mathbb{R}^2$ is connected.

3.7.9 Let (X, d) be a metric space. Let $A \subseteq X$. Prove that if A is connected and $A \subseteq B \subseteq \bar{A}$, then B is connected.

3.7. Compactness, Connectedness, and Completeness in Metric Spaces

3.7.10 Let (X,d) be a metric space, and let x be a point in X. Define the **connected component** of x to be the union of all the connected sets containing x. Prove that the connected component of x is the largest connected set containing x.

3.7.11 Let (X,d) be a metric space. Define X to be **totally disconnected** if the connected component of points is just the point itself. Prove that this agrees with the definition of "totally disconnected" for subsets of $\mathbb{R}$.

3.7.12 Let (X,d) be a metric space, and let $A \subseteq X$. Prove that A is disconnected if and only if there exist closed sets E, F such that $A \cap E \cap F = \emptyset$, $A \subseteq E \cup F$, $A \cap E \neq \emptyset$, and $A \cap F \neq \emptyset$.

3.7.13 Prove that $\mathbb{R}^d$ with the Euclidean metric is Cauchy complete.

3.7.14 Prove Lemma 3.7.14.

3.7.15 Let C_i be a nonempty compact set in $\mathbb{R}^d$ for each $i \in \mathbb{N}$, and suppose that $C_1 \supset C_2 \supset C_3 \cdots$. Prove that $\bigcap_i C_i \neq \emptyset$. (The case for $d = 1$ was Exercise 3.4.6.)

3.7.16 Let (X,d) be a metric space. Prove that a sequence (x_n) converges to x in (X,d) if and only if for every open set G such that $x \in G$ there exists $N \in \mathbb{N}$ such that $x_n \in G$ for all $n \geq N$.

3.7.17 Let (X,d) be a metric space, and let d' be a metric on X topologically equivalent to d. Prove that if (x_n) is a sequence in X that converges with respect to the metric d, then it also converges with respect to the metric d', and that if this property holds for every sequence, then d' is topologically equivalent to d.

3.7.18 Let (X,d) be a complete metric space. Define the diameter of a set A in X to be $\dim(A) = \sup\{d(x,y) : x, y \in A\}$. It is well-defined though it may be ∞. Let (A_n) be a nested sequence of closed sets in X, i.e., each A_n is closed and $A_{n+1} \subseteq A_n$. Prove that if $\dim(A_n) \to 0$, then $\bigcap_n A_n$ consists of a single point p in X.

3.7.19 Prove that if d' is equivalent to d, then d' is topologically equivalent to d.

3.7.20 Let (X,d) be a metric space, and let d' be a metric on X that is equivalent to d. Prove that a sequence is Cauchy with respect to d if and only if it is Cauchy with respect to d'. Use Exercise 3.7.3 to show that this does not hold for topologically equivalent metrics. (See also Exercise 4.5.12) Prove that if (X,d) is a complete metric space and d' is equivalent to d, then (X,d') is a complete metric space.

Chapter 4

Continuous Functions

4.1. Continuous Functions on $\mathbb{R}$

Continuity is now one of the basic concepts of analysis. It was not given a formal definition until the early 1800s by Bolzano (1781–1848) and Cauchy (1789–1857).

Let A be a subset of $\mathbb{R}$, and let $f : A \to \mathbb{R}$ be a function. The first notion that we consider is continuity at a point x in the domain A of f. The idea of continuity at x is that f should not vary too much around x. In other words, we want to think of f as being "locally almost constant" at x ($f(y)$ should be "just about" $f(x)$ for y suitably chosen around x). We can write this as

$$|f(x) - f(y)| < \varepsilon, \tag{4.1}$$

for some $\varepsilon > 0$ that we shall think of as small. Note that (4.1) is equivalent to

$$f(x) - \varepsilon < f(y) < f(x) + \varepsilon,$$

which informally we may think of as saying that $f(y)$ is ε-*close* to $f(x)$.

So by locally almost constant we mean that $f(y)$ is ε-close to $f(x)$ for all y sufficiently close to x. Finally, we need to formalize the term "sufficiently close to x". By this we mean that $|x - y| < \delta$, for some positive number δ that has been chosen in advance of y.

Definition 4.1.1. Let $f : A \to \mathbb{R}$ be a function, and let x be a point in A. We say that f is **continuous at** x if

> for every $\varepsilon > 0$ there exists $\delta > 0$ such that
> when $y \in A$ and $|x - y| < \delta$, then $|f(x) - f(y)| < \varepsilon$. ◊

It is important to note that the number δ generally depends on x and ε.

Definition 4.1.2. The function f is said to be **continuous on** A if it is continuous at x for each x in A. If f is not continuous at x, we say that f is **discontinuous at** x. It is **discontinuous on** A if it is discontinuous at some point of A. ◊

Question 4.1.3. Let $f : A \to \mathbb{R}$ be a function, and let $x \in A$ be an isolated point of A. Prove that f is continuous at x.

Example 4.1.4. The simplest example of a continuous function is a constant function: let $c \in \mathbb{R}$, and for any $A \subseteq \mathbb{R}$ define the function $C : A \to \mathbb{R}$ by $C(x) = c$ for all $x \in A$. Fix $x \in A$. We note that $|C(x) - C(y)| = |c - c| = 0$ for any $y \in A$. So given $\varepsilon > 0$, we can choose any $\delta > 0$. This is because for all $y \in A$ such that $|x - y| < \delta$ it follows that $|C(x) - C(y)| = 0 < \varepsilon$. Therefore, C is continuous at x for all $x \in A$.

Example 4.1.5. For the next simple example let $f : \mathbb{R} \to \mathbb{R}$ be defined by $f(x) = kx$, where $k \neq 0$ is a constant (the $k = 0$ case has already been considered in Example 4.1.4). We show that f is continuous at x for all $x \in \mathbb{R}$. We start by examining in detail the condition that needs to be satisfied for a fixed $x \in \mathbb{R}$ and a given $\varepsilon > 0$. We call this step the **development**. Fix $x \in \mathbb{R}$. We need to find a condition on y so that $|f(x) - f(y)| < \varepsilon$, which using the definition of f becomes $|kx - ky| < \varepsilon$, or $|k||x - y| < \varepsilon$. So we need to choose $\delta > 0$ such that y satisfies $|k||x - y| < \varepsilon$ when $|x - y| < \delta$. This suggests that we should let $\delta = \varepsilon/|k|$. This ends the development part, and now we are ready to write the proof.

Let $x \in \mathbb{R}$. Given $\varepsilon > 0$, choose $\delta = \varepsilon/|k|$. Then $\delta > 0$. If $|x - y| < \delta$, then $|x - y| < \varepsilon/|k|$, or $|kx - ky| < \varepsilon$ for all y in $\mathbb{R}$. Therefore, f is continuous at x. (The development part is often needed to find the proof, but logically only the proof part is necessary and typically only the proof is published.)

Remark 4.1.6. A function $f : A \to \mathbb{R}$ is not continuous at $x \in A$ if and only if

there exists $\varepsilon > 0$ such that for every $\delta > 0$

there is some $y \in A$ such that $|x - y| < \delta$ and $|f(x) - f(y)| \geq \varepsilon$.

Example 4.1.7. Define a function $D : \mathbb{R} \to \mathbb{R}$ by

$$(4.2) \qquad D(x) = \begin{cases} 1 & \text{if } x \in \mathbb{Q}, \\ 0 & \text{if } x \in \mathbb{R} \setminus \mathbb{Q}. \end{cases}$$

This function is called **Dirichlet's function**. We will show that D is discontinuous at every $x \in \mathbb{R}$.

To be discontinuous at $x \in \mathbb{R}$ means that there exists $\varepsilon > 0$ so that for all $\delta > 0$, for some $y \in A$ with $|x - y| < \delta$ we have $|D(x) - D(y)| \geq \epsilon$.

First assume that x is rational, so $D(x) = 1$. Choose any ε such that $0 < \varepsilon < 1$. For any $\delta > 0$ there exists $y \in \mathbb{R} \setminus \mathbb{Q}$ such that $|x - y| < \delta$. For this y, $D(y) = 0$, so $|D(x) - D(y)| = 1 > \varepsilon$; therefore, D is not continuous at $x \in \mathbb{Q}$. A similar argument shows that D is not continuous at $x \in \mathbb{R} \setminus \mathbb{Q}$. (We shall show in Chapter 6 that f is not Riemann integrable.)

We conclude this example by remarking that D restricted to $\mathbb{Q}$ is continuous. In fact, define a function $g : \mathbb{Q} \to \mathbb{R}$ by $g(x) = D(x)$ for all $x \in \mathbb{Q}$, the **restriction** of D to $\mathbb{Q}$. Of course g is just the constant function 1 on $\mathbb{Q}$ and is continuous at every $x \in \mathbb{Q}$ as a function whose domain is $\mathbb{Q}$. This example and Question 4.1.3 show that even for continuity at a point, it is relevant what the domain of the function is.

4.1. Continuous Functions on $\mathbb{R}$

Example 4.1.8. Let $f : (0, \infty) \to \mathbb{R}$ be defined by $f(x) = \frac{1}{x}$. We show that for all $x \in (0, \infty)$ f is continuous at x. We start with the development, where we want to study the choice of δ given $\varepsilon > 0$. For this we calculate

$$|f(x) - f(y)| = \left|\frac{1}{x} - \frac{1}{y}\right| = \frac{|x-y|}{|xy|} = \frac{|x-y|}{xy}.$$

As x is fixed, we see that we need to control y so that

$$\frac{|x-y|}{xy} < \varepsilon.$$

We know we can require $|x - y| < \delta$, which is equivalent to

$$x - \delta < y < x + \delta.$$

To find an upper bound for $1/y$, we look for a lower bound for y. If we had $\delta < \frac{x}{2}$ (any other choice of δ that is less than x also works), we would have

$$\frac{x}{2} < y,$$

which implies

$$\frac{1}{y} < \frac{2}{x}.$$

Now we have $1/y$ bounded above by $2/x$. Next we continue the calculation

$$|f(x) - f(y)| = \frac{|x-y|}{xy} < \frac{\delta}{x}\frac{2}{x}.$$

From here we see that the second condition we need δ to satisfy is $2\delta/x^2 < \varepsilon$, which we can write as

$$\delta < \frac{\varepsilon x^2}{2}.$$

Now we start the proof. Let $x \in (0, \infty)$. Given $\varepsilon > 0$, choose $\delta > 0$ so that

$$0 < \delta < \min\left\{\frac{x}{2}, \frac{\varepsilon x^2}{2}\right\}.$$

If $|x - y| < \delta$ and $y \in (0, \infty)$, then

$$\frac{1}{y} < \frac{2}{x}.$$

Then

$$|f(x) - f(y)| = \left|\frac{1}{x} - \frac{1}{y}\right| = \frac{|x-y|}{xy} < \frac{\delta}{x}\frac{2}{x} < \varepsilon.$$

This shows that f is continuous at x.

The following characterization is very useful.

Proposition 4.1.9 below is trivially true when x is an isolated point of A, since we have seen that a function is always continuous at its isolated points, and if p is an isolated point of A, then there is no sequence of points in A converging to p. So the main content of the proposition is when x is an accumulation point of A. If the domain of the function is an interval, which is the case in most applications, then every point of the domain is an accumulation point.

Proposition 4.1.9. *Let $A \subseteq \mathbb{R}$, let $f : A \to \mathbb{R}$ be a function, and let $x \in A$. Then f is continuous at x if and only if for every sequence (a_n) in A such that $\lim_{n \to \infty} a_n = x$ it is the case that $\lim_{n \to \infty} f(a_n) = f(x)$.*

Proof. Suppose that f is continuous at $x \in A$, and let (a_n) be a sequence in A converging to x. Let $\varepsilon > 0$. There exists $\delta > 0$ so that
$$|f(x) - f(y)| < \varepsilon \text{ whenever } y \in A \text{ and } |x - y| < \delta.$$
For this $\delta > 0$ there exists $N \in \mathbb{N}$ such that
$$|a_n - x| < \delta \text{ for all } n \geq N.$$
Now let $n \geq N$. So $|a_n - x| < \delta$, which implies that $|f(a_n) - f(x)| < \varepsilon$, showing that $(f(a_n))$ converges to $f(x)$.

For the converse suppose that for every sequence (a_n) in A that converges to x, the sequence $(f(a_n))$ converges to $f(x)$. If f is not continuous at $x \in A$, then there exists $\varepsilon > 0$ so that for every $\delta > 0$ there exists $y \in A$ such that
$$|x - y| < \delta \quad \text{but } |f(x) - f(y)| \geq \varepsilon.$$
So if for each $n \in \mathbb{N}$ we set $\delta_n = 1/n$, then there exists $y_n \in B(x, 1/n) \cap A$ such that
$$|f(x) - f(y_n)| \geq \varepsilon.$$
Since $y_n \in B(x, 1/n)$, it follows that (y_n) converges to x (Exercise 4.1.6), but $(f(y_n))$ cannot converge to $f(x)$, a contradiction. Therefore, f is continuous at x. $\square$

Proposition 4.1.10. *Let $A \subseteq \mathbb{R}$, and let $f, g : A \to \mathbb{R}$ be continuous functions. Then the following hold:*

(1) *$c \cdot f : A \to \mathbb{R}$ is continuous for each constant $c \in \mathbb{R}$.*

(2) *$f + g : A \to \mathbb{R}$ is continuous.*

(3) *$f \cdot g : A \to \mathbb{R}$ is continuous.*

(4) *f/g is continuous provided that $g(x) \neq 0$ for all $x \in A$.*

Proof. We only prove part (2) as the proofs of the other parts are similar. Let x be an element of A. If x is an isolated point, we are done. Now let (a_n) be a sequence in A converging to x. Since f and g are continuous, the sequences $(f(a_n))$ and $(g(a_n))$ converge to $f(x)$ and $g(x)$, respectively. Then $(f + g)(a_n) = f(a_n) + g(a_n)$ converges to $f(x) + g(x) = (f + g)(x)$. Therefore, $f + g$ is continuous at x. $\square$

The following proposition is useful to prove continuity for new functions that are the composition of two given continuous functions.

Proposition 4.1.11. *Let $A, B \subseteq \mathbb{R}$, and let $f : A \to \mathbb{R}$ and $g : B \to \mathbb{R}$ be continuous functions. If $f(A) \subseteq B$, then*
$$g \circ f : A \to \mathbb{R}$$
is a continuous function.

Proof. Let $x \in A$. Assume that x is an accumulation point of A, and let (a_n) be a sequence in A converging to x. As f is continuous at x, $(f(a_n))$ is a sequence converging to $f(x) \in B$. Since g is continuous at $f(x)$, then $(g(f(a_n)))$ converges to $g(f(x))$. $\square$

4.1.1. Continuity Using Inverse Images. There is another characterization of continuity that does not use the distance between points explicitly, just the properties of open sets. We will discuss it only in the setting of functions defined on $\mathbb{R}$. The case of more general domains is left for the exercises and is covered later in the context of metric spaces (Section 4.5).

We start with the standard definition. We recall that the condition $|x - y| < \delta$ can be written equivalently as $y \in B(x, \delta)$, and the condition $|f(x) - f(y)| < \varepsilon$ can be written as $f(y) \in B(f(x), \varepsilon)$. The function $f : \mathbb{R} \to \mathbb{R}$ is continuous at $x \in \mathbb{R}$ if and only if

for every $\varepsilon > 0$ there exists $\delta > 0$ such that
$$y \in B(x, \delta) \text{ implies } f(y) \in B(f(x), \varepsilon).$$

But we know that $f(y) \in B(f(x), \varepsilon)$ is the same as saying that $y \in f^{-1}(B(f(x), \varepsilon)$. Then the condition that $y \in B(x, \delta)$ implies $y \in f^{-1}(B(f(x), \varepsilon)$ can be written as $B(x, \delta) \subseteq f^{-1}(B(f(x), \varepsilon))$.

So we have that f is continuous at $x \in \mathbb{R}$ if and only if

(4.3) \qquad for every $\varepsilon > 0$ there exists $\delta > 0$ such that
$$B(x, \delta) \subseteq f^{-1}(B(f(x), \varepsilon)).$$

Then we are ready to prove the following characterization of continuity; it is in terms of inverse images of sets.

Lemma 4.1.12. *A function $f : \mathbb{R} \to \mathbb{R}$ is continuous if and only if for every open set G in $\mathbb{R}$, its inverse image $f^{-1}(G)$ is open.*

Proof. Assume that f is continuous on $\mathbb{R}$, and let G be an open set in $\mathbb{R}$. Recall that the definition of the inverse image of a set does not need the existence of the inverse function of f; it is defined in terms of f as $f^{-1}(G) = \{x \in \mathbb{R} : f(x) \in G\}$. Now, if $f^{-1}(G)$ is empty, we are done, so let $x \in f^{-1}(G)$. Then $f(x) \in G$ and as G is open, there exists $\varepsilon > 0$ so that the open ball $B(f(x), \varepsilon)$ is contained in G. Using the continuity of f at x, for this ε there exists $\delta > 0$ so that $B(x, \delta) \subseteq f^{-1}(B(f(x), \varepsilon))$. This shows that $f^{-1}(G)$ is open. This is illustrated in Figure 4.1.

For the converse, we desire to show continuity of f at each $x \in \mathbb{R}$. Given $\varepsilon > 0$, the ball $B(f(x), \varepsilon)$ is an open set, so by assumption, the set $f^{-1}(B(f(x), \varepsilon))$ is open. As we know that x is a point in $f^{-1}(B(f(x), \varepsilon))$, there exists $\delta > 0$ so that $B(x, \delta) \subseteq f^{-1}(B(f(x), \varepsilon))$. This shows that f is continuous at x. $\square$

4.1.2. Continuity of the Inverse Function. We study conditions for when an invertible continuous function has an inverse that is continuous. We start with an example.

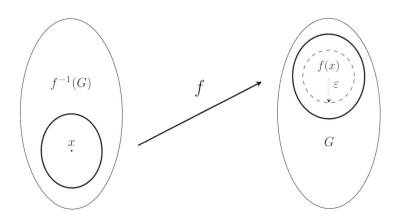

Figure 4.1. The inverse image of G under f.

Example 4.1.13. Let $A = [1,2] \cup (3,4]$, and let $f : A \to \mathbb{R}$ be defined by $f(x) = x$ when $x \in [1,2]$ and $f(x) = x - 1$ when $x \in (3,4]$. One can verify that f is injective on A and so it has an inverse $g : [1,3] \to A$ given by

$$g(y) = \begin{cases} y & \text{if } 1 \leq y \leq 2, \\ y+1 & \text{if } 2 < y \leq 3. \end{cases}$$

However, while f is continuous, g is not continuous at $y = 2$ since $\lim_{y \to 2^-} g(y) = 2$ but $\lim_{y \to 2^+} g(y) = 3$.

Proposition 4.1.14. *Let A be a compact subset of $\mathbb{R}$, and let $f : A \to f(A)$ be a continuous invertible function. Then its inverse f^{-1} is continuous.*

Proof. To simplify notation, let g stand for f^{-1}. We use Proposition 4.1.9 to show that g is continuous at $y \in f(A)$. Let (b_n) be a sequence in $f(A)$ such that

$$\lim_{n \to \infty} b_n = y.$$

If $(g(b_n))$ does not converge to $g(y)$, then there exists $\varepsilon > 0$ such that for all $N \in \mathbb{N}$ there is $n \geq N$ such that $|g(b_n) - g(y)| \geq \varepsilon$. So we can choose an increasing subsequence (n_i) so that

(4.4) $$|g(b_{n_i}) - g(y)| \geq \varepsilon.$$

As A is compact, $(g(b_{n_i}))_i$ has a convergent subsequence $(g(b_{n_{i_j}}))_j$,

(4.5) $$\lim_{j \to \infty} g(b_{n_{i_j}}) = p \in A.$$

By continuity of f,

$$f(p) = f(\lim_{j \to \infty} g(b_{n_{i_j}})) = \lim_{j \to \infty} f(g(b_{n_{i_j}}))$$
$$= \lim_{j \to \infty} b_{n_{i_j}}$$
$$= y,$$

showing $p = g(y)$. So (4.4) contradicts (4.5) and f^{-1} must be continuous. □

Corollary 4.1.15. *Let (a,b) be an interval, and let $f : (a,b) \to \mathbb{R}$ be continuous and invertible. Then its inverse f^{-1} is continuous.*

Proof. This follows from Proposition 4.1.14 since continuity is a local property. Let $(c,d) = f((a,b))$ and $y \in (c,d)$. Choose $c < c' < y < d' < d$, and choose a', b' such that $f(a') = c', f(b') = d'$. Then we apply Proposition 4.1.14 to $f : [a', b'] \to [c', d']$. □

Exercises: Continuous Functions on $\mathbb{R}$

4.1.1 Show that if $|x - y| < \alpha$ for $x, y \in \mathbb{R}$, and $\alpha \in (0, \infty)$, then $|y| < |x| + \alpha$. Is the converse true?

4.1.2 Show that all polynomials $p : \mathbb{R} \to \mathbb{R}$ of the form $p(x) = a_n x^n + \cdots + a_0$, for $a_i \in \mathbb{R}, i \in \{0, \ldots, n\}$ are continuous.

4.1.3 Let f be defined by
$$f(x) = \frac{x^2 - 9}{x + 3}$$
for $x \neq -3$.
(a) Define $f(-3)$ so that $f : \mathbb{R} \to \mathbb{R}$ is continuous, and prove your claim.
(b) Define $f(-3)$ so that f is not continuous at -3 and prove your claim.

4.1.4 Let $f(x) = 1/x$. Show that $f : (-\infty, 0) \to \mathbb{R}$ is continuous on $(-\infty, 0)$.

4.1.5 Find an example of a function $f : \mathbb{R} \to \mathbb{R}$ such that there exists only one $x \in \mathbb{R}$ for which f is continuous at x. Can you extend this so that f is continuous at only k points for arbitrary k? What about an example that is continuous on $\mathbb{Z}$ but discontinuous on $\mathbb{R} \setminus \mathbb{Z}$?

4.1.6 Complete the proof that (y_n) converges to x in Proposition 4.1.9.

4.1.7 Prove parts $(1), (3)$, and (4) of Proposition 4.1.10.

4.1.8 Prove, using the ε, δ definition, that if $f : A \to [0, \infty)$ is continuous, then $\sqrt{f} : A \to [0, \infty)$ is continuous.

4.1.9 Let $f : \mathbb{R} \to \mathbb{R}$ be a continuous function such that $f(x + y) = f(x) + f(y)$ for all $x, y \in \mathbb{R}$. Show that f is of the form $f(x) = xf(1)$ for all $x \in \mathbb{R}$.

4.1.10 Let $f : \mathbb{R} \to \mathbb{R}$ be a function. Prove that f is continuous if and only if for all $a, b \in \mathbb{Q}$ the set $\{x \in \mathbb{R} : a < f(x) < b\}$ is open.

4.1.11 Let $A \subseteq \mathbb{R}$, and let $f : A \to \mathbb{R}$ be a function. Prove that f is continuous on A if and only if for every $G \subseteq \mathbb{R}$ that is open, the set $f^{-1}(G)$ is of the form $A \cap H$ for some open set $H \subseteq \mathbb{R}$.

4.1.12 Let $f : \mathbb{R} \to \mathbb{R}$ be a function.
(a) Show that if f is continuous, then for every $F \subseteq \mathbb{R}$ that is closed, the set $f^{-1}(F)$ is closed.
(b) Is the converse true?

4.1.13 Let $A \subseteq \mathbb{R}$, and let $f : A \to \mathbb{R}$ be continuous. Is $|f| : A \to \mathbb{R}$ continuous?

4.1.14 Let $f : \mathbb{R} \to \mathbb{R}$ be continuous.

(a) If $F \subseteq \mathbb{R}$ is closed, is $f(F)$ closed?

(b) If $G \subseteq \mathbb{R}$ is open, is $f(G)$ open?

4.1.15 Let $A \subseteq \mathbb{R}$, and let $f : A \to \mathbb{R}$ be a function. We say f is **increasing** or **monotone increasing** if $f(x) \leq f(y)$ whenever $x \leq y$ in A. Show that if f is monotone increasing, then there exists at most a countable number of points $x \in A$ such that f is not continuous at x. (We say f is **strictly increasing** if $f(x) < f(y)$ whenever $x < y$ in A.) Show a similar result holds for **decreasing** or **monotone decreasing** functions: $f(x) \geq f(y)$ whenever $x \leq y$ in A. (A function is **monotone** if it is increasing or decreasing.)

4.1.16 Let $f : [a,b] \to \mathbb{R}$ be a function. We say that f is **Lipschitz** if there is a constant $C \in \mathbb{R}$ such that
$$|f(x) - f(y)| \leq C|x - y|$$
for all $x, y \in [a,b]$. Show that a Lipschitz function is continuous. Is the converse true?

4.1.17 *Modified Dirichlet's function or Thomae's function*: Define the function f on $[0,1]$ by
$$f(x) = \begin{cases} \frac{1}{q} & \text{if } x = \frac{p}{q}, \text{ where } p, q \in \mathbb{N} \text{ are in lowest terms,} \\ 1 & \text{if } x \in \mathbb{Q}^c \text{ or } x = 0. \end{cases}$$
Prove that f is continuous at the irrationals but not continuous at the rationals. (It can be shown that there is no function that is continuous on the rationals and discontinuous on the irrationals.)

* 4.1.18 Let $f : \mathbb{R} \to \mathbb{R}$ be continuous, and let $A \subseteq \mathbb{R}$. Is $f(\bar{A}) \subseteq \overline{f(A)}$?

4.2. Intermediate Value and Extreme Value Theorems

The first theorem we study in this section has several interesting applications. We start with the statement of the theorem.

Theorem 4.2.1 (Intermediate value theorem). *Let $f : [a,b] \to \mathbb{R}$ be a continuous function on an interval $[a,b]$ such that $f(a) \neq f(b)$. Let d be a number that is between $f(a)$ and $f(b)$. Then there exists $c \in (a,b)$ such that*
$$f(c) = d.$$

Question 4.2.2. Show that the intermediate value theorem is not true when the function is not continuous or when the domain is not an interval.

Before proving Theorem 4.2.1 (which is illustrated in Figure 4.2), we study some of its consequences. We start with a useful corollary (see Figure 4.3).

Corollary 4.2.3 (Zero theorem). *Let f be a continuous function on an interval $[a,b]$. Suppose that $f(a)f(b) < 0$ (i.e., they are nonzero and have different signs). Then there exists $c \in (a,b)$ such that*
$$f(c) = 0.$$

4.2. Intermediate Value and Extreme Value Theorems

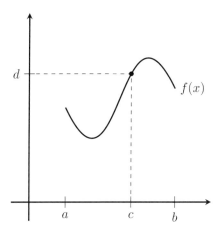

Figure 4.2. Intermediate value theorem illustration.

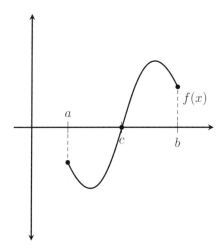

Figure 4.3. Illustration of Corollary 5.2.2 to the intermediate value theorem.

Proof. This is a direct consequence of the intermediate value theorem. As $f(a)f(b) < 0$, if $d = 0$, then d is between $f(a)$ and $f(b)$. Then Theorem 4.2.1 implies that there exists a $c \in (a, b)$ such that $f(c) = 0$. $\square$

Remark 4.2.4. It is interesting to note that Theorem 4.2.1 can be proved if one assumes that Corollary 4.2.3 is true. In fact, let f be a continuous function on an interval $[a, b]$ such that $f(a) \neq f(b)$, and let d be a number that is between $f(a)$ and $f(b)$. Define $h : [a, b] \to \mathbb{R}$ by

$$h(x) = f(x) - d.$$

If $f(a) < d < f(b)$, then $h(a) < 0$ and $h(b) > 0$. If $f(b) < d < f(a)$, then $h(a) > 0$ and $h(b) < 0$. In either case $h(a)h(b) < 0$, so there exists $c \in (a, b)$ such that $h(c) = 0$. This implies $f(c) = d$, completing the proof of Theorem 4.2.1.

Example 4.2.5. One can use Corollary 4.2.3 to show the existence of real roots of polynomials of odd degree. (A **real root** of a polynomial $p(x)$ is a number $c \in \mathbb{R}$ such that $p(c) = 0$.) For example, let $p(x) = x^3 + x^2 + 1$. We note that $p(1) = 3$ and $p(-2) = -3$. So by Corollary 4.2.3 there exists $c \in (-2, 1)$ such that $p(c) = 0$. (Not all polynomials have real roots, for example the even-degree polynomial $x^2 + 1$ does not have real roots.) Now we give an argument for a general monic polynomial of degree 3, i.e., a polynomial of the form $p(x) = x^3 + a_2 x^2 + a_1 x + a_0$. (*Monic* means that $a_3 = 1$, and for an n degree polynomial that is $a_n = 1$.) Write

$$p(x) = x^3(1 + a_2 \frac{1}{x} + a_1 \frac{1}{x^2} + a_0 \frac{1}{x^3}).$$

Now we show there exists $n \in \mathbb{N}$ such that $p(n) > 0$ and $p(-n) < 0$. The idea is that, by the Archimedean principle, we can choose n large enough so that

$$1 + a_2 \frac{1}{n} + a_1 \frac{1}{n^2} + a_0 \frac{1}{n^3}$$

is essentially 1, i.e., it is between 1.1 and 0.9 (say). Then $p(n)$ behaves essentially like n^3. To find such an n, first set

$$a = \max\{|a_0|, |a_1|, |a_2|\}.$$

Choose $n \in \mathbb{N}$ so that

$$\frac{1}{n} a < \frac{1}{30}.$$

One can verify that this n satisfies $p(n) > 0$ and $p(-n) < 0$. Corollary 4.2.3 then implies that there exists c between $-n$ and n with $p(c) = 0$.

Proof of Theorem 4.2.1. By Remark 4.2.4 we know it suffices to prove Corollary 4.2.3. Let $f : [a, b] \to \mathbb{R}$ be a continuous function. Assume that $f(a) < 0 < f(b)$ (the case when $f(b) < 0 < f(a)$ is similar). Let

$$A = \{x \in [a, b] : f(x) < 0\}.$$

Clearly, $a \in A$ since $f(a) < 0$. Also A is bounded by definition. So we may consider

$$\beta = \sup A.$$

Then $a \leq \beta \leq b$. We will show that $f(\beta) = 0$.

First we show that $f(\beta) \leq 0$. By Lemma 1.2.10, for each $\varepsilon_n = 1/n$ there exists $a_n \in A$ such that $|\beta - a_n| < 1/n$. It follows that (a_n) is a sequence in A converging to β. By continuity of f, $\lim_{n \to \infty} f(a_n) = f(\beta)$. But each $f(a_n) < 0$, so $f(\beta) \leq 0$. As $f(b) > 0$, this also shows that $\beta < b$.

Next we show that assuming $f(\beta) < 0$ leads to a contradiction. We first outline the idea. As $f(\beta)$ is negative, using the continuity of f at β, we can find a point arbitrarily close to β, which we can choose to be to the right of β, and such that f is still negative at this point. This would contradict that β is the supremum of A, so it must be that $f(\beta) = 0$. Now we formalize this argument. Since $f(\beta) < 0$ we can choose ε such that $0 < \varepsilon < -f(\beta)$. By continuity at β there exists $\delta > 0$ so that $|f(x) - f(\beta)| < \varepsilon$ whenever $|x - \beta| < \delta$ and $x \in [a, b]$. So there exists x with $b > x > \beta$ and such that $f(x) < 0$, contradicting that β is an upper bound of A. Therefore, we can set $c = \beta$, and we have that $f(c) = 0$. $\square$

4.2. Intermediate Value and Extreme Value Theorems

Remark 4.2.6. We outline another proof of Theorem 4.2.1 that is also of interest, but the details are left to the reader. As in the first proof, we again prove Corollary 4.2.3. We generate a nested sequence of closed intervals whose intersection is a unique point that turns out to be c. The interesting part about this proof is that it gives an algorithm that generates a sequence of points that converges to a zero of f. For concreteness suppose $f(a)$ is negative and $f(b)$ is positive. Start by calculating the midpoint between a and b and call it m_1. If m_1 is a zero of f we are done. Otherwise, it must have a different sign from $f(a)$ or from $f(b)$. If $f(m_1)$ is positive, as $f(a)$ is negative, we next consider the interval $[a, m_1]$, and if $f(m_1)$ is negative, we consider the interval $[m_1, b]$. Say $f(m_1) > 0$. Then let m_2 be the midpoint between a and m_1. If m_2 is a zero of f, we are done, else the sign of $f(m_2)$ differs from the sign of $f(a)$ or the sign of $f(m_1)$. Say $f(m_2) < 0$. Then let m_3 be the midpoint between m_2 and m_1. In this way we generate a sequence of closed nested intervals that by the strong nested intervals theorem must intersect in a unique point c, and at the same time the sequence of midpoints must converge to c. It remains to show that $f(c) = 0$. This is done by showing that both $f(c) > 0$ and $f(c) < 0$ lead to a contradiction as f is continuous at c.

Remark 4.2.7. If f is continuous on $[a, b]$, it follows from the intermediate value theorem that $f([a, b])$ is a connected set. More generally, it can be shown, using the characterization of continuity in Exercise 4.1.11, that if f is a continuous function on a connected set A, then $f(A)$ is also connected. In turn this would give a third proof of the intermediate value theorem (see Exercises 4.2.9 and 4.2.8).

We conclude with another theorem that also has many applications. Let $f : A \to \mathbb{R}$ be a function.

Definition 4.2.8. We say that f has a **minimum point**, or a **global minimum point**, on A if there is a point $z \in A$ such that

$$f(z) \leq f(x) \text{ for all } x \in A.$$

In this case we also say that f attains its minimum or has a minimum on A. We say $f(z)$ is its **global minimum value**. Similarly, we say that f has a **maximum point**, or a **global maximum point**, on A if there is a point $w \in A$ such that

$$f(w) \geq f(x) \text{ for all } x \in A.$$

We also say that f attains its maximum on A. We say $f(w)$ is its **global maximum value**. ◇

In general, a function need not have a minimum or maximum on a set. For example, if $A = (0, 1)$ and $f(x) = 1/x$, then f does not have a minimum on A: for each $x \in A$ there exists $x' \in A$ such that $f(x') < f(x)$. The following theorem gives conditions for when a minimum and maximum exist.

Theorem 4.2.9 (Extreme value theorem). *Let A be a compact subset of $\mathbb{R}$ and let $f : A \to \mathbb{R}$ be a continuous function. Then f has a maximum and a minimum value on A.*

Proof. Recall that the image of f, $f(A)$, consists of all points in $\mathbb{R}$ of the form $f(x)$ for some x in A. We first show that $f(A)$ is a bounded set. We proceed by

contradiction and suppose first that $f(A)$ is not bounded above. Then for each integer $N \in \mathbb{N}$ there exists an element of $f(A)$, which we may write as $f(a_N)$, such that
$$f(a_N) > N.$$
This gives a sequence $(a_N)_{N \in A}$. As A is compact (sequentially compact), the sequence (a_N) has a subsequence (a_{N_k}) converging to a point p in A. The continuity of f implies
$$\lim_{k \to \infty} f(a_{N_k}) = f(p).$$
This means that, for $\varepsilon = 1$, there exists $K \in \mathbb{N}$ such that
$$|f(p) - f(a_{N_k})| < 1 \text{ for all } k \geq K.$$
So
$$f(p) > f(a_{N_k}) - 1 > N_k - 1 \text{ for all } k \geq K.$$
But this is a contradiction since it gives a bound for the infinite increasing sequence $N_k - 1$. Therefore, $f(A)$ is bounded above. A similar argument shows that $f(A)$ is bounded below (Exercise 4.2.10).

Now let
$$\alpha = \inf f(A) \quad \text{and} \quad \beta = \sup f(A).$$

By Example 2.1.8, there exist sequences (x_n) and (y_n) in $f(A)$ converging to α and β, respectively. Then x_n must be of the form $f(c_n)$ for some $c_n \in A$, and $y_n = f(d_n)$ for some $d_n \in A$. As A is sequentially compact, there exists (c_{n_k}), a subsequence of (c_n), converging to a point in A that we denote by z, and (d_{n_k}) a subsequence of (d_n) converging to a point in A that we denote by w. By continuity of f,
$$\alpha = \lim_{k \to \infty} f(c_{n_k}) = f(z) \quad \text{and} \quad \beta = \lim_{k \to \infty} f(d_{n_k}) = f(w).$$
This means that z is a global minimum and w is a global maximum. □

We conclude with a theorem whose proof is similar to that of Theorem 4.2.9.

Theorem 4.2.10. *Let A be a compact subset of $\mathbb{R}$, and let $f : A \to \mathbb{R}$ be a continuous function. Then $f(A)$ is compact.*

Proof. One proof could follow the ideas in the proof of Theorem 4.2.9 to show that $f(A)$ is bounded and closed, so compact. Instead we give a direct proof that $f(A)$ is sequentially compact. Let (x_n) be a sequence in $f(A)$. Then x_n must be of the form $x_n = f(a_n)$ for some $a_n \in A$. As A is compact, the sequence (a_n) has a convergent subsequence a_{n_k} to a point a in A. The continuity of f implies that $f(a_{n_k})$ converges to $f(a) \in f(A)$. But $f(a_{n_k})$ is a subsequence of (x_n). Thus $f(A)$ is sequentially compact. □

Theorem 4.2.10 can be used to give another proof that a continuous function attains a minimum and a maximum on a compact set; since $f(A)$ is compact one can use Lemma 3.4.3 to obtain the maximum and minimum values.

Exercises: Intermediate Value and Extreme Value Theorems

4.2.1 Prove Theorem 4.2.1 in the case when $f(b) < f(a)$.

4.2.2 *Fixed point theorem*: Let $f : [a,b] \to [a,b]$ be continuous. Show that there exists a point $p \in [a,b]$ such that $f(p) = p$. The point p that satisfies this equation is called a **fixed point** of f.

4.2.3 Complete the details of the second proof of Theorem 4.2.1 outlined in Remark 4.2.6.

4.2.4 Let $f : \mathbb{R} \to \mathbb{R}$ be continuous. Suppose there is a point $q \in \mathbb{R}$ such that $f(f(q)) = q$ but $f(q) \neq q$ (such a point is said to be a point of *proper period* 2). Prove that then there is a fixed point $p \in \mathbb{R}$, i.e., $f(p) = p$.

4.2.5 Give an example of a closed set $A \subseteq \mathbb{R}$ and a continuous function $f : A \to \mathbb{R}$ such that $f(A)$ is not closed.

4.2.6 Give an example of a bounded set $A \subseteq \mathbb{R}$ and a continuous function $f : A \to \mathbb{R}$ such that $f(A)$ is not bounded.

4.2.7 Let $p(x)$ be a polynomial of odd degree with coefficients in $\mathbb{R}$. Show that there exists $c \in \mathbb{R}$ such that $p(c) = 0$.

4.2.8 Let $A \subseteq \mathbb{R}$ and $f : \mathbb{R} \to \mathbb{R}$ be continuous. Prove that if A is connected, then $f(A)$ is connected. Give one proof using Theorem 3.6.3 and another without using the theorem.

4.2.9 Use Exercise 4.2.8 to give another proof of Theorem 4.2.1.

4.2.10 Complete the proof that $f(A)$ is bounded below in Theorem 4.2.9.

4.2.11 Give an example to show that the converse of the intermediate value theorem does not hold. (*Hint*: Use a trigonometric function.)

4.2.12 Use the intermediate value theorem to give another proof that there is a positive real number c such that $c^2 = 2$.

4.2.13 Use the intermediate value theorem to prove that a nonconstant polynomial of odd degree is surjective on $\mathbb{R}$. What about even degree?

* 4.2.14 Use the intermediate value theorem to prove the supremum property, i.e., show that if an ordered field satisfies the intermediate value theorem, then it is complete.

4.3. Limits

Let $A \subseteq \mathbb{R}$, and let $f : A \to \mathbb{R}$ be a function. Let $a \in \mathbb{R}$ be an accumulation point of A. (Note that a is not required to be in the domain A of f, but if it is not an accumulation point of A, the notion of limit does not make sense since one could not approach a from points in A.) The idea is that the limit of $f(x)$, as x approaches a, should be a number L such that $f(x)$ is arbitrarily close to L whenever x is close enough to a (but different from a). We note that f may not be defined at a, or if it is, its value at a is not important. Therefore, in the definition of limit, there is no requirement when $x = a$.

Definition 4.3.1. Let $A \subseteq \mathbb{R}$, let $f : A \to \mathbb{R}$ be a function, and let a be an accumulation point of A. We define a number $L \in \mathbb{R}$ to be a **limit** of f at the point a, and we write
$$\lim_{x \to a} f(x) = L \text{ if}$$

(4.6) \qquad for every $\varepsilon > 0$ there exists $\delta > 0$ such that
$$x \in A \text{ and } 0 < |x - a| < \delta \text{ imply } |f(x) - L| < \varepsilon. \qquad \diamond$$

In Proposition 4.3.2 we prove that if a limit exists there is only one number L satisfying the properties of a limit, justifying the notation in Definition 4.3.1.

Proposition 4.3.2. *Limits of functions are unique: if L_1 and L_2 are limits of $f(x)$ as x approaches a, then $L_1 = L_2$.*

Proof. Let $\varepsilon > 0$. There exists $\delta_1 > 0$ such that $|f(x) - L_1| < \varepsilon/2$ whenever $x \in A$ and $0 < |x - a| < \delta_1$. Similarly, there exists $\delta_2 > 0$ so that $|f(x) - L_2| < \varepsilon/2$ whenever $x \in A$ and $0 < |x - a| < \delta_2$. If x is a point such that $0 < |x - a| < \min\{\delta_1, \delta_2\}$ (which we can choose as a is an accumulation point and δ_1 and δ_2 are positive), then
$$\begin{aligned} |L_1 - L_2| &= |L_1 - f(x) + f(x) - L_2| \\ &\leq |L_1 - f(x)| + |f(x) - L_2| \\ &< \frac{\varepsilon}{2} + \frac{\varepsilon}{2} \\ &= \varepsilon. \end{aligned}$$
Since this is true for all $\varepsilon > 0$, $L_1 = L_2$. $\qquad \square$

As we shall see, it may occur that there is no number L satisfying the conditions of the limit, in which case we say the limit does not exist. If there is a number L satisfying the properties of the limit, we say the limit exists.

Limits are easiest to calculate in the case of continuous functions for points in the domain of the function. The reader has probably noticed the similarities in the definitions of limit and continuity. These similarities make the proof of the following proposition short. However, it is an important proposition that is sometimes used as the definition of continuity when one knows in advance the definition of a limit.

Proposition 4.3.3. *Let $f : A \to \mathbb{R}$ be a function, and let $a \in A$ be an accumulation point of A. The function f is continuous at a if and only if*
$$\lim_{x \to a} f(x) = f(a).$$

Proof. Let f be continuous at a, and let $\varepsilon > 0$. There exists $\delta > 0$ such that
$$|f(x) - f(a)| < \varepsilon \text{ whenever } x \in A \text{ and } |a - x| < \delta.$$
This means that $\lim_{x \to a} f(x) = f(a)$. The converse is similar. $\qquad \square$

4.3. Limits

Example 4.3.4. We have already seen that if $p(x)$ is a polynomial with real coefficients, then $p(x)$ is continuous on $\mathbb{R}$. It follows that for every $a \in \mathbb{R}$, $\lim_{x \to a} p(x) = p(a)$.

The next proposition gives another useful way to think about limits and can be used to translate properties about limits of functions into properties about limits of sequences.

Proposition 4.3.5. *Let $f : A \to \mathbb{R}$ be a function, and let $a \in \mathbb{R}$ be an accumulation point of A. Then*
$$\lim_{x \to a} f(x) = L$$
if and only if for every sequence (a_n) in $A \setminus \{a\}$ converging to a,
$$\lim_{n \to \infty} f(a_n) = L.$$

Proof. Suppose $\lim_{x \to a} f(x) = L$, and let (a_n) be a sequence in $A \setminus \{a\}$ converging to a. Then for $\varepsilon > 0$ there exists $\delta > 0$ such that
$$|f(x) - L| < \varepsilon \text{ whenever } x \in A \text{ and } 0 < |x - a| < \delta.$$
For $\delta > 0$, there exists $N \in \mathbb{N}$ such that
$$0 < |a_n - a| < \delta \text{ if } n \geq N.$$
Then $|f(a_n) - L| < \varepsilon$ for all $n \geq N$. Therefore, the sequence $(f(a_n))$ converges to L.

For the converse suppose that L does not satisfy the definition of being the limit of $f(x)$ as x approaches a. The negation of condition (4.6) is that there exists $\varepsilon > 0$ so that for each $\delta > 0$, some x with $0 < |x - a| < \delta$ satisfies
$$|f(x) - L| \geq \varepsilon.$$
Since x depends on δ, by choosing $\delta = 1/n$ for each $n \in \mathbb{N}$, we obtain a point x_n such that $0 < |x_n - a| < 1/n$. This means that the sequence (x_n) converges to a and also $|f(x_n) - L| \geq \varepsilon$. But this means that the sequence $(f(x_n))$ cannot converge to L, a contradiction. Therefore $\lim_{x \to a} f(x) = L$. $\square$

Example 4.3.6. Let $A = (0, 1]$, and let $f : A \to \mathbb{R}$ be defined by
$$f(x) = \begin{cases} \frac{1}{x} & \text{if } x \in (0, 1), \\ 2 & \text{if } x = 1. \end{cases}$$
We show that
$$\lim_{x \to 1} f(x) = 1.$$
We start with the development. Given $\varepsilon > 0$, we would like to have
$$\left| \frac{1}{x} - 1 \right| < \varepsilon$$
for x suitably chosen. So
$$\left| \frac{1 - x}{x} \right| < \varepsilon.$$

As x is chosen close to 1, we can assume $x > 1/2$ (for example), so $1/x < 2$. Then we have
$$\left|\frac{1}{x} - 1\right| = \left|\frac{1-x}{x}\right| < 2|1-x|.$$
This shows that if we make $2|1-x|$ small, then we obtain that $|\frac{1}{x} - 1|$ is small. We know that we can control $|1-x|$ since we are evaluating a limit at $x = 1$.

Now we start the proof. Let $\varepsilon > 0$. Choose $0 < \delta = \min\{\varepsilon/2, 1/2\}$. If $x \in (0,1)$ and $|x-1| < \delta$, then $x > 1/2$. So
$$\left|\frac{1}{x} - 1\right| < 2|1-x| < 2\delta \le \varepsilon.$$
This completes the proof of the limit. (Note that the value of the function f at $x = 1$ did not play a role in the proof.)

Example 4.3.7. Let $A = (0,1) \cup (1,2)$, and let $f : A \to \mathbb{R}$ be defined by
$$f(x) = \begin{cases} \frac{1}{x} & \text{if } x \in (0,1), \\ -\frac{1}{x} & \text{if } x \in (1,2). \end{cases}$$
We show that $\lim_{x \to 1} f(x)$ does not exist. Consider the sequences $a_n = 1 - (1/n)$ and $b_n = 1 + (1/n)$, both converging to 1. But $(f(a_n))$ converges to 1 and $(f(b_n))$ converges to -1. Proposition 4.3.5 implies the limit cannot exist.

Remark 4.3.8. The function in Example 4.3.6 is not continuous. We can consider, however, the function $g(x) = 1/x$ on $(0,1]$, which is continuous and agrees with f at all points of its domain but at the point of discontinuity $a = 1$. If we know g is continuous on A, we can find $\lim_{x \to 1} g(x) = 1$ which is the same as $\lim_{x \to 1} f(x)$. This idea is explored in more detail in Exercise 4.3.8.

Remark 4.3.9. The function in Example 4.3.7 is continuous on its domain, but the limit does not exist at a point that is an accumulation point but is not in the domain (namely at $x = 1$). We could consider the function $h(x) = 1/x$ defined on $(0,1)$. For this function the limit exists, and one can show $\lim_{x \to 1} h(x) = 1$. This is similar to what one can call a one-sided limit of $f(x)$, or a limit as x approaches 1 from the left. These ideas are developed in the Exercises.

Example 4.3.10. We show that $\lim_{x \to 2} \frac{x^3 - 8}{x - 2} = 12$. We start with the development. First note that $x^3 - 8 = (x-2)(x^2 + 2x + 4)$. Then

(4.7)
$$\left|\frac{x^3 - 8}{x - 2} - 12\right| = \left|\frac{(x-2)(x^2 + 2x + 4)}{x - 2} - 12\right|$$
$$= |x^2 + 2x - 8| = |x - 2||x + 4|.$$

We know we can choose $\delta > 0$ so that $|x - 2| < \delta$. If δ is restricted so that $\delta \le 1$, then $-1 < x - 2 < 1$, so $5 < x + 4 < 7$. Thus $|x-2||x+4| < 7\delta$. Therefore, we wish to choose δ so that in addition to satisfying $\delta \le 1$ we have $7\delta \le \varepsilon$. Now we are ready to start the proof.

4.3. Limits

Let $\varepsilon > 0$. Choose δ so that $\delta = \min\{1, \varepsilon/7\} > 0$. If $|x - 2| < \delta$, then from (4.7) we have that

$$\left|\frac{x^3 - 8}{x - 2} - 12\right| = |x - 2||x + 4| < 7\delta \leq \varepsilon.$$

Therefore, $\lim_{x \to 2} \frac{x^3 - 8}{x - 2} = 12$.

The following is a basic proposition for computing limits of sums, products, and quotients.

Proposition 4.3.11. *Let $A \subseteq \mathbb{R}$ be a set. Let $f, g : A \to \mathbb{R}$ be functions, and let a be an accumulation point of A. Suppose that*

$$\lim_{x \to a} f(x) = L_1 \quad \text{and} \quad \lim_{x \to a} g(x) = L_2.$$

Then

(1)
$$\lim_{x \to a} (f + g)(x) = L_1 + L_2,$$

(2)
$$\lim_{x \to a} (f \cdot g)(x) = L_1 \cdot L_2,$$

(3)
$$\lim_{x \to a} \frac{f}{g}(x) = \frac{L_1}{L_2} \quad \text{provided} \ L_2 \neq 0.$$

Proof. All three cases are a consequence of Propositions 4.3.5 and 2.1.11. We illustrate the ideas by showing part (3); the other parts are similar and are left to the reader. First note that as $L_2 \neq 0$, there exists an open interval around a so that for every $x \in A$ in this interval, $g(x) \neq 0$ (this follows from the fact that we can force $g(x)$ to be sufficiently close to L_2 on this interval). Recall that f/g is the function defined by

$$\frac{f}{g}(x) = \frac{f(x)}{g(x)}.$$

Let (a_n) be a sequence in $A \setminus \{a\}$ converging to a. By hypothesis, $(f(a_n))$ and $(g(a_n))$ converge to L_1 and L_2, respectively. As $g(a_n) \neq 0$ for all sufficiently large n, Proposition 2.1.11 completes the proof. $\square$

4.3.1. Infinite Limits. Recall that $\lim_{x \to a} f(x) = L$ of a function $f : A \to \mathbb{R}$ was defined for the case when a and L are in $\mathbb{R}$. Now we consider the cases when a can be $\pm\infty$ and L can be $\pm\infty$.

Definition 4.3.12. Let $f : A \to \mathbb{R}$ be a function, and let $a \in \mathbb{R}$ be an accumulation point of A. We define

$$\lim_{x \to a} f(x) = \infty$$

if for all $\alpha > 0$ there exists $\delta > 0$ such that

$$x \in A \text{ and } 0 < |x - a| < \delta \text{ imply } f(x) > \alpha. \qquad \diamond$$

Definition 4.3.13. Similarly, we define
$$\lim_{x \to a} f(x) = -\infty$$
if for all $\beta < 0$ there exists $\delta > 0$ such that
$$x \in A \text{ and } 0 < |x - a| < \delta \text{ imply } f(x) < \beta. \qquad \diamond$$

Example 4.3.14. Let $f : (0, \infty) \to (0, \infty)$ be defined by $f(x) = \frac{1}{x^2}$. We show that $\lim_{x \to 0} f(x) = \infty$. Let $\alpha > 0$. Choose $\delta > 0$ so that
$$\delta < 1/\sqrt{\alpha}.$$
If $0 < x < \delta$, then $0 < x < 1/\sqrt{\alpha}$. So $0 < x^2 < 1/\alpha$, which implies that $1/x^2 > \alpha$. Therefore, $1/x^2$ diverges to ∞ as $x \to 0$. (Note that as the domain of f is $(0, \infty)$, when x approaches 0 it has to be from the right.)

Example 4.3.15. Define $f : (0, \infty) \to \mathbb{R}$ by $f(x) = -\frac{1}{x}$. We show that
$$\lim_{x \to 0} f(x) = -\infty.$$
Let $\beta < 0$. Choose $0 < \delta < -1/\beta$. If $0 < x < \delta$, then $0 < x < -1/\beta$. Hence, $-1/x < \beta$. Therefore, $-1/x$ diverges to $-\infty$ as $x \to 0$ and $x \in (0, \infty)$.

Remark 4.3.16. We can consider a function $h : \mathbb{R} \setminus \{0\}$ defined by $h(x) = -\frac{1}{x}$. Then the limit of h as x approaches 0 from the right, written as $\lim_{x \to 0^+} h(x)$, is $-\infty$, and the limit as x approaches 0 from the left, $\lim_{x \to 0^-} h(x) = +\infty$, so $\lim_{x \to 0} h(x)$ does not exist (see Exercise 4.3.6). On the other hand if we define $g : (0, \infty) \to \mathbb{R}$ by $g(x) = -\frac{1}{x}$, then in this case $\lim_{x \to 0} g(x)$ exists and equals $-\infty$.

Definition 4.3.17. Let $f : A \to \mathbb{R}$ be a function, suppose ∞ is an accumulation point of A (i.e., there is a sequence (a_n) in A such that $a_n \to \infty$), and let $L \in \mathbb{R}$. Define
$$\lim_{x \to \infty} f(x) = L \text{ if}$$
for every $\varepsilon > 0$ there exists $\alpha > 0$ such that
$$x \in A \text{ and } x > \alpha \text{ imply } |f(x) - L| < \varepsilon.$$
In a similar way we define $\lim_{x \to -\infty} f(x) = L$ when $-\infty$ is an accumulation point of A. $\qquad \diamond$

Definition 4.3.18. Now we consider the case when a and L are ∞. Let $f : A \to \mathbb{R}$ be a function, and suppose ∞ is an accumulation point of A. Define
$$\lim_{x \to \infty} f(x) = \infty \text{ if}$$
for all $\alpha > 0$ there exists $\gamma > 0$ such that
$$x \in A \text{ and } x > \gamma \text{ imply } f(x) > \alpha. \qquad \diamond$$

Example 4.3.19. Let $f(x) = x^3$. We show that $\lim_{x \to \infty} f(x) = \infty$. Let $\alpha > 0$. Choose $\gamma = \sqrt[3]{\alpha}$. If $x > \gamma$, then $x > \sqrt[3]{\alpha}$, so $x^3 > \alpha$. Therefore, x^3 diverges to ∞ as $x \to \infty$.

The definitions of the remaining cases (i.e., $\lim_{x \to \infty} f(x) = -\infty$, $\lim_{x \to -\infty} f(x) = \infty$, and $\lim_{x \to -\infty} f(x) = -\infty$) are similar and are left to the reader.

4.3. Limits

Exercises: Limits

4.3.1 Find the following limits if they exist, and justify your answers.
  (a) $\lim_{x \to -3} x^3 + 2x^2 + 1$
  (b) $\lim_{x \to 2} \frac{x^4 - 3}{x - 1}$
  (c) $\lim_{x \to 2} x^3 + 1 + \frac{1}{x}$
  (d) $\lim_{x \to 2} \frac{x^4 - 16}{x^2 - 4}$

4.3.2 Let $A = [0, 4]$, and define $f : A \to \mathbb{R}$ by $f(x) = x^2 + 3$. Prove that $\lim_{x \to 1} f(x) = 4$ using the definition (do not use Proposition 4.3.3). First give the development and then the proof.

4.3.3 Let $A = (0, 3)$, and define $f : A \to \mathbb{R}$ by

$$f(x) = \begin{cases} x^2 + 2 & \text{if } x \in (0, 3) \setminus \{1\}, \\ 10 & \text{if } x = 1. \end{cases}$$

Find $\lim_{x \to 1} f(x)$, and prove your answer using the definition. First give the development and then the proof.

4.3.4 Let $A = (-2, 2)$, and define $f : A \to \mathbb{R}$ by

$$f(x) = \begin{cases} x^2 + 5 & \text{if } x \in (-2, 0], \\ x^3 + 3 & \text{if } x \in (0, 2). \end{cases}$$

Find $\lim_{x \to 0} f(x)$, if it exists, and prove your answer.

4.3.5 Let $A \subseteq \mathbb{R}$, and let $f, g : A \to \mathbb{R}$ be functions. Let $a \in \mathbb{R}$ be an accumulation point of A. Suppose f is bounded on A and $\lim_{x \to a} g(x) = 0$. Prove that $\lim_{x \to a} f(x)g(x) = 0$.

4.3.6 Let $A \subseteq \mathbb{R}$, and let $f : A \to \mathbb{R}$ be a function. Let $a \in \mathbb{R}$ be an accumulation point of $A \cap (c, a)$ for some $c < a$. Define a number $L \in \mathbb{R}$ to be the **left limit** of f as x approaches a from the left, and write

$$\lim_{x \to a^-} f(x) = L \text{ if}$$

for every $\varepsilon > 0$ there exists $\delta > 0$ such that

$$x \in A \text{ and } 0 < a - x < \delta \text{ imply } |f(x) - L| < \varepsilon.$$

If $a \in \mathbb{R}$ is an accumulation point of $A \cap (a, d)$ for some $d > a$, define a number $L \in \mathbb{R}$ to be the **right limit** of f as x approaches a from the right, and write

$$\lim_{x \to a^+} f(x) = L \text{ if}$$

for every $\varepsilon > 0$ there exists $\delta > 0$ such that

$$x \in A \text{ and } 0 < x - a < \delta \text{ imply } |f(x) - L| < \varepsilon.$$

  (a) Prove that the left and right limits, when they exist, are unique.
  (b) Suppose a is an accumulation point of A from both the left and the right (as above). Prove that $\lim_{x \to a} f(x)$ exists and equals L if and only if both $\lim_{x \to a^-} f(x)$ and $\lim_{x \to a^+} f(x)$ exist and equal L.

4.3.7 Let $A = [0,4]$, and define $f : A \to \mathbb{R}$ by
$$f(x) = \begin{cases} x+5 & \text{if } x \in [0,2), \\ x^2+5 & \text{if } x \in [2,4]. \end{cases}$$
Find $\lim_{x \to 2^-} f(x)$ and $\lim_{x \to 2^+} f(x)$, and prove your answer using the definition.

4.3.8 Let $A \subseteq \mathbb{R}$, and let $f : A \to \mathbb{R}$ be a function. Let $a \in \mathbb{R}$ be an accumulation point of A. Suppose there exists a set $B \subseteq \mathbb{R}$ and a continuous function $g : B \to \mathbb{R}$ such that $A \subseteq B$, $a \in B$, and $f(x) = g(x)$ for all $x \in A \setminus \{a\}$. (We think of g as an extension of f except at the point a.) Prove that $\lim_{x \to a} f(x)$ exists and equals $g(a)$.

4.3.9 Prove parts (1) and (2) of Proposition 4.3.11.

4.3.10 Let $f : A \to \mathbb{R}$ be a function, and let $a \in \mathbb{R}$ be an accumulation point of A. Extend the definition we have seen to the cases $\lim_{x \to a^-} f(x) = \infty$, $\lim_{x \to a^-} f(x) = -\infty$, $\lim_{x \to a^+} f(x) = \infty$, and $\lim_{x \to a^+} f(x) = -\infty$. Find $\lim_{x \to 0^-} \frac{1}{x}$ and $\lim_{x \to 0^+} \frac{1}{x}$, and prove all your claims. (Here we understand the domain of $\frac{1}{x}$ to be $\mathbb{R} \setminus \{0\}$.)

4.3.11 Find $\lim_{x \to 1^+} \frac{x^2+4}{x-1}$, and prove all your claims.

4.3.12 Find $\lim_{x \to \infty} \frac{x^3+5}{x^2+1}$, and prove all your claims.

4.4. Uniform Continuity

Uniform continuity is not a local concept, that is, there is no definition of uniform continuity at a point; it is a concept that applies to a function, and the domain of the function plays a crucial role in the definition. For example, the function $f(x) = 1/x$, when considered as a function with domain $(0,1)$, $f : (0,1) \to \mathbb{R}$, is not uniformly continuous, while when considered as a function with domain, the interval $(2,3)$, $f : (2,3) \to \mathbb{R}$, is uniformly continuous. As we shall see in Theorem 4.4.5, a continuous function is uniformly continuous on any compact interval in its domain.

Definition 4.4.1. Let A be a subset of $\mathbb{R}$, and let $f : A \to \mathbb{R}$ be a function. The function f is said to be **uniformly continuous,** or **uniformly continuous** on A, if

for all $\varepsilon > 0$ there exists $\delta > 0$ such that

$x, y \in A$ and $|x-y| < \delta$ imply $|f(x) - f(y)| < \varepsilon$. ◇

Note that there is no point that is chosen in advance of ε. So in the definition, while δ may depend on ε (and in general it does), there is no dependence of δ on any point of the domain A (i.e., given ε, the same δ works for all points of A). Sometimes to emphasize the difference with (standard) continuity we may write that f is uniformly continuous on A if and only if

for all $\varepsilon > 0$ there exists $\delta > 0$ such that

$|f(x) - f(y)| < \varepsilon$ whenever $x, y \in A$ and $|x-y| < \delta$.

4.4. Uniform Continuity

From the definition it follows directly that if $f : A \to \mathbb{R}$ is uniformly continuous on A, then it is continuous at x for each x in A. (The converse is not true, as can be seen by Example 4.4.4.)

Also, another direct consequence of the definition is that if $f : A \to \mathbb{R}$ is a uniformly continuous function and $B \subseteq A$, then $f : B \to \mathbb{R}$ is uniformly continuous (given $\varepsilon > 0$, the same δ that is obtained for A works for B).

Example 4.4.2. Let $f : (0, 1) \to \mathbb{R}$ be defined by $f(x) = x^2$. Then one can calculate
$$|f(x) - f(y)| = |x^2 - y^2| = |x - y||x + y| < 2|x - y|,$$
as $x, y < 1$. Therefore, for any $\varepsilon > 0$, if we choose $\delta = \varepsilon/2$, then
$$|f(x) - f(y)| < 2|x - y| < \varepsilon \text{ whenever } |x - y| < \delta.$$
This shows that x^2 is uniformly continuous on $(0, 1)$. (The same proof also shows that it is uniformly continuous on $[0, 1]$, but we wanted to use $(0, 1)$ since we use the same interval in the next example. Also, as we shall see later in Theorem 4.4.5, on a closed bounded interval, continuity implies uniform continuity, and then one has uniform continuity on the open subinterval.)

Remark 4.4.3. A function $f : A \to \mathbb{R}$ is not uniformly continuous on A if and only if

> there exists $\varepsilon > 0$ such that for every $\delta > 0$
> there is some $x, y \in A$ such that $|x - y| < \delta$ and $|f(x) - f(y)| \geq \varepsilon$.

Example 4.4.4. Let $f : (0, 1) \to \mathbb{R}$ be defined by $f(x) = 1/x$. In this case we show f is not uniformly continuous. We calculate
$$|f(x) - f(y)| = \left|\frac{1}{x} - \frac{1}{y}\right| = \frac{|x - y|}{xy}.$$
We observe that x and y may be very close to 0, independently of how small $|x - y|$ is. So the difference $|f(x) - f(y)|$ cannot be made arbitrarily small by simply choosing x and y sufficiently close. In fact, we show that for every $\delta > 0$ we can choose x and y that are δ-close but such that $|f(x) - f(y)|$ cannot be made smaller than $1/2$. Let $\varepsilon = 1/2$. For any $1 > \delta > 0$ we can choose x and y so that
$$0 < x, y < \delta \quad \text{and} \quad |x - y| = \frac{\delta}{2}.$$
(For example, choose $y = x + \frac{\delta}{2}$ and $x = \frac{\delta}{4}$.) Then
$$|f(x) - f(y)| = \frac{|x - y|}{|xy|} > \frac{\delta/2}{\delta^2} = \frac{1}{2\delta} > \frac{1}{2} = \varepsilon.$$
It follows that f is not uniformly continuous on $(0, 1)$. (The reader should compare this with the proof of continuity in Example 4.1.8.) It follows that $f(x) = 1/x$ cannot be extended to a continuous function on $[0, 1]$—if it could be extended, then that extension would be uniformly continuous on $[0, 1]$ (by Theorem 4.4.5), which would imply $1/x$ is uniformly continuous on $(0, 1)$.

Under some conditions, namely compactness, continuity implies uniform continuity as the following theorem shows.

Theorem 4.4.5. *Let $A \subseteq \mathbb{R}$, and let $f : A \to \mathbb{R}$ be a continuous function. If A is compact, then f is uniformly continuous.*

Proof. We give two proofs. The first one uses the characterization of compactness by sequential compactness. The second uses the characterization of compactness by open covers. While the second proof is longer, it illustrates the ideas of open covers.

We know that A is sequentially compact. We proceed by contradiction and assume that f is not uniformly continuous. So there exists $\varepsilon_0 > 0$ so that for all $\delta > 0$,

$$\text{there exists } x, y \in A \text{ such that } |x - y| < \delta \text{ and } |f(x) - f(y)| \geq \varepsilon_0.$$

For each $n \in \mathbb{N}$, choose $\delta = 1/n$ to obtain points $x_n, y_n \in A$ with $|x_n - y_n| < \frac{1}{n}$. By sequential compactness, there is (x_{n_i}), a subsequence of (x_n), that converges to a point x^* in A. It follows that (y_{n_i}) also converges to x^*. Now choose N so that when $i \geq N$, we have $|x_{n_i} - y_{n_i}| < \delta$. Then $|f(x_{n_i}) - f(y_{n_i})| \geq \varepsilon_0$. But by continuity of f, $\lim_{i \to \infty} |f(x_{n_i}) - f(y_{n_i})| = |f(x^*) - f(x^*)| = 0$, a contradiction.

The second proof is not by contradiction. Let $\varepsilon > 0$. Since f is continuous at every point of A, for each $x \in A$, there exists $\delta_x > 0$ so that

$$(4.8) \qquad |f(y) - f(x)| < \frac{\varepsilon}{2} \text{ for all } y \in A \text{ with } |y - x| < \delta_x.$$

To find a δ that works for all x one might at first be tempted to take the smallest, i.e., the infimum, of all δ_x. However, in general this is an infinite collection of numbers and their infimum may be 0. Here is where we use compactness to find a finite number of appropriate δ_x.

We observe that the collection of open balls $\{B(x, \delta_x/2)\}_{x \in A}$ forms an open cover of A. (We see below that it is makes a difference to take the radii of these balls to be $\delta_x/2$ and not simply δ_x.) By compactness it has a finite subcover, which we may denote by

$$B(x_1, \delta_1/2), \ldots, B(x_n, \delta_n/2),$$

for some $x_1, \ldots, x_n$ and where $\delta_i = \delta_{x_i}$, $i \in \{1, \ldots, n\}$. Set

$$\delta = \min\{\frac{\delta_1}{2}, \ldots, \frac{\delta_n}{2}\}.$$

Now let $x, y \in A$ with $|y - x| < \delta$. Then x must be contained in some element of the finite subcover: $x \in B(x_i, \delta_i/2)$ for some $i \in \{1, \ldots, n\}$; so $|x - x_i| < \delta_i/2$. Next we find that

$$|y - x_i| \leq |y - x| + |x - x_i| < \frac{\delta_i}{2} + \frac{\delta_i}{2} = \delta_i.$$

Using (4.8), we obtain

$$|f(y) - f(x)| \leq |f(y) - f(x_i)| + |f(x_i) - f(x)| < \frac{\varepsilon}{2} + \frac{\varepsilon}{2} = \varepsilon.$$

This completes that proof that f is uniformly continuous on A. $\square$

Exercises: Uniform Continuity

4.4.1 Let $a, b \in \mathbb{R}$ be constants, and define $f(x) = ax + b$ for $x \in [0, \infty)$. Show that $f : [0, \infty) \to \mathbb{R}$ is uniformly continuous. What about $f : \mathbb{R} \to \mathbb{R}$?

4.4.2 Let $A = [0, 2] \setminus \{1\}$, and define $f : A \to \mathbb{R}$ by
$$f(x) = \frac{1}{x-1}.$$
Is f uniformly continuous on A? Prove your claim.

4.4.3 Give an example to show that the converse of Theorem 4.4.5 does not hold.

4.4.4 Let $0 < a < 1$ and $f(x) = 1/x$. Give two proofs, one using Theorem 4.4.5 and another without using the theorem, that f is uniformly continuous on $[a, 1)$.

4.4.5 Prove that $f(x) = 1/x^k$, $k \in \mathbb{N}$, is not uniformly continuous on $(0, 1)$.

4.4.6 Let $A \subseteq \mathbb{R}$, and let $f : A \to \mathbb{R}$ be uniformly continuous function on A. Prove that if (a_n) is a Cauchy sequence in A, then $(f(a_n))$ is Cauchy.

4.4.7 Let $A \subseteq \mathbb{R}$, and let $f : A \to \mathbb{R}$ be a uniformly continuous function on A. Prove that if A is bounded, then $f(A)$ is bounded.

4.4.8 Let $g(x) = x^2$. Show that $g : [0, \infty) \to \mathbb{R}$ is not uniformly continuous. (*Hint*: It suffices to show that uniform continuity fails for a specific $\varepsilon > 0$. You can choose $\varepsilon = 1$. Then show that for any $\delta > 0$ you can choose $x, y \in [0, \infty)$ so that $|x - y| < \delta$ but $|g(x) - g(y)| \geq 1$.)

4.4.9 Is $f(x) = \frac{\sin x}{x}$ uniformly continuous on $(0, 1)$?

4.4.10 Prove that $f(x) = x^2 + 1$ is not uniformly continuous on $\mathbb{R}$.

4.5. Continuous Functions on Metric Spaces

Throughout this section (X, d) and (Y, ρ) will be metric spaces.

Definition 4.5.1. A function $f : X \to Y$ is **continuous** at a point $x \in X$ if for all $\varepsilon > 0$ there exists $\delta > 0$ so that
$$d(x, y) < \delta \text{ implies } \rho(f(x), f(y)) < \varepsilon.$$
$\diamond$

Equivalently, $f : X \to Y$ is continuous at x if
$$y \in B(x, \delta) \text{ implies } f(y) \in B(f(x), \varepsilon).$$
While the notation does not show this, $B(x, \delta)$ is an open ball in X and $B(f(x), \delta)$ is an open ball in Y. Furthermore, this last condition is equivalent to
$$B(x, \delta) \subseteq f^{-1}(B(f(x), \varepsilon)).$$

Definition 4.5.2. The function $f : X \to Y$ is said to be **continuous on** X if it is continuous at every point of X. $\diamond$

We note that this definition agrees with the definition of continuity for functions $f : A \to \mathbb{R}$.

We are ready for the following characterization.

Lemma 4.5.3. *Let (X, d) and (Y, ρ) be metric spaces, and let $f : X \to Y$ be a function. Then f is continuous on X if and only if for all open sets G in Y, the set $f^{-1}(G)$ is open in X.*

Proof. Suppose f is continuous, and let G be an open set in Y. To show that $f^{-1}(G)$ is open, let $x \in f^{-1}(G)$. So $f(x) \in G$ and as G is open, there exists $\varepsilon > 0$ so that $B(f(x), \varepsilon) \subseteq G$. Since f is continuous at x there exists $\delta > 0$ so that
$$B(x, \delta) \subseteq f^{-1}(B(f(x), \varepsilon)) \subseteq f^{-1}(G).$$
Therefore, $f^{-1}(G)$ is open.

For the converse, suppose that for every open set G in Y, the set $f^{-1}(G)$ is open. Let $x \in X$, and let $\varepsilon > 0$. Consider the open set $G = B(f(x), \varepsilon)$. Since $f^{-1}(B(f(x), \varepsilon))$ is open and x is a point in $f^{-1}(B(f(x), \varepsilon))$, there exists $\delta > 0$ so that $B(x, \delta) \subseteq f^{-1}(B(f(x), \varepsilon))$, which means that f is continuous at x. □

The characterization of continuity in terms of sequences also holds; see Exercise 4.5.13.

In the definition of continuity, it is implicit that the ball $B(x, \delta)$ refers to all points y in X that satisfy $d(x, y) < \delta$. For example if X is the metric space $[1, 2)$ with the induced metric from $\mathbb{R}$, then when testing continuity at $x = 1$, the balls $B(1, \delta)$ will be of the form $[1, 1 + \delta)$. To clarify these ideas, we recall the notion of relatively open sets from Section 3.3. Let (X, d) be a metric space, and let $Z \subseteq X$. We will use that $A \subseteq Z$ is relatively open and may say it is open relative to Z if A is of the form $A = Z \cap G$, where G is open in X. For example, the set $[1, 2)$ is open relative to $[1, 3]$ even though it is clearly not open in $\mathbb{R}$. Now we can state an extension of Lemma 4.5.3. The proof is similar to the proof of Lemma 4.5.3, and it is left to the reader.

Lemma 4.5.4. *Let (X, d) and (Y, ρ) be metric spaces, let $Z \subseteq X$, $W \subseteq Y$, and let $f : Z \to W$ be a function. Then f is continuous on Z if and only if for all relatively open sets G in W, the set $f^{-1}(G)$ is relatively open in Z.*

We define a useful class of continuous functions.

Definition 4.5.5. A function $f : X \to Y$ is said to be an **isometry** if for all $x, y \in X$,
$$\rho(f(x), f(y)) = d(x, y). \qquad \diamond$$

Definition 4.5.6. We say f is a **contraction** if there is a constant $0 < r < 1$ such that
$$\rho(f(x), f(y)) \leq r d(x, y). \qquad \diamond$$

Isometries and contractions are continuous; this follows directly from the definition of continuity. Translations, and more generally rigid motions in $\mathbb{R}^n$, are isometries.

Our next theorem has many applications. We need a definition first. If $f : X \to X$ is a function, we can consider its composition iterates (that is, we can consider the function $f \circ f$ that is denoted by f^2, called the **second iterate** and defined by $f^2(x) = f(f(x)))$, and in this way we define the nth **iterate** of f by

$$f^n(x) = f(f^{n-1}(x)).$$

(We note that the nth derivative will be denoted $f^{(n)}$.)

Theorem 4.5.7 (Banach fixed point theorem). *Let (X, d) be a complete metric space, and let $f : X \to X$ be a contraction. For every $x \in X$, the sequence $(f^n(x))$ converges to a point p that is fixed under f: $f(p) = p$. Furthermore, the fixed point is unique.*

Proof. We first show that for each $x \in X$, the sequence $(f^n(x))$ is a Cauchy sequence. Let $n, \ell \in \mathbb{N}$. We obtain the following inequality by first applying the property of a contraction n times, then using the triangle inequality, and finally the geometric series:

$$\begin{aligned} d(f^n(x), f^{n+\ell}(x)) &\leq r^n d(x, f^\ell(x)) \\ &\leq r^n[d(x, f(x)) + d(f(x), f^2(x)) + \cdots + d(f^{\ell-1}(x), f^\ell(x))] \\ &\leq r^n d(x, f(x))[1 + r + \cdots + r^{\ell-1}] \\ &< r^n \cdot d(x, f(x)) \cdot \frac{1}{1-r}. \end{aligned}$$

Given $\varepsilon > 0$, as $r < 1$, we can choose $N \in \mathbb{N}$ so that

$$r^N \cdot d(x, f(x)) \cdot \frac{1}{1-r} < \varepsilon.$$

Then it follows that for all $n \geq N$ and all $\ell \in \mathbb{N}$,

$$d(f^n(x), f^{n+\ell}(x)) < \varepsilon.$$

Thus the sequence $(f_n(x))$ is Cauchy, and so it converges to a point $p \in X$. The continuity of f implies that p is a fixed point since

$$f(p) = f(\lim_{n \to \infty} f^n(x)) = \lim_{n \to \infty} f^{n+1}(x) = p.$$

If q is another fixed point, then by the contraction property we have

$$d(p, q) = d(f(p), f(q)) \leq r d(p, q),$$

which is only possible when $p = q$ as $r < 1$. $\square$

Remark 4.5.8. We illustrate one of the many applications of Theorem 4.5.7. Several of the interesting applications of this theorem occur when the space X has additional structure. Here we discuss the case when X consists of all nonempty compact subsets of $\mathbb{R}$ (or $\mathbb{R}^n$). So each point of X is now a nonempty compact subset of $\mathbb{R}$. For concreteness we will consider just compact subsets of $[0, 1]$, but the theory extends in a natural way to compact subsets of $\mathbb{R}^n$. We need to define a metric on X, which intuitively is supposed to capture the distance between any two compact sets A and B in $[0, 1]$. First, define the distance from a point x to the set B as the shortest distance to B, or

(4.9) $$d(x, B) = \inf\{d(x, y) : y \in B\}.$$

(One can show that it is possible to use the minimum instead of the infimum as B is compact.) Then define

(4.10) $$d_1(A, B) = \sup\{d(x, B) : x \in A\}.$$

It turns out that d_1 is not symmetric (it may happen that $d_1(A, B) \neq d_1(B, A)$), so d_1 is not a metric, but one can finally define a distance between compact sets, called the **Hausdorff metric** by

(4.11) $$H(A, B) = \max\{d_1(A, B), d_1(B, A)\}.$$

It can be shown that H is indeed a metric on X (Exercise 4.5.5). It is harder to show, and we do have all the results that are needed for the proof, that with this metric, X is a complete metric space, and we just ask for an outline of the argument in Exercise 4.5.6.

We now define a particular contraction on X that we denote by T. First define $f_1(x) = \frac{1}{3}x$ and $f_2(x) = \frac{1}{3}x + \frac{2}{3}$. Now define, for each $A \subseteq X$, the map $T : X \to X$ by

(4.12) $$T(A) = f_1(A) \cup f_2(A).$$

Note that $f_1(A)$ is another set, so $f([0, 1]) = [0, \frac{1}{3}]$. Thus, for example,

$$T([0, 1]) = [0, \frac{1}{3}] \cup [\frac{2}{3}, 1],$$

and if we apply T again, noting that $f_1([0, \frac{1}{3}]) = [0, \frac{1}{9}] \cup [\frac{2}{3}, \frac{7}{9}]$, we obtain

$$T^2([0, 1]) = [0, \frac{1}{9}] \cup [\frac{2}{9}, \frac{3}{9}] \cup [\frac{2}{3}, \frac{7}{9}] \cup [\frac{8}{9}, 1].$$

One can see that this generates the sets F_n in the construction of the Cantor set in Figure 3.5 by $F_n = T^n([0, 1])$.

It remains to verify that T is a contraction on (X, d) (Exercise 4.5.7). Then Theorem 4.5.7 implies that the sequence of sets $T^n([0, 1])$ converges to a unique fixed point, K, i.e., $T(K) = K$. It should be clear that the Cantor set is the only set that after iterating by T remains fixed: $f_1(K)$ gives us the first third of K and $f_2(K)$ its second third, so that after applying T to K we get back K. Amazingly, the theorem states that one can start with any compact set A, and eventually $T^n(A)$ also converges to K. For example, if $A = \{0\}$, we get $T(A) = \{0, \frac{2}{3}\}, T^2(A) = \{0, \frac{2}{9}, \frac{2}{3}, \frac{8}{9}\}$, and as n increases one can see that the pattern of K starts to emerge.

These ideas can be generalized to nonempty compact subsets of $\mathbb{R}^n$, in particular of $\mathbb{R}^2$. Then one can obtain sequences that generate many of the well-known fractals of the plane.

4.5.1. Metric Completion. We outline a proof that any metric space (X, d) can be extended to a complete metric space in the sense that there is a complete metric space (X', d') and an isometry $\phi : X \to X'$. We recall that an isometry is injective, and it preserves distances, so we can regard $\phi(X)$ as a copy of X. The idea of copy here is informal, but let us just mention that ϕ is a bijection from X to $\phi(X)$, and all metric-dependent properties on X get carried over to $\phi(X)$ using

4.5. Continuous Functions on Metric Spaces

the identity $d'(\phi(x), \phi(y)) = d(x,y)$. It can further be shown that $\phi(X)$ is dense in X'.

Cauchy sequences in an arbitrary metric space X may not converge. To construct a complete metric space extending X, the idea is to add new points to X that would be the limits of Cauchy sequences in X. The starting idea of the construction is that the point that the Cauchy sequence is supposed to converge to should be represented by the Cauchy sequence itself. But different Cauchy sequences might converge to the same point, so we need to identify all Cauchy sequences that would converge to the same point. We do this by creating equivalence classes of Cauchy sequences. The notion of equivalence classes is a powerful notion in mathematics when creating new structures; it groups into one entity all objects that share a given property (in our case, all Cauchy sequences that converge to the same point). We have used that notion when constructing $\mathbb{Z}$ and $\mathbb{Q}$ in Subsection 0.6.2, but this can be read independently of those sections. Equivalence classes are defined in terms of equivalence relations, which are discussed in Section 0.4. We mention here that the word "class" in this context means the same as set.

First we define a relation among Cauchy sequences. Let (a_n) and (b_n) be Cauchy sequences. We say they are related by R and write

$$(a_n) R (b_n) \text{ if and only if } d(a_n, b_n) \to 0.$$

(We use this notation instead of writing $((a_n), (b_n)) \in R$.) For example, if (a_n) is a Cauchy sequence that converges to a, it is identified by this relation with all Cauchy sequences that converge to a.

We prove that this relation satisfies three properties.

- A sequence (a_n) is related to itself, i.e., $(a_n) R (a_n)$.
- If (a_n) is related to (b_n), then (b_n) is related to (a_n).
- If (a_n) is related to (b_n), and (b_n) is related to (c_n), then (a_n) is related (c_n).

These are the properties that define an equivalence relation. To show the first property, we note that $d(a_n, a_n) = 0$, so clearly $(a_n) R (a_n)$. The second property follows from the fact that $d(a_n, b_n) = d(b_n, a_n)$, so $(a_n) R (b_n)$ implies $(b_n) R (a_n)$. The third property follows from the triangle inequality

$$d(a_n, c_n) \leq d(a_n, b_n) + d(b_n, c_n),$$

so if $d(a_n, b_n) \to 0$ and $d(b_n, c_n) \to 0$, then $d(a_n, c_n) \to 0$.

Now, given a Cauchy sequence (a_n), define the equivalence class of (a_n) by

$$[(a_n)] = \{(b_n) : d(b_n, a_n) \to 0\};$$

i.e., $[(a_n)]$ consists of all Cauchy sequences either related to (a_n) or equivalent to (a_n) by this relation. By the first property, the class of (a_n) contains the sequence (a_n), or $(a_n) \in [(a_n)]$. Also, if (a_n) and (b_n) are related, then the class of (a_n) is the same as the class of (b_n). For if (c_n) is in $[(a_n)]$, then $(c_n) R (a_n)$, and as $(a_n) R (b_n)$ then $(c_n) R (b_n)$, which means that $[(a_n)] \subseteq [(b_n)]$. Similarly, one shows $[(b_n)] \subseteq [(a_n)]$.

For example, if x is an element in X, we can define the constant sequence $a_n = x$ for all $n \in \mathbb{N}$; we shall denote this constant sequence (a_n) by $(x)_n$. It is clear that the constant sequence $(x)_n$ converges to x and is a Cauchy sequence in X. If $y \neq x$, then the equivalence classes $[(y)_n]$ and $[(x)_n]$ are different (and do not intersect). But the sequence $(x + 1/n)$ is equivalent to $(x)_n$, so it is in the class $[(x)_n]$. Another sequence in this class is $(x - 1/n^2)$.

So far our discussion has been in the context of a general metric space. To give an example of sequences when the metric space is the set of rational numbers $\mathbb{Q}$, we can let $a_n = (1 + 1/n)^n$. This is a sequence of rationals that we know converges to e, which is irrational. Without knowing in advance about $\mathbb{R}$, one can show that (a_n) is a Cauchy sequence in $\mathbb{Q}$ that does not converge in $\mathbb{Q}$. One can still consider its equivalence class $[(a_n)]$, and one would like to think of this class as representing the number e.

We are ready to define the notion of a metric completion. It is useful to have a notation for the set of all Cauchy sequences in a given metric space (X, d), which we will denote by X^*. So, for example, for each $x \in X$, the sequence $(x)_n$ is an element of X^*. Also, in Question 4.5.9 the reader is asked to show that if (a_n) and (b_n) are Cauchy sequences, then $(d(a_n, b_n)_n)$ is a Cauchy sequence in $\mathbb{R}$, so it converges. Thus the limit $\lim_{n \to \infty} d(a_n, b_n)$ exists in $\mathbb{R}$.

Question 4.5.9. Let (a_n) and (b_n) be in X^*. Prove that $(d(a_n, b_n))$ is a Cauchy sequence.

Definition 4.5.10. Given a metric space (X, d), we define its **metric completion** to be the space (X', d'), where X' is the space of equivalence classes of Cauchy sequences in X (i.e., equivalence classes of elements of X^*), and d' is defined by

$$(4.13) \qquad d'([(a_n)], [(b_n)]) = \lim_{n \to \infty} d(a_n, b_n),$$

where (a_n) and (b_n) are Cauchy sequences.

We have already remarked that we can define the limit in (4.13). Before showing that d' is a metric, in Question 4.5.11 the reader is asked to show it is well-defined, i.e., that the value of distance between two equivalence classes does not depend on the representative of the equivalence classes that is used.

Question 4.5.11. Prove that the value of d' does not depend on the representative of the class, that is, if $[(a'_n)] = [(a_n)]$ and $[(b'_n)] = [(b_n)]$, then $\lim_{n \to \infty} d(a'_n, b'_n) = \lim_{n \to \infty} d(a_n, b_n)$.

The following lemma is the one that says X' is a nice space, and it also says that equivalence classes are disjoint and partition the space X^*. While we state it on our specific context, it is true for all equivalence classes.

Lemma 4.5.12. *The set X' of equivalence classes of Cauchy sequences forms a partition of X^* in the sense that*

- *every Cauchy sequence is an element of some class;*
- *each Cauchy sequence is in only one class, i.e., if two classes share the same Cauchy sequence, then they must be equal.*

Proof. First, a Cauchy sequence is in its own class as it is related to itself, so each element of X^* is in some element of X': if (a_n) is a Cauchy sequence, then $(a_n) \in [(a_n)] \in X'$.

4.5. Continuous Functions on Metric Spaces

For the second part, we start by noting that if (c_n) is in $[(b_n)]$, then $[(c_n)] \subseteq [(b_n)]$; this is because any sequence that is related to (c_n), by transitivity, is related to (b_n). Now suppose classes $[(b_n)]$ and $[(c_n)]$ have the element (a_n) in common. So (b_n) is related to (a_n), which is related to (c_n), so by transitivity (b_n) is related to (c_n). It follows that $[(c_n)] \subseteq [(b_n)]$, and similarly one shows $[(b_n)] \subseteq [(c_n)]$. $\square$

To show d' is a metric we verify the first property: suppose $d([(a_n)],[(b_n)]) = 0$. Then $\lim_{n\to\infty} d(a_n, b_n) = 0$. This means that (a_n) and (b_n) are related, so $[(a_n)] = [(b_n)]$. The fact that d' is symmetric is a direct result of the definition. The last property, the triangle inequality, is left as an exercise.

It remains to show that (X', d') is Cauchy complete. Let (α_k) be a Cauchy sequence in X'. Then each α_k is of the form $\alpha_k = [(a_{k,n})_n]$, where for each k, $(a_{k,n})_n$ is a Cauchy sequence (in n). Let $\varepsilon > 0$. There exists $K \in \mathbb{N}$ such that

(4.14) $\qquad$ if $k \geq K, \ell \in \mathbb{N}$, then $d'([(a_{k,n})_n], [(a_{k+\ell,n})_n]) < \varepsilon$.

Equivalently,

(4.15) $\qquad$ if $k \geq K, \ell \in \mathbb{N}$, then $\lim_{n\to\infty} d(a_{k,n}, a_{k+\ell,n}) < \varepsilon$.

Thus, there exists $N \in \mathbb{N}$ so that, for $k \geq K, \ell \in \mathbb{N}$,

$$\text{if } n \geq N, \text{ then } d(a_{k,n}, a_{k+\ell,n}) < \varepsilon.$$

Then we can choose, for fixed K, the sequence $b_n = a_{K,n}$. It follows that

$$\text{if } n \geq N, \ell \in \mathbb{N}, \text{ then } d(b_n, b_{n+\ell}) < \varepsilon.$$

Thus (b_n) is a Cauchy sequence, so $\alpha = [(b_n)]$ is an element of X'. Now we have to show (α_k) converges to α in (X', d'). We observe that from (4.15), when $k = K$, we have

$$\text{if } \ell > K, \text{ then } \lim_{n\to\infty} d(a_{K,n}, a_{\ell,n}) < \varepsilon$$

or

$$\text{if } \ell > K, \text{ then } d'([(b_n)], [(a_{\ell,n})]) < \varepsilon.$$

This shows that α_ℓ converges to α, completing the proof that the metric is complete.

Now we define the isometry $\phi : X \to X'$. For $x \in X$, we consider as before the constant sequence $(x)_n$, which is clearly a Cauchy sequence in X. The equivalence class $[(x)_n]$ consists of all Cauchy sequences in X that converge to x. Then define

$$\phi(x) = [(x)_n].$$

We verify that ϕ is an isometry,

$$d'(\phi(x), \phi(y)) = d'([(x)_n], [(y)_n]) = \lim_{n\to\infty} d(x,y) = d(x,y).$$

Therefore, ϕ is an isometry from X to X'. This finishes the construction of the completion (X', d') of a metric space (X, d).

Remark 4.5.13. The reader may be thinking at this moment that one could apply this completion process to the rational numbers $\mathbb{Q}$ with the Euclidean distance, and then obtain a complete metric space that would be a model for $\mathbb{R}$. This is in fact the construction of the real numbers obtained by Cantor in 1872. There is one thing we have to modify, though. In our definition of the metric d', we used that the real numbers are Cauchy complete (the metric is defined as the limit of a Cauchy sequence in $\mathbb{R}$). If one is going to use this process to construct the real numbers, one cannot make this assumption. The modification is not complicated and is left as a challenge to the interested reader.

Exercises: Continuous Functions on Metric Spaces

Exercises

4.5.1 Prove Lemma 4.5.4.

4.5.2 Let $f : (X, d) \to (Y, \rho)$ be continuous. Prove that if $A \subseteq X$ is compact, then $f(A) \subseteq Y$ is compact.

4.5.3 Prove that in (4.9) one can replace inf with min when B is compact, and sup with max when A and B are compact in (4.10).

4.5.4 Give examples of compact sets A and B in $\mathbb{R}$ to show that $d_1(A, B) \neq d_1(B, A)$, where d_1 is as in (4.10).

4.5.5 Prove that H as defined in Remark 4.5.8 is a metric.

* 4.5.6 Prove that H as defined in Remark 4.5.8 is a complete metric when defined on the collection of all nonempty compact subsets of $\mathbb{R}$.

4.5.7 Prove that the map T defined in (4.12) is a contraction under H.

4.5.8 Prove that if (a_n) and (b_n) are Cauchy sequences, then $\lim_{n \to \infty} d(a_n, b_n)$ exists.

4.5.9 Let X be the symbolic space $\{0, 1\}^{\mathbb{N}}$ of Example 3.1.6. Define the *shift map* $\sigma : X \to X$ by $\sigma(a_1 a_2 a_3 \cdots) = (a_2 a_3 \cdots)$, i.e., it shifts coordinates to the left. Prove that σ is continuous.

4.5.10 Prove that the metric d' in (4.13) satisfies the triangle inequality.

4.5.11 Let (X, d) be a metric space, and let $f : X \to X$ be a bijection such that both f and f^{-1} are continuous. Define $d'(x, y) = d(f(x), f(y))$ for $x, y \in X$. Prove that d' is a metric on X that is topologically equivalent to d.

4.5.12 Let $X = (0, \infty)$, let d be the Euclidean metric (i.e., $d(x, y) = |x - y|$), and let $f(x) = 1/x$ for $x \in X$. Let d' be defined by $d'(x, y) = d(f(x), f(y))$ as in Exercise 4.5.11. Find a sequence in X that is Cauchy with respect to d but not with respect to d' and another that is Cauchy with respect to d' and not with respect to d.

4.5.13 Let $f : X \to Y$ be a function, and let $x \in X$. Show that f is continuous at x if and only if for all sequences (x_n) that converge to x, the sequence $(f(x_n))$ converges to $f(x)$.

* 4.5.14 Use the metric space completion process to give another construction of the real numbers. You need to define the operations of addition and multiplication and prove all their properties.

Chapter 5

Differentiable Functions

5.1. Differentiable Functions on $\mathbb{R}$

The notion of the derivative of a function is due to Newton (1642–1727) and Leibniz (1646–1716). The definition is based on the notion of a limit. Differentiation, like continuity, is a local concept.

We will consider a function f defined on an open interval (a,b) (with $a < b$). Fix a point x in (a,b). For each $z \in (a,b)$, consider the ratio

$$\frac{f(z) - f(x)}{z - x}$$

which represents the slope of the chord (i.e., line) from $(z, f(z))$ to $(x, f(x))$ (see Figure 5.1). As z approaches x, this chord approaches the tangent line to the graph of the function at the point $(x, f(x))$ if there is a tangent at this point.

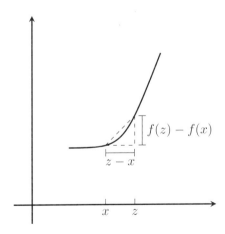

Figure 5.1. Slope of the chord from $(x, f(x))$ to $(z, f(z))$.

Definition 5.1.1. We say that the function $f : (a,b) \to \mathbb{R}$ is **differentiable at** $x \in (a,b)$ if the limit

$$(5.1) \qquad \lim_{z \to x} \frac{f(z) - f(x)}{z - x}$$

exists. In this case we write

$$f'(x) = \lim_{z \to x} \frac{f(z) - f(x)}{z - x}$$

and say that $f'(x)$ is the **derivative** of f at x. ◊

As x is an interior point of the domain (a,b), we can talk about the ratio in the limit for all z that are sufficiently close to x. From our geometric interpretation of this limit, the derivative will be the slope of the tangent line, if the tangent line exits. There is no tangent line if and only if the limit does not exist.

Example 5.1.2. Constant functions are differentiable. Let $c \in \mathbb{R}$, and let $C : \mathbb{R} \to \mathbb{R}$ be defined by $C(x) = c$ for all $x \in \mathbb{R}$. Then

$$C'(x) = \lim_{z \to x} \frac{C(z) - C(x)}{z - x} = \lim_{z \to x} \frac{c - c}{z - x} = 0.$$

Remark 5.1.3. By thinking of z as $x+h$ for $h \in \mathbb{R}$, we can write $f'(x)$ equivalently as

$$f'(x) = \lim_{h \to 0} \frac{f(x+h) - f(x)}{h}.$$

Remark 5.1.4. In most applications we consider functions defined on an open interval (a,b). More generally, if a function f is defined on a set A, $f : A \to \mathbb{R}$, this definition applies to any point in the interior of A: if x is in the interior of A, we can consider x in an interval inside A, and we say f is differentiable at x if condition (5.1) is satisfied. While we consider functions defined on an open interval, for the theorems in this section we could just as well have considered the functions defined on an open set. In Remark 5.1.7 we discuss the case of functions defined on a closed interval.

Example 5.1.5. Let $n \in \mathbb{N}$, and let $f : \mathbb{R} \to \mathbb{R}$ be defined by $f(x) = x^n$ for all $x \in \mathbb{R}$. We show that f is differentiable at every $x \in \mathbb{R}$ and

$$f'(x) = nx^{n-1}.$$

We do a couple of special cases first since this is one of our first examples. If $n = 1$, $f(x) = x$. So

$$f'(x) = \lim_{z \to x} \frac{f(z) - f(x)}{z - x} = \lim_{z \to x} \frac{z - x}{z - x} = 1.$$

Now when $n = 2$, $f(x) = x^2$, then

$$f'(x) = \lim_{z \to x} \frac{f(z) - f(x)}{z - x} = \lim_{z \to x} \frac{z^2 - x^2}{z - x} = \lim_{z \to x} z + x = 2x.$$

5.1. Differentiable Functions on $\mathbb{R}$

For the general case we use the binomial theorem and calculate

$$f'(x) = \lim_{h \to 0} \frac{f(x+h) - f(x)}{h} = \lim_{h \to 0} \frac{(x+h)^n - x^n}{h}$$

$$= \lim_{h \to 0} \frac{\sum_{i=0}^{n} \binom{n}{i} x^{n-i} h^i - x^n}{h}$$

$$= \lim_{h \to 0} \frac{\sum_{i=1}^{n} \binom{n}{i} x^{n-i} h^i}{h} = \lim_{h \to 0} \sum_{i=1}^{n} \binom{n}{i} x^{n-i} h^{i-1}$$

$$= n x^{n-1} \text{ (as the only term in the sum without } h \text{ corresponds to } i=1).$$

Example 5.1.6. Let $f : \mathbb{R} \to \mathbb{R}$ be defined by $f(x) = \frac{1}{x}$ for all $x \in \mathbb{R}, x \neq 0$. Then f is differentiable at every $x \in \mathbb{R} \setminus \{0\}$ since

$$f'(x) = \lim_{z \to x} \frac{f(z) - f(x)}{z - x} = \lim_{z \to x} \frac{\frac{1}{z} - \frac{1}{x}}{z - x} = \lim_{x \to z} \frac{x - z}{zx(z - x)} = -\frac{1}{x^2}.$$

Remark 5.1.7. In the case when a function is defined on a closed interval $A = [a, b]$, one could define a one-sided derivative at the endpoint $x = a$ as the one-sided limit as z approaches a from the right. The *left-sided derivative* $f'^{-}(b)$ is defined in a similar way but as z approaches b from the left. We note that the one-sided derivatives can also be defined at any point of the domain, not just the endpoints. For points where the function is differentiable, the right-sided and left-sided derivatives must exist and be equal. However, we will not emphasize one-sided derivatives. One could, though, rephrase many of the theorems below in this context, and some of the exercises may ask the reader to do this.

Definition 5.1.8. Let $f : [a, b] \to \mathbb{R}$ be a function. Define the **right-sided derivative** of f at a by

$$f'^{+}(a) = \lim_{z \to a^+} \frac{f(z) - f(a)}{z - a},$$

and the **left-sided derivative** $f'^{-}(b)$ is defined as z approaches b from the left by

$$f'^{-}(b) = \lim_{z \to b^-} \frac{f(z) - f(b)}{z - b}. \qquad \diamond$$

The proposition below shows that differentiation implies continuity; its converse is not true as discussed in Remark 5.1.11.

Proposition 5.1.9. *Let $f : (a, b) \to \mathbb{R}$ be a function, and let $x \in (a, b)$. If f is differentiable at x, then f is continuous at x.*

Proof. We give two proofs. By Proposition 4.3.3 it suffices to show that

$$\lim_{z \to x} f(z) - f(x) = 0.$$

Now, using Proposition 4.3.11,

$$\lim_{z \to x} f(z) - f(x) = \lim_{z \to x} \frac{f(z) - f(x)}{z - x} \cdot (z - x)$$

$$= \lim_{z \to x} \frac{f(z) - f(x)}{z - x} \cdot \lim_{z \to x} (z - x)$$

$$= f'(x) \cdot 0 = 0.$$

For the second proof we use the definition of continuity. Let $\varepsilon > 0$. We need to find $\delta > 0$ so that if $|x - z| < \delta$ and $x \in A$, then $|f(x) - f(z)| < \varepsilon$. As f is differentiable at x, the limit
$$\lim_{z \to x} \frac{f(z) - f(x)}{z - x}$$
exists and equals $f'(x)$. Therefore, there exists $\delta_1 > 0$ so that
$$\left| \frac{f(z) - f(x)}{z - x} - f'(x) \right| < 1,$$
for all $z \in A$ such that $0 < |z - x| < \delta_1$. Then
$$|f(z) - f(x) - f'(x)(z - x)| < |z - x|.$$
Therefore,
$$\begin{aligned} |f(z) - f(x)| &= |f(z) - f(x) - f'(x)(z - x) + f'(x)(z - x)| \\ &\leq |f(z) - f(x) - f'(x)(z - x)| + |f'(x)(z - x)| \\ &< |z - x| + |f'(x)(z - x)| \\ &= |z - x|(1 + |f'(x)|). \end{aligned}$$
Set
$$\delta = \min\left\{ \delta_1, \frac{\varepsilon}{1 + |f'(x)|} \right\}.$$
Thus if $z \in (a, b)$ and $|z - x| < \delta$, then $|f(z) - f(x)| < \varepsilon$ and f is continuous at x. □

Question 5.1.10. Let $f : [a, b] \to \mathbb{R}$ be a function such that the left-sided derivative exists at b and the right-sided derivative exists at a. Show, from the definitions, that f is continuous at a and at b.

Remark 5.1.11. A continuous function need not be differentiable, so the converse of Proposition 5.1.9 does not hold. In fact let $f(x) = |x|$. We know f is continuous at 0. However, considering the sequence $(1/n)$ which converges to 0, we observe that
$$\lim_{n \to \infty} \frac{f(\frac{1}{n}) - f(0)}{\frac{1}{n} - 0} = \frac{\frac{1}{n} - f(0)}{\frac{1}{n} - 0} = 1.$$
But if we consider the sequence $(-1/n)$, which also converges to 0, we have
$$\lim_{n \to \infty} \frac{f(-\frac{1}{n}) - f(0)}{-\frac{1}{n} - 0} = \frac{\frac{1}{n} - f(0)}{-\frac{1}{n} - 0} = -1.$$
As these limits are different, it follows that f is not differentiable at 0. We will see in Subsection 8.3.1 an example of a continuous function that is differentiable at no point.

Remark 5.1.12. The absolute value function $|x|$ is probably the best known example of a continuous function that is not differentiable, but the reader should not think that such "sharp corners" are necessary for the counterexamples. The property that was observed for $|x|$ at $x = 0$ is that the *one-sided* derivatives are different. As we have remarked, if f is differentiable at x, then the right-sided and left-sided derivatives must exist and be equal. In the case of absolute value, one computes the right-sided derivative at $x = 0$ to be $+1$ and the left-sided derivative

5.1. Differentiable Functions on $\mathbb{R}$

to be -1. But one can have examples where the one-sided derivatives are different and there is no visible "corner". For example, consider $g : (0,2) \to \mathbb{R}$ defined by

$$g(x) = \begin{cases} x^2 & \text{if } 0 < x \leq 1, \\ x^4 & \text{if } 1 < x < 2. \end{cases}$$

Then one can calculate that $g'^+(1) = 4$ and $g'^-(1) = 2$ (see Exercise 5.1.12). So g is not differentiable at $x = 1$. There are also examples where the function is continuous but the one-sided derivatives do not exist (see Exercise 5.1.4).

Theorem 5.1.13. *Let $f, g : (a, b) \to \mathbb{R}$ be functions. Suppose $x \in (a, b)$ and f, g are differentiable at $x \in (a, b)$.*

(1) *If c is a constant, then cf is differentiable at x and*
$$(cf)'(x) = cf'(x).$$

(2) *The function $f + g$ is differentiable at x and*
$$(f + g)'(x) = f'(x) + g'(x).$$

(3) *Product rule: The function $f \cdot g$ is differentiable at x and*
$$(f \cdot g)'(x) = f'(x) \cdot g(x) + f(x) \cdot g'(x).$$

(4) *Quotient rule: If $g(x) \neq 0$, then f/g is differentiable at x and*
$$\left(\frac{f}{g}\right)'(x) = \frac{g(x) \cdot f'(x) - f(x) \cdot g'(x)}{g^2(x)}.$$

Proof. For the first property we note that

$$(5.2) \qquad \frac{(cf)(z) - (cf)(x)}{z - x} = c \cdot \frac{f(z) - f(x)}{z - x}.$$

Since $\lim_{z \to x} \frac{f(z) - f(x)}{z - x}$ exists and equals $f'(x)$, by taking limits as z approaches x in (5.2), we obtain that cf is differentiable at x and $(cf)'(x) = cf'(x)$.

The second property also follows directly from the definition. We observe that

$$(5.3) \qquad \frac{(f + g)(z) - (f + g)(x)}{z - x} = \frac{f(z) - f(x)}{z - x} + \frac{g(z) - g(x)}{z - x}.$$

Since $\lim_{z \to x} \frac{f(z) - f(x)}{z - x}$ and $\lim_{z \to x} \frac{g(z) - g(x)}{z - x}$ exist and equal $f'(x)$ and $g'(x)$, respectively, by taking limits as z approaches x in (5.3), we obtain that $f + g$ is differentiable at x and $(f + g)'(x) = f'(x) + g'(x)$.

We show the product rule; the last part is left as an exercise. Note that for all x and z,

$$f(z)g(z) - f(x)g(x) = f(z)g(z) - f(x)g(z) + f(x)g(z) - f(x)g(x)$$
$$= [f(z) - f(x)]g(z) + f(x)[g(z) - g(x)].$$

Then,

$$(5.4) \qquad \frac{f(z)g(z) - f(x)g(x)}{z - x} = \frac{f(z) - f(x)}{z - x} \cdot g(z) + f(x) \cdot \frac{g(z) - g(x)}{z - x}.$$

After taking limits as $z \to x$ in both sides of (5.4) and using that
$$f'(x) = \lim_{z \to x} \frac{f(z) - f(x)}{z - x}, \qquad g'(x) = \lim_{z \to x} \frac{g(z) - g(x)}{z - x},$$
and that since g is continuous at x, $\lim_{z \to x} g(z) = g(x)$, we obtain that the product is differentiable at x and satisfies the desired formula. $\square$

Theorem 5.1.14 (Chain rule)**.** *Let $g : (a, b) \to \mathbb{R}$ and $f : (c, d) \to \mathbb{R}$ be functions such that $g(a, b) \subseteq (c, d)$. If g is differentiable at $x \in (a, b)$ and f is differentiable at $g(x)$, then $f \circ g$ is differentiable at x and*
$$(f \circ g)'(x) = f'(g(x))g'(x).$$

Proof. We need to show that
$$\lim_{z \to x} \frac{f(g(z)) - f(g(x))}{z - x}$$
exists and equals $f'(g(x))g'(x)$ for x in (a, b). This suggests that we write
$$(5.5) \qquad \frac{f(g(z)) - f(g(x))}{z - x} = \frac{f(g(z)) - f(g(x))}{g(z) - g(x)} \cdot \frac{g(z) - g(x)}{z - x}.$$
When we consider the limit as z approaches x, we know we can assume $z \neq x$ in (5.5). It may be the case, however, that $g(z) = g(x)$ even when $z \neq x$. So we cannot take $\lim_{z \to x}$ in both expressions in (5.5). To address this, we define a new function for fixed x. For $z \in (a, b)$, let
$$h(z) = \begin{cases} \frac{f(g(z)) - f(g(x))}{g(z) - g(x)} & \text{if } g(z) \neq g(x), \\ f'(g(x)) & \text{if } g(z) = g(x). \end{cases}$$
Then for all $z \in (a, b)$ such that $z \neq x$, we have that
$$(5.6) \qquad \frac{f(g(z)) - f(g(x))}{z - x} = h(z) \cdot \frac{g(z) - g(x)}{z - x}.$$
This is the case since when $g(z) = g(x)$ we obtain 0 on both sides of (5.6), and when $g(z) \neq g(x)$ then (5.6) is nothing but (5.5), which holds in this case.

We claim that $h(z)$ is continuous at $z = x$ (see Exercise 5.1.9). Then $\lim_{z \to x} h(z) = f'(g(x))$. Then taking limits as $z \to x$ on both sides of (5.6) completes the proof. $\square$

Example 5.1.15. Let $n \in \mathbb{N}$ and $f : \mathbb{R} \to \mathbb{R}$ be defined by
$$f(x) = \frac{1}{x^n} \text{ for all } x \in \mathbb{R}, x \neq 0.$$
We show that f is differentiable at every $x \in \mathbb{R} \setminus \{0\}$ and
$$f'(x) = -\frac{n}{x^{n+1}}.$$
To see this, first let $g(x) = 1/x$ and $h(x) = x^n$. Then $f(x) = g(h(x))$. By Examples 5.1.6 and 5.1.5, g is differentiable for $x \neq 0$, $g'(x) = -1/x^2$, h is differentiable for all x, and $h'(x) = nx^{n-1}$. By Theorem 5.1.14, $g \circ h$ is differentiable for $x \neq 0$ and
$$f'(x) = g'(h(x))h'(x) = -\frac{1}{(x^n)^2}(nx^{n-1}) = -\frac{n}{x^{n+1}}.$$

5.1. Differentiable Functions on ℝ

The following question goes over the definition of differentiation, and will have some applications.

Question 5.1.16. Let $f : (a, b) \to \mathbb{R}$ be a function, and let $x \in (a, b)$. Let (p_n) and (q_n) be sequences such that $a < p_n \leq x < q_n < b$, and assume $\lim_{n \to \infty} p_n = \lim_{n \to \infty} q_n = x$. Prove that if f is differentiable at x, then

$$\lim_{n \to \infty} \frac{f(q_n) - f(p_n)}{q_n - p_n} = f'(x).$$

Exercises: Differentiable Functions on ℝ

5.1.1 Let $f : (a, b) \to \mathbb{R}$ be a function. Suppose f is Lipschitz. Is f differentiable on (a, b)?

5.1.2 Complete the proof of Theorem 5.1.13.

5.1.3 Let $n, m \in \mathbb{N}$ and $f : (0, \infty) \to \mathbb{R}$ be defined by $f(x) = x^{n/m}$. Prove that f is differentiable and

$$f'(x) = (n/m)x^{n/m-1}.$$

5.1.4 Let $f : [-1, 1] \to \mathbb{R}$ be defined by

$$f(x) = \begin{cases} x \sin(1/x) & \text{if } x \neq 0, \\ 0 & \text{if } x = 0. \end{cases}$$

Prove that f is continuous on $[-1, 1]$, but f is not differentiable at $x = 0$; see Figure 5.2. (Assume the standard properties of $\sin x$, i.e., that it is continuous and differentiable and its derivative is $\cos x$.)

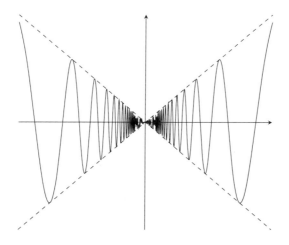

Figure 5.2. The function $x \sin(1/x), x \neq 0$.

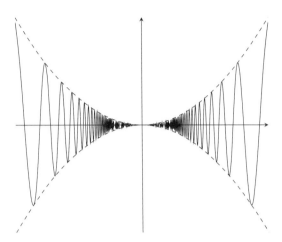

Figure 5.3. The function $x^2 \sin(1/x), x \neq 0$.

5.1.5 Let $g : [-1, 1] \to \mathbb{R}$ be defined by

$$g(x) = \begin{cases} x^2 \sin(1/x) & \text{if } x \neq 0, \\ 0 & \text{if } x = 0. \end{cases}$$

Prove that g is differentiable on $(-1, 1)$, but g' is not continuous at $x = 0$. See Figure 5.3. (Assume the standard properties of $\sin x$, i.e., that it is continuous and differentiable and its derivative is $\cos x$.)

5.1.6 Give an example of a continuous invertible function on $(0, 1)$ that is not differentiable.

5.1.7 Prove that a continuous invertible function defined on a closed or open interval is strictly monotone.

5.1.8 Let $n \in \mathbb{N}$. Construct a function that is n times differentiable on an interval but fails to be $n + 1$ times differentiable at a point in the interval.

5.1.9 Prove that the function h defined in the proof of Theorem 5.1.14 is continuous at $z = x$.

5.1.10 Prove that the function $f(x) = |x|x$ has one-sided derivatives at $x = 0$ and find their values.

5.1.11 Let $f : (a, b) \to \mathbb{R}$ be a function, and let $x \in (a, b)$. Prove that f is differentiable at x if and only if the left-sided and right-sided derivatives at x exist and are equal.

5.1.12 Let g be the function of Remark 5.1.12. Compute its one-sided derivatives at $x = 1$.

5.1.13 Extend Proposition 5.1.9 to the case of functions defined on closed intervals $f : [a, b] \to \mathbb{R}$.

5.2. Mean Value Theorem

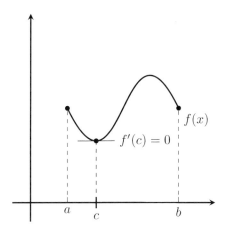

Figure 5.4. Rolle's theorem illustration.

5.2. Mean Value Theorem

The mean value theorem is one of the most important theorems in calculus. It applies to functions that are differentiable inside an interval. It tells us that for a differentiable function on an interval, there is a point in the graph of the function at which the tangent is parallel to the line joining the two endpoints of the graph (see Figure 5.5). In particular, if the line joining the two endpoints of the graph is horizontal, then this point is a local minimum or a local maximum. This consequence is called Rolle's theorem and is illustrated in Figure 5.4.

Our proof of the mean value theorem uses the idea of local extrema. Throughout this section we only consider functions whose domains are closed bounded intervals.

Definition 5.2.1. Let $f : [a, b] \to \mathbb{R}$ be a function. A point $c \in (a, b)$ is said to be a **local minimum** if for some $\delta > 0$,
$$f(x) \geq f(c) \text{ for all } x \in (c - \delta, c + \delta).$$
A point $d \in (a, b)$ is said to be a **local maximum** if for some $\delta > 0$,
$$f(x) \leq f(d) \text{ for all } x \in (d - \delta, d + \delta).$$
A **local extremum** is a point that is either a local minimum or a local maximum.
◊

The derivative plays an important role when looking for local extrema, as shown by the following theorem, which was formulated by Fermat.

Theorem 5.2.2. *Let $f : [a, b] \to \mathbb{R}$ be a function. Suppose that f is continuous on $[a, b]$ and differentiable on (a, b). If $c \in (a, b)$ is a local extremum for f, then*
$$f'(c) = 0.$$

Proof. Suppose first that c is a local maximum. Then there exists $\delta > 0$ such that $(c - \delta, c + \delta) \subseteq [a, b]$ and for all $z \in (c - \delta, c + \delta)$
$$f(z) \leq f(c).$$
We know that
$$f'(c) = \lim_{z \to c} \frac{f(z) - f(c)}{z - c}.$$
Choose a sequence (a_n) converging to c in $(c - \delta, c)$. As $a_n < c$ and $f(a_n) \leq f(c)$, using Proposition 4.3.5 we obtain
$$f'(c) = \lim_{n \to \infty} \frac{f(a_n) - f(c)}{a_n - c} \geq 0.$$
By taking a sequence (b_n) converging to c in $(c, c + \delta)$, we get
$$f'(c) \leq 0.$$
This implies that $f'(c) = 0$.

Now if c is a local minimum, then it is a local maximum for $(-f)$, and by the part we have shown $(-f)'(c) = 0$, which implies $f'(c) = 0$. □

Remark 5.2.3. The function $f : \mathbb{R} \to \mathbb{R}$ defined by $f(x) = x^3$ is differentiable and satisfies $f'(0) = 0$, but 0 is not a local extremum. Therefore, the converse of Theorem 5.2.2 is not true. Also, in Theorem 5.2.2 we in fact only need f to be differentiable at c.

We are now ready to prove Rolle's theorem. Its statement is illustrated in Figure 5.4 which says that a differentiable function on an interval that has the same height (value) at the endpoints must have a point with a horizontal tangent.

Theorem 5.2.4 (Rolle's theorem). *Let $f : [a, b] \to \mathbb{R}$ be a function. Suppose that f is continuous on $[a, b]$ and differentiable on (a, b). If $f(a) = f(b)$, then there exists a point $c \in (a, b)$ such that*
$$f'(c) = 0.$$

Proof. Let f be a function satisfying the hypotheses of Rolle's theorem. By Theorem 4.2.9 there exists $c_1, c_2 \in [a, b]$ such that $f(c_1)$ is a (global) minimum value and $f(c_2)$ is a (global) maximum value. If $c_1, c_2 \in \{a, b\}$, as $f(a) = f(b)$, f is a constant function on $[a, b]$ and therefore $f'(c) = 0$ for all $c \in (a, b)$.

So assume that either $c_1 \in (a, b)$ or $c_2 \in (a, b)$. Therefore f has a local extremum at a point c in (a, b) and Theorem 5.2.2 implies that $f'(c) = 0$. □

The following is the main theorem of this section and is illustrated in Figure 5.5.

Theorem 5.2.5 (Mean value theorem). *Let $f : [a, b] \to \mathbb{R}$ be a function. Suppose that f is continuous on $[a, b]$ and differentiable on (a, b). Then there exists a point $c \in (a, b)$ such that*
$$f'(c) = \frac{f(b) - f(a)}{b - a}.$$

5.2. Mean Value Theorem

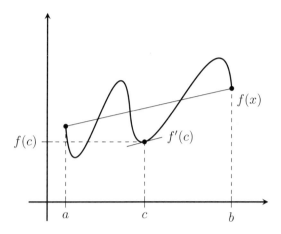

Figure 5.5. Mean value theorem illustration.

Proof. We will modify the function f so that we can apply Rolle's theorem. Let f be a function as in the mean value theorem. We may assume that $f(a) \neq f(b)$, since if they are equal, the conclusion is just Rolle's theorem. We want to translate f down so that we can apply Rolle's theorem to the translated function. The idea is to translate the graph of f so that the value of the translated function at a, instead of being $f(a)$, is 0, and the value of the translated function at b is also 0. To accomplish this, we first find the line that goes from the point $(a, f(a))$ to the point $(b, f(b))$. This line has equation

$$y = f(b) + \frac{f(b) - f(a)}{b - a}(x - b).$$

The new translated function will be obtained from f by subtracting this line:

$$g(x) = f(x) - [f(b) + \frac{f(b) - f(a)}{b - a}(x - b)].$$

One can verify that $g(a) = 0 = g(b)$ and that g is continuous on $[a, b]$ and differentiable on (a, b). Therefore Rolle's theorem applied to g implies that there exists $c \in (a, b)$ such that $g'(c) = 0$. Note that

$$g'(x) = f'(x) - \frac{f(b) - f(a)}{b - a},$$

so

$$0 = f'(c) - \frac{f(b) - f(a)}{b - a},$$

which shows that the mean value theorem holds for f. $\square$

Corollary 5.2.6. *Let $f : [a, b] \to \mathbb{R}$ be a function. Suppose that f is continuous on $[a, b]$ and differentiable on (a, b) and $f'(x) = 0$ for all $x \in (a, b)$. Then $f(x) = f(a)$ for all $x \in [a, b]$, i.e., f is constant on $[a, b]$.*

Proof. We proceed by contradiction. Suppose there exists $b_1 \in [a,b]$ with $a < b_1$ and such that $f(a) \neq f(b_1)$. Then applying the mean value theorem to f on $[a,b_1]$, we obtain that there exists a point $c \in (a,b)$ such that
$$f'(c) = \frac{f(b_1) - f(a)}{b-a}.$$
But as $f'(c) = 0$, this implies that $f(b_1) = f(a)$, a contradiction. $\square$

Remark 5.2.7. Suppose we have a function f that is defined on an open interval (a,b) instead of the closed interval $[a,b]$. We can obtain the same conclusion as in Corollary 5.2.6 if $f: (a,b) \to \mathbb{R}$ is differentiable on (a,b), and $f'(x) = 0$ for all $x \in (a,b)$. To show that f must be constant on (a,b), apply Corollary 5.2.6 to f: choose c the midpoint of the interval, and apply it for f on $[a', c]$ and on $[c, b']$ for arbitrary $a' < c < b'$.

Question 5.2.8. Let $G = (0,1) \cup (1,2)$. Define f on the open set G by $f(x) = 1$ if $x \in (0,1)$ and $f(x) = 2$ if $x \in (1,2)$. Show that $f'(x) = 0$ for all $x \in G$, but f is not constant on G. Discuss.

Question 5.2.9. Let (a,b) be an open interval with $0 \in (a,b)$, and let $f: (a,b) \to \mathbb{R}$ be differentiable on (a,b). If $f'(x) = 1$ for all $x \in (a,b)$ and $f(0) = 0$, show that then $f(x) = x$ for all $x \in (a,b)$.

Example 5.2.10. Let G be an open set in $\mathbb{R}$, and let $f: G \to \mathbb{R}$ be a function that is differentiable on G. Suppose there exists a constant $M > 0$ so that $|f'(x)| < M$ for all $x \in G$. We show that then f is uniformly continuous on G. Let $\varepsilon > 0$, and choose $\delta = \varepsilon/M$. Suppose $x, y \in G$ and $|x - y| < \delta$. Without loss of generality we may assume $x < y$. By the mean value theorem applied to f on $[x, y]$, there exists $z \in (x, y)$ such that
$$\frac{f(y) - f(x)}{y - x} = f'(z).$$
Therefore,
$$|f(y) - f(x)| = |f'(z)||y - x| < M\delta = M\frac{\varepsilon}{M} = \varepsilon,$$
showing that f is uniformly continuous on G.

Remark 5.2.11. If $f: [a,b] \to \mathbb{R}$ is continuous and differentiable on (a,b) with $|f'(x)| \leq M$ for some $0 < M < 1$, then $f: [a,b] \to \mathbb{R}$ is a contraction. This is a consequence of the mean value theorem.

The following is an interesting generalization that has several applications. The new idea here, and the reason it is not an immediate consequence of Theorem 5.2.5, is that the same point c is in both sides of equation (5.7).

Theorem 5.2.12 (Generalized mean value theorem). *Let $f, g: [a,b] \to \mathbb{R}$ be functions. Suppose that f, g are continuous on $[a,b]$ and differentiable on (a,b). Then there exists a point $c \in (a,b)$ such that*

(5.7) $$(f(b) - f(a))g'(c) = (g(b) - g(a))f'(c).$$

Proof. Define a new function ϕ by
$$\phi(x) = (f(b) - f(a))g(x) - (g(b) - g(a))f(x).$$

5.2. Mean Value Theorem

Then clearly ϕ is continuous on $[a,b]$ and differentiable on (a,b) as it is defined in terms of continuous and differentiable functions. Furthermore, one has that $\phi(a) = \phi(b)$. By Rolle's theorem there exists $c \in (a,b)$ such that $\phi'(c) = 0$, which completes the proof. $\square$

We end this section with a useful result for calculating the derivative of the inverse function without necessarily having a formula for the inverse.

First a few comments about the hypotheses in the theorem below. The function f we will consider will be differentiable, so continuous on an interval (a,b). Thus the image $f(a,b)$ has to be an interval, which we will write as (c,d). In addition, f needs to be invertible. Now, for example, the function $h : (-1,1) \to (-1,1)$, given by $h(x) = x^3$, is differentiable, invertible, and strictly monotone, but its inverse, given by $g(x) = \sqrt[3]{x}$, while continuous, is not differentiable at 0 (Exercise 5.2.9). We will need to assume that the derivative of f is nonzero at the points where we want its inverse to be differentiable. If $f'(x) \neq 0$ on (a,b), then it must be that $f'(x) > 0$ on (a,b), or $f'(x) < 0$ on (a,b) (Exercise 5.2.11). Finally, if $f'(x) > 0$ or $f'(x) < 0$ on (a,b), then f is already invertible on (a,b); so we will assume this condition, and invertibility is assumed implicitly.

Theorem 5.2.13 (Derivative of the inverse function). *Let $f : (a,b) \to (c,d)$ be differentiable and such that for all $x \in (a,b)$, $f'(x) > 0$ or $f'(x) < 0$. Let $g : (c,d) \to (a,b)$ be the inverse of f. Then g is differentiable and*

$$g'(y) = \frac{1}{f'(g(y))}.$$

Proof. We first show that g is differentiable at $y \in (c,d)$. For this we consider the limit

$$\lim_{z \to y} \frac{g(z) - g(y)}{z - y}.$$

Since g is the inverse of f, we have $y = f(g(y))$ and $z = f(g(z))$. Letting $u = g(z)$ and $w = g(y)$, this suggests that we consider the following limit

(5.8) $$\lim_{w \to u} \frac{w - u}{f(w) - f(u)}.$$

In the limit we may assume $w \neq u$, and since f is injective, we have $f(w) \neq f(u)$. Since f is differentiable and $f'(x) \neq 0$ for all $x \in (a,b)$, using Proposition 4.3.11(3) we have that the limit in (5.8) exists and is equal to $1/f'(u)$.

Therefore, for $\varepsilon > 0$ there exists $\delta_1 > 0$ such that

$$0 < |w - u| < \delta_1 \implies \left| \frac{w - u}{f(w) - f(u)} - \frac{1}{f'(u)} \right| < \varepsilon.$$

Using that g is continuous, for this $\delta_1 > 0$ there exists $\delta_2 > 0$ such that

$$0 < |z - y| < \delta_2 \implies |g(z) - g(y)| < \delta_1.$$

Therefore,

$$0 < |z - y| < \delta_2 \implies \left| \frac{g(z) - g(y)}{f(g(z)) - f(g(y))} - \frac{1}{f'(g(y))} \right| < \varepsilon,$$

which is equivalent to
$$\left| \frac{g(z) - g(y)}{z - y} - \frac{1}{f'(g(y))} \right| < \varepsilon,$$
showing that g is differentiable and satisfies the equation. □

Example 5.2.14. Let $f(x) = 3x^3 + x$. Then $f'(x) = 9x^2 + 1$. So $f'(x) > 0$ for all x. Let us consider f on $(0, 2)$. We have $f((0, 2)) = (0, 26)$, and f has an inverse $g : (0, 26) \to (0, 2)$. We want to calculate $g'(4)$. We note that $f(1) = 4$, so
$$g'(4) = \frac{1}{f'(g(4))} = \frac{1}{f'(1)} = \frac{1}{10}.$$

5.2.1. L'Hôpital's Rule. We cover a very useful rule for calculating limits. There are several versions of this rule; we will cover one and leave the others for the Exercises.

Theorem 5.2.15 (L'Hôpital's rule). *Let $a < c < b$ and $f, g : (a, c) \cup (c, b) \to \mathbb{R}$ be differentiable functions. Suppose that*

(1) $\lim_{x \to c} f(x) = \lim_{x \to c} g(x) = 0$,
(2) $g'(x) \neq 0$ for all $x \in (a, c) \cup (c, b)$,
(3) $\lim_{x \to c} \frac{f'(x)}{g'(x)} = L \in \mathbb{R}$.

Then

(5.9) $$\lim_{x \to c} \frac{f(x)}{g(x)} = L.$$

Proof. Extend the definitions of f and g by setting $f(c) = g(c) = 0$. Then by condition (1), $\lim_{x \to c} f(x) = f(c)$, so f is continuous at c, and similarly, g is continuous at c. Let $x \in (a, c) \cup (c, b)$. We will consider two cases: $x < c$ and $x > c$. For each case apply the generalized mean value theorem to $[x, c]$ or $[c, x]$ to obtain
$$\frac{f(x) - f(c)}{g(x) - g(c)} = \frac{f'(c_x)}{g'(c_x)},$$
for some c_x between x and c. We note that $g(x) \neq g(c)$ by condition (2), and Rolle's theorem is applied to g on the interval $[x, c]$ or $[c, x]$. Then,

(5.10) $$\frac{f(x)}{g(x)} = \frac{f'(c_x)}{g'(c_x)},$$

for all $x \in (a, c) \cup (c, b)$. Now let $\varepsilon > 0$. By (3) there exists $\delta_1 > 0$ so that if $|y - c| < \delta_1$ and $y \in (a, c) \cup (c, b)$, then
$$\left| \frac{f'(y)}{g'(y)} - L \right| < \varepsilon.$$
If we choose x so that $|x - c| < \delta_1$, then the corresponding c_x satisfies $|c_x - c| < \delta_1$, so
$$\left| \frac{f'(c_x)}{g'(c_x)} - L \right| < \varepsilon.$$
Then using (5.10), we have
$$\left| \frac{f(x)}{g(x)} - L \right| < \varepsilon,$$

which implies (5.9) and completes the proof. □

Example 5.2.16. We prove that the following limit exists and find its value:
$$\lim_{x \to 0} \frac{\sin(3x^2)}{x^2}.$$
Let $f(x) = \sin(3x^2)$ and $g(x) = x^2$. Then f and g are defined and differentiable on $(-1, 1)$ and $g'(x) = 2x \neq 0$ for $x \in (-1, 0) \cup (0, 1)$. Also, $\lim_{x \to 0} f(x) = \lim_{x \to 0} g(x) = 0$. To apply L'Hôpital's rule, we wish to compute the limit
$$\lim_{x \to 0} \frac{f'(x)}{g'(x)} = \lim_{x \to 0} \frac{6x \cos(3x^2)}{2x} = \lim_{x \to 0} \frac{6 \cos(3x^2)}{2} = 3,$$
since $\lim_{x \to 0} \cos(3x^2) = 1$ by continuity of the cosine function. Then
$$\lim_{x \to 0} \frac{\sin(3x^2)}{x^2} = 3.$$

Exercises: Mean Value Theorem

5.2.1 Let $f : [a, b] \to \mathbb{R}$ be a function. Suppose that f is continuous on $[a, b]$ and differentiable on (a, b). Show that if $f'(x) \neq 0$ for all $x \in (a, b)$, then f is one-to-one.

5.2.2 Let $f, g : [a, b] \to \mathbb{R}$ be functions that are continuous on $[a, b]$ and differentiable on (a, b). Show that if $f'(x) = g'(x)$ for all $x \in (a, b)$, then there exists $c \in \mathbb{R}$ such that $f(x) = g(x) + c$ for all $x \in [a, b]$.

5.2.3 Let $f, g : [a, b] \to \mathbb{R}$ be functions that are continuous on $[a, b]$ and differentiable on (a, b) such that $f(a) = g(a)$. Show that if $f'(x) \leq g'(x)$ for all $x \in (a, b)$, then $f(x) \leq g(x)$ for all $x \in (a, b)$.

5.2.4 *Darboux's intermediate value theorem for the derivative*: Let $f : (a, b) \to \mathbb{R}$ be differentiable on (a, b). Let $x, y \in (a, b), x < y$. Show that if there is a number c such that $f'(x) < c < f'(y)$ or $f'(x) > c > f'(y)$, then there exists $z \in (x, y)$ such that $f'(z) = c$. (*Hint*: Define a function $\phi(x) = f(x) - cx$.)

5.2.5 *L'Hôpital's rule—case 2*: Let $a < c < b$ and $f, g : (a, c) \cup (c, b) \to \mathbb{R}$ be differentiable functions. Suppose that the following hold:
(a) $\lim_{x \to c} g(x) = \infty$ and $\lim_{x \to c} f(x) = \infty$;
(b) $g'(x) \neq 0$ for all $x \in (a, c) \cup (c, b)$;
(c) $\lim_{x \to c} \frac{f'(x)}{g'(x)} = L \in \mathbb{R}$.
Show that
$$\lim_{x \to c} \frac{f(x)}{g(x)} = L.$$

5.2.6 Discuss and prove a version of Theorem 5.2.15 in the case when $c = \infty$ or $c = -\infty$.

5.2.7 Determine if the following limits exit, and if so find their value.
(a) $\lim_{x \to 0} \frac{\sin^3(2x)}{x^3}$

(b) $\lim_{x\to 0} \frac{1-\cos 4x}{x^2}$

5.2.8 Let $f : (a,b) \to \mathbb{R}$ be a differentiable function. Prove that if $f'(x) \geq 0$ for all $x \in (a,b)$, then f is an increasing function on (a,b). Furthermore, if $f'(x) > 0$ for all $x \in (a,b)$, then f is a strictly increasing function on (a,b).

5.2.9 Prove that the function $g : (-1,1) \to \mathbb{R}$ given by $g(x) = \sqrt[3]{x}$ is not differentiable at 0.

5.2.10 Prove that $f(x) = x^{1/5}$ is uniformly continuous on $[0, \infty)$.

5.2.11 Let $f : (a,b) \to \mathbb{R}$ be differentiable. Prove that if $f'(x) \neq 0$ for all $x \in (a,b)$, then it must be that $f'(x) > 0$ for all $x \in (a,b)$ or $f'(x) < 0$ for all $x \in (a,b)$.

5.2.12 Let $f : (a,b) \to \mathbb{R}$ be a differentiable function. Show that if f' is bounded on (a,b), then f is bounded on (a,b). (*Hint*: Choose a fixed point in (a,b) and apply the mean value theorem.) Give an example to show this does not hold when the interval (a,b) is infinite.

* 5.2.13 Let $f : (a,b) \to \mathbb{R}$ be a function. We say f is **convex** on (a,b) if $\ell(x) > f(x)$ for all $x \in (a,b)$, where $\ell(x) = m(x-a) + f(a)$, with $m = (f(b)-f(a))/(b-a)$, is the line from $(a, f(a))$ to $(b, f(b))$. Prove that if f is differentiable and f' is increasing on (a,b), then f is convex on (a,b).

5.3. Taylor's Theorem

Polynomials form the class of functions we are most familiar with, and they are well understood in terms of differentiation and integration. Surprisingly, sufficiently differentiable functions can be approximated, locally, by their Taylor polynomials. This is the content of Taylor's theorem that we study in this section; it is also the foundation for Taylor series, which will be covered in Section 8.5.

We start by first defining the *higher order derivatives* of a function.

Definition 5.3.1. Let f be a function whose domain contains an open interval (a,b), and let x be a point in that interval. Write $f^{(0)}(x)$ for the function $f(x)$ and $f^{(1)}(x)$ for its derivative $f'(x)$. Suppose $f^{(n-1)}(x)$ has been defined and is differentiable for x in (a,b). We let $f^{(n)}(x)$ be the derivative of $f^{(n-1)}(x)$ and call it the n**th order derivative** of f. ◇

Definition 5.3.2. Let $f : (a,b) \to \mathbb{R}$ be a continuous function and suppose it is differentiable of order n on the interval (a,b). The n**th degree Taylor polynomial centered at** $c \in (a,b)$, **or about** c, of f is defined to be the polynomial

$$P_n(x) = f(c) + f^{(1)}(c)(x-c) + \cdots + \frac{f^{(n)}(c)}{n!}(x-c)^n = \sum_{i=0}^{n} \frac{f^{(i)}(c)}{i!}(x-c)^i.$$ ◇

5.3. Taylor's Theorem

Example 5.3.3. Let $f(x) = \frac{1}{1+x}$. We find its Taylor polynomial of degree n centered at $c = 0$. We have

$$f'(x) = -\frac{1}{(1+x)^2}; \qquad f''(x) = \frac{2}{(1+x)^3};$$

$$f^{(3)}(x) = -\frac{6}{(1+x)^4} = -3!(1+x)^{-4};$$

$$f^{(4)}(x) = \frac{24}{(1+x)^5} = 4!(1+x)^{-5}.$$

By induction it follows that the nth derivative is given by

$$f^{(n)}(x) = (-1)^n n! (1+x)^{-(n+1)}.$$

Therefore,

$$f(0) = 1, f'(0) = -1, f''(0) = 2, \ldots, f^{(n)}(0) = (-1)^n n!.$$

We find

$$P_n(x) = 1 - x + \frac{2}{2!}x^2 - \frac{6}{3!}x^3 + \cdots = \sum_{i=0}^{n} (-1)^n x^n.$$

The following is Taylor's theorem with the Lagrange remainder.

Theorem 5.3.4 (Taylor's theorem with Lagrange remainder). *Let $f : (a,b) \to \mathbb{R}$ be $n+1$ times differentiable, for $n \in \mathbb{N}_0$. Let $c \in (a,b)$. Then, for every $x \in (a,b)$, we can write*

$$f(x) = P_n(x) + R_n(x),$$

*where $P_n(x)$ is the nth degree Taylor polynomial of f centered at c, and $R_n(x)$, called the **Lagrange remainder**, is given by*

$$R_n(x) = \frac{f^{(n+1)}(\zeta)}{(n+1)!}(x-c)^{n+1} \text{ for some } \zeta \text{ between } x \text{ and } c.$$

Proof. For $n = 0$ the theorem follows from the mean value theorem. In fact, let $x \in (a,b)$. Assume first that $x > c$ (the proof for the case when $x < c$ is similar). Considering f on $[c,x]$ (when $x > c$), by Theorem 5.2.5, there is a ζ so that

$$\frac{f(x) - f(c)}{x - c} = f'(\zeta).$$

Hence,

$$f(x) = f(c) + f'(\zeta)(x - c),$$

where $f(c)$ is the 0th degree Taylor polynomial and $f'(\zeta)(x-c)$ is the Lagrange remainder.

For the induction part we use some auxiliary functions before applying the generalized mean value theorem. Define

$$\phi(x) = f(x) - P_n(x) \quad \text{and} \quad g(x) = (x-c)^{n+1} \text{ for } x \in (a,b).$$

We claim that several applications of the generalized mean value theorem will yield that there is a ζ between x and c such that

(5.11) $$\phi(x) \, g^{(n+1)}(\zeta) = g(x) \phi^{(n+1)}(\zeta).$$

We give the proof for $n = 1$ as it illustrates the main ideas and the inductive step. Assume that $x > c$ (again a similar argument can be given when $x < c$). Apply the generalized mean value theorem to the functions $\phi(x)$ and $g(x)$ on the interval $[c, x]$. Then there exists ζ_1 in the interval (c, x) such that

$$(\phi(x) - \phi(c))g^{(1)}(\zeta_1) = (g(x) - g(c))\phi^{(1)}(\zeta_1).$$

We next apply the generalized mean value theorem to the functions $\phi^{(1)}(x)$ and $g^{(1)}(x)$ on the interval $[c, \zeta_1]$ to obtain a ζ_2 in (c, ζ_1) such that

$$(\phi^{(1)}(\zeta_1) - \phi^{(1)}(c))g^{(2)}(\zeta_2) = (g^{(1)}(\zeta_1) - g^{(1)}(c))\phi^{(2)}(\zeta_2).$$

But using that $0 = \phi(c) = \phi^{(1)}(c) = g(c) = g^{(1)}(c)$, we obtain

$$\phi(x)g^{(1)}(\zeta_1) = g(x)\phi^{(1)}(\zeta_1) \quad \text{and} \quad \phi^{(1)}(\zeta_1)g^{(2)}(\zeta_2) = g^{(1)}(\zeta_1)\phi^{(2)}(\zeta_2).$$

Therefore,

$$\phi(x)g^{(1)}(\zeta_1)g^{(2)}(\zeta_2) = g(x)\phi^{(1)}(\zeta_1)g^{(2)}(\zeta_2) = g(x)g^{(1)}(\zeta_1)\phi^{(2)}(\zeta_2).$$

After simplifying (since $g^{(1)}(\zeta_1) \neq 0$), we obtain

$$\phi(x)\, g^{(2)}(\zeta_2) = g(x)\, \phi^{(2)}(\zeta_2),$$

which completes the case for $n = 1$ of (5.11) by letting $\zeta = \zeta_2$. The inductive step follows the same idea.

Now using $g^{(n+1)}(x) = (n+1)!$ and $\phi^{n+1}(x) = f^{(n+1)}(x)$ in (5.11), we get

$$\phi(x)(n+1)! = (x - c)^{n+1}\phi^{n+1}(\zeta).$$

This translates to

$$(f(x) - P_n(x))(n+1)! = (x - c)^{n+1}f^{(n+1)}(\zeta) \text{ or}$$

$$f(x) = P_n(x) + \frac{(x - c)^{n+1}}{(n+1)!}f^{(n+1)}(\zeta). \qquad \square$$

Remark 5.3.5. The remainder term in Taylor's theorem can be used to estimate the error of the approximating polynomials. Let $A = \max\{c - a, b - c\}$. If we know that

$$|f^{(n+1)}(x)| \leq K \text{ for all } x \in [a, b]$$

for some constant K, then

$$|R_n(x)| = \left|\frac{(x-c)^{n+1}}{(n+1)!}f^{n+1}(\zeta)\right| \leq \frac{|x-c|^{n+1}K}{(n+1)!} \leq \frac{A^{n+1}K}{(n+1)!}.$$

Example 5.3.6. We find the Taylor polynomial of degree n for $f(x) = \sin x$ around $c = 0$ with remainder. We note that $f'(x) = \cos x$, $f^{(2)}(x) = -\sin x$, $f^{(3)}(x) = -\cos x$, $f^{(4)}(x) = \sin x, \ldots$. So we see a pattern where $f^{(2i)}(0) = 0$ and $f^{(2i+1)}(0) = (-1)^i$, for $i = 0, 1, \ldots$. Then, for $n = 2k + 1$,

$$P_{2k+1}(x) = \sum_{i=0}^{k} \frac{(-1)^i}{(2i+1)!}x^{2i+1}.$$

We now observe that $P_{2k+2}(x) = P_{2k+1}(x)$ since $f^{(2k+2)}(0) = 0$. This has consequences when estimating the error. For example, one can write $\sin x = x + R_1(x)$,

5.3. Taylor's Theorem

but a better estimate is given by $\sin x = x + R_2(x)$. So when calculating the remainder for P_{2k+1}, we might as well calculate the remainder for P_{2k+2}. Then

$$R_{2k+2}(x) = \frac{(-1)^{k+1}}{(2k+3)!} \cos \zeta (x-c)^{2k+3}.$$

A bound for the remainder of P_{2k+1} when $x \in [-1, 1]$ is given by

$$|R_{2k+2}(x)| \leq \frac{1}{(2k+3)!}.$$

In most applications we just need an estimate for the remainder. There is a simpler argument for estimating the remainder as pointed out recently in [4] that we now present.

Lemma 5.3.7 (Taylor's remainder estimate). *Let $f : (a,b) \to \mathbb{R}$ be $n+1$ times differentiable, for $n \geq 0$. Let $c \in (a,b)$. Assume that for some $m, M \in \mathbb{R}$,*

$$m \leq f^{(n+1)}(x) \leq M \text{ for all } x \in (a,b).$$

Then

(5.12) $$\frac{m}{(n+1)!}(x-c)^{n+1} \leq R_n(x) \leq \frac{M}{(n+1)!}(x-c)^{n+1} \text{ for } x \in (a,b),$$

where $R_n(x)$ is the nth remainder of f.

Proof. The proof is by induction. The case $n = 0$ is as in Theorem 5.3.4 and is left to the reader to verify. Assume (5.12) it is true for all functions that are n times differentiable, and let f be $n+1$ times differentiable. Write

$$f(x) = P_n(x) + R_n(x).$$

Then

$$f'(x) = P_n'(x) + R_n'(x) \text{ for } x \in (a,b).$$

We next observe that P_n' is the Taylor polynomial of degree $n-1$ of f', and f' is n times differentiable. Then by the inductive hypothesis

$$\frac{m}{n!}(x-c)^n \leq R_n'(x) \leq \frac{m}{n!}(x-c)^n.$$

An application of Exercise 5.2.3 completes the inductive step. $\square$

It possible to use Lemma 5.3.7 to give another proof of Taylor's theorem with the Lagrange remainder (a special case of this is in Exercise 5.3.10).

The following little "oh" notation is used in asymptotic analysis and number theory, and it is useful when thinking of approximations by Taylor polynomials.

Definition 5.3.8. We say that a function $f(x)$ is **little "oh"** of another function $g(x)$ and write

$$f(x) = o(g(x))$$

if

$$\lim_{x \to 0} \frac{f(x)}{g(x)} = 0. \qquad \diamond$$

So, for example, from the estimate in Example 5.3.6, we can write
$$\frac{\sin x}{x} = 1 + o(x^2),$$
since $R_2(x) = \frac{-\cos(\zeta)}{6}x^3$.

Example 5.3.9. We evaluate the following limit.
$$\lim_{x \to 0} \frac{x \sin^3(2x)}{(1 - \cos x)^2}.$$

First note that, by Taylor's theorem, we can write $\sin(2x) = 2x + o(x^3)$. It follows that $\lim_{x \to 0} \frac{\sin 2x}{x} = 2$. Then by Proposition 4.3.3,
$$\lim_{x \to 0} \frac{\sin^3 2x}{x^3} = \lim_{x \to 0} \left(\frac{\sin 2x}{x}\right)^3 = 8.$$

Similarly, by Exercise 5.3.5, $\cos x = 1 - \frac{x^2}{2!} + o(x^4)$, so $\lim_{x \to 0} \frac{1 - \cos x}{x^2} = \frac{1}{2}$. Then,
$$\lim_{x \to 0} \frac{x^4}{(1 - \cos x)^2} = 4.$$

It follows that
$$\lim_{x \to 0} \frac{x \sin^3(2x)}{(1 - \cos x)^2} = \lim_{x \to 0} \frac{\sin^3(2x)}{x^3} \cdot \frac{x^4}{(1 - \cos x)^2} = 32.$$

Exercises: Taylor's Theorem

5.3.1 Use Taylor's theorem to determine if the following limits exist and if so find their values.
  (a) $\lim_{x \to 0} \frac{\sin^3(2x)}{x^3}$
  (b) $\lim_{x \to 0} \frac{1 - \cos 4x}{x^2}$

5.3.2 Let $f(x) = e^x$ for $x \in \mathbb{R}$. Determine the Taylor polynomial P_n centered at $c = 0$ and the form of the remainder $R_n(x)$ for this function.

5.3.3 Let $f(x) = \frac{1}{1 - x^3}$ for $x \in \mathbb{R}$. Determine the Taylor polynomial P_n centered at $c = 0$ and the form of the remainder $R_n(x)$ for this function.

5.3.4 Let $f(x) = \ln x$ for $x \in (0, \infty)$. Determine the Taylor polynomial P_n centered at $c = 1$ for this function.

5.3.5 Let $f(x) = \cos x$ for $x \in \mathbb{R}$. Determine its Taylor polynomial P_n centered at $c = 0$ and the form of the remainder $R_n(x)$.

5.3.6 Let $f(x) = \sin 2x$ for $x \in \mathbb{R}$. Determine its Taylor polynomial P_n centered at $c = \pi/4$ and the form of the remainder $R_n(x)$.

5.3.7 Let $f(x) = \sin 2x$ for $x \in \mathbb{R}$. Determine its Taylor polynomial P_n centered at $c = \pi/4$ and the form of the remainder $R_n(x)$.

5.3.8 Let $f : [a, b] \to \mathbb{R}$ be continuous on $[a, b]$ and $n + 1$ times differentiable on (a, b) for $n \in \mathbb{N}$, and let $c \in (a, b)$. Prove that $f^{(i)}(c) = P_n^{(i)}(c)$ for $i = 0, 1, \ldots, n$, and that if Q is any polynomial of degree n such that $f^{(i)}(c) = P_n^{(i)}(c)$ for $i = 0, 1, \ldots, n$, then $Q = P$.

5.3.9 Justify the calculations in (6.13).

5.3.10 Use Lemma 5.3.7 to give another proof of Theorem 5.3.4 in the special case where $f^{(n+1)}$ is assumed to be continuous. (*Hint*: Use the intermediate value theorem for $f^{(n+1)}$.)

5.3.11 Prove that a function $f : \mathbb{R} \to \mathbb{R}$ is differentiable at $a \in \mathbb{R}$ if and only if
$$f(x) = f(a) + M(x-a) + o(x-a)$$
for some $M \in \mathbb{R}$.

Chapter 6

Integration

6.1. The Riemann Integral

In this section we develop the theory of Riemann integration. The theory is developed for functions defined on closed bounded intervals. In Section 6.3 we discuss the extension of the Riemann integral to unbounded intervals, which leads us to consider what are called improper Riemann integrals.

For concreteness, let us first consider a function $f : [a, b] \to \mathbb{R}$ that is positive. We are interested in defining the notion of the area under the graph of the function f that is over the interval $[a, b]$, i.e., the area of the region

$$\{(x, y) : a \leq x \leq b, 0 \leq y \leq f(x)\}.$$

For example, if $f(x) = 2$, then this region is a rectangle and its area is $2(b-a)$. But, for example, in the case of $f(x) = x^2$, we do not have a readily available formula for the area. In this and more general cases our technique will be to approximate the region by rectangles and to use the sum of the areas of the rectangles as an approximation for the area of the region. The base of each rectangle will be a subinterval of $[a, b]$, and the height will be given by the value of the function f at a point in this subinterval.

To construct the subintervals of $[a, b]$, we introduce the notion of a partition.

Definition 6.1.1. A **partition** of an interval $[a, b]$ consists of a finite sequence of points

$$\mathcal{P} = \{x_0, \ldots, x_n\}$$

such that

$$a = x_0 < x_1 < \cdots < x_n = b. \qquad \diamond$$

A partition $\mathcal{P} = \{x_0, \ldots, x_n\}$ of $[a, b]$ with $n + 1$ points gives a partition of the interval into the following n subintervals

$$[x_0, x_1], [x_1, x_2], \ldots, [x_{n-1}, x_n].$$

The ith subinterval of the partition is
$$[x_{i-1}, x_i],$$
where i ranges from 1 to n, so the first subinterval ($i = 1$) is $[x_0, x_1]$. In applications it is often convenient to use equally spaced points.

Definition 6.1.2. The **uniform partition** of $[a, b]$ consists of points $\{x_i : i = 0, \ldots, n\}$ given by
$$x_i = a + \frac{b-a}{n} i.$$
In this case each subinterval has the same length, and we write
$$\Delta x = \frac{b-a}{n}. \qquad \diamond$$

Example 6.1.3. Given an interval $[0, 1]$, we may consider partitioning it into equally spaced points. So for each n we can use a uniform partition of $[0, 1]$ consisting of the $n + 1$ points given by
$$x_i = \frac{i}{n} \text{ for } i = 0, 1, \ldots, n.$$
These points determine the n subintervals
$$[0, \frac{1}{n}], [\frac{1}{n}, \frac{2}{n}], \ldots, [\frac{n-1}{n}, 1].$$

For a given partition, we need to specify the points that are used to calculate the heights of the rectangles.

Definition 6.1.4. Given a partition $\mathcal{P}$ we say $\mathcal{P}^*$ is a sequence of **corresponding points** to $\mathcal{P}$ if
$$\mathcal{P}^* = \{x_1^*, \ldots, x_n^*\}, \quad \text{where } x_i^* \in [x_{i-1}, x_i] \text{ for } i = 1, \ldots, n. \qquad \diamond$$

In practice, there are three typical choices for the points in $\mathcal{P}^*$:

the left endpoint, $x_i^* = x_{i-1}$.

the right endpoint, $x_i^* = x_i$.

the midpoint, $x_i^* = \dfrac{x_{i-1} + x_i}{2}$.

Definition 6.1.5. For each partition $\mathcal{P}$ and choice of corresponding points $\mathcal{P}^*$, we define the **Riemann sum** of f corresponding to $\mathcal{P}$ and $\mathcal{P}^*$:
$$S(f, \mathcal{P}, \mathcal{P}^*) = \sum_{i=1}^{n} f(x_i^*)(x_i - x_{i-1}). \qquad \diamond$$

Figure 6.1 illustrates the calculation of $S(f, \mathcal{P}, \mathcal{P}^*)$, where $\mathcal{P}^*$ consists of the left endpoints. In this case we see the Riemann sum giving an approximation to the "area" under the function. (We put area in quotation marks because, while we have an intuitive notion of area, it has not been formally defined for arbitrary regions. We are using the fact that we know what the area of a rectangle is and will use that to define the integral for more general regions, thus giving a notion of area for these regions.)

6.1. The Riemann Integral

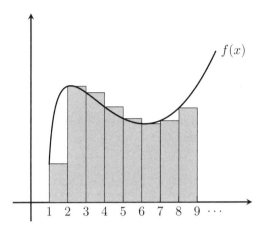

Figure 6.1. A Riemann sum with the left endpoints.

Example 6.1.6. Let $f : [0,1] \to \mathbb{R}$ be defined by $f(x) = x^2$. Using the uniform partition and choosing $\mathcal{P}^*$ to be the right endpoints, we obtain

$$S(f, \mathcal{P}, \mathcal{P}^*) = \sum_{i=1}^{n} f(x_i^*)(x_i - x_{i-1})$$

$$= \sum_{i=1}^{n} f(\frac{i}{n})\frac{1}{n} = \sum_{i=1}^{n} \frac{i^2}{n^2}\frac{1}{n}$$

$$= \frac{1}{n^3} \sum_{i=1}^{n} i^2 = \frac{1}{n^3} \frac{n(n+1)(2n+1)}{6}$$

$$= \frac{1}{6}(1 + \frac{1}{n})(2 + \frac{1}{n}).$$

Remark 6.1.7. A partition $\mathcal{P}$ with corresponding points $\mathcal{P}^*$ is called a *tagged partition* in the theory of the Henstock–Kurzweil integral that generalizes the Riemann integral (see the Bibliographical Notes for references).

We are interested in considering Riemann sums for sequences of increasingly finer partitions, as they give increasingly better approximations. To make this more precise, we introduce the following definition.

Definition 6.1.8. We define the **mesh** of a partition $\mathcal{P}$ by

$$||\mathcal{P}|| = \max\{x_i - x_{i-1} : i = 1, \ldots, n\}. \qquad \diamond$$

A sequence of partitions becomes finer as the mesh of the partitions decreases to 0. For example, in the case of the uniform partition $\mathcal{P}_n$ of an interval $[a,b]$ into $n+1$ points, its mesh is

$$||\mathcal{P}_n|| = \frac{b-a}{n}.$$

In this case we see that as the number of points in the partition increases to infinity, the mesh $||\mathcal{P}_n|| \to 0$.

We define a useful notion to compare partitions.

Definition 6.1.9. A partition $\mathcal{Q}$ of $[a, b]$ is said to be **finer** than a partition $\mathcal{P}$ of $[a, b]$ if every partition point of $\mathcal{P}$ is a partition point of $\mathcal{Q}$. ◇

For example, $\mathcal{Q} = \{0, 1/4, 1/2, 3/4, 1\}$ is a finer partition of $[0, 1]$ than $\mathcal{P} = \{0, 1/2, 1\}$. We also introduce the notion of the common refinement of two partitions.

Definition 6.1.10. Given partitions $\mathcal{P}$ and $\mathcal{Q}$ of $[a, b]$, the **common refinement**, denoted by $\mathcal{P} \vee \mathcal{Q}$, is the partition consisting of all points of both $\mathcal{P}$ and $\mathcal{Q}$ written in increasing order. ◇

So the common refinement $\mathcal{P} \vee \mathcal{Q}$ is finer than both $\mathcal{P}$ and $\mathcal{Q}$. For example, if $\mathcal{P} = \{0, 1/4, 1/2, 3/4, 1\}$ and $\mathcal{Q} = \{0, 1/3, 1/2, 7/8, 1\}$, their common refinement is $\mathcal{P} \vee \mathcal{Q} = \{0, 1/4, 1/3, 1/2, 3/4, 7/8, 1\}$.

We are ready for the main definition of this chapter.

Definition 6.1.11. Let f be a function defined on an interval $[a, b]$. We define f to be **Riemann integrable** if there exists a real number denoted by

$$\int_a^b f \quad \text{or} \quad \int_a^b f(x)\, dx$$

such that for all $\varepsilon > 0$ there exists $\delta > 0$ so that

(6.1) $$\left| \int_a^b f(x)\, dx - S(f, \mathcal{P}, \mathcal{P}^*) \right| < \varepsilon$$

whenever $||\mathcal{P}|| < \delta$ and $\mathcal{P}^*$ is any sequence of corresponding points to $\mathcal{P}$. In this case we say that $\int_a^b f(x)\, dx$ is the **Riemann integral** of f and write

$$\lim_{||\mathcal{P}|| \to 0} S(f, \mathcal{P}, \mathcal{P}^*) = \int_a^b f(x)\, dx.$$ ◇

It is important to note that condition (6.1) needs to hold for all sequences of corresponding points $\mathcal{P}^*$.

Example 6.1.12. The first example of a Riemann integrable function is a constant function. Let $f(x) = c$ for all $x \in [a, b]$ and some $c \in \mathbb{R}$. For any partition $\mathcal{P}$ of $[a, b]$ and corresponding points $\mathcal{P}^*$, we have

$$S(f, \mathcal{P}, \mathcal{P}^*) = \sum_{i=1}^n f(x_i^*)(x_i - x_{i-1}) = c \sum_{i=1}^n (x_i - x_{i-1}) = c(b - a).$$

So in this case the number $\int_a^b f$ is $c(b - a)$, and we see that for all $\varepsilon > 0$ if we let $\delta = 1$ (in fact, any value works), then for all partitions $\mathcal{P}$ with $||\mathcal{P}|| < \delta$ we have

$$\left| \int_a^b f - S(f, \mathcal{P}, \mathcal{P}^*) \right| = 0 < \varepsilon,$$

for all corresponding $\mathcal{P}^*$.

Example 6.1.13. Let f be the Dirichlet function (defined in Example 4.1.7) on any interval $[a, b]$. We will show that f is not Riemann integrable. We observe that for any partition $\mathcal{P}$ of $[a, b]$, we can choose a sequence of corresponding points $\mathcal{P}_1^*$

6.1. The Riemann Integral

which are not rational, and then $S(f, \mathcal{P}, \mathcal{P}_1^*) = 0$. However, we can also choose a sequence of corresponding points $\mathcal{P}_2^*$ that are rational so that $S(f, \mathcal{P}, \mathcal{P}_2^*) = b - a$. This implies that f is not Riemann integrable when $b > a$.

Lemma 6.1.14. *A Riemann integrable function on an interval $[a, b]$ is bounded on $[a, b]$.*

Proof. Let f be Riemann integrable on $[a, b]$, and write

$$T = \int_a^b f.$$

As f is integrable, for $\varepsilon = 1$, there exists $\delta > 0$ so that if $\mathcal{P}$ is a partition with $\|\mathcal{P}\| < \delta$, then

$$|T - S(f, \mathcal{P}, \mathcal{P}^*)| < 1$$

for any sequence of corresponding points $\mathcal{P}^*$. Choose a uniform partition $\mathcal{P}_n = \{x_0, \ldots, x_n\}$ of $n + 1$ points such that $\|\mathcal{P}_n\| < \delta$. So

$$(6.2) \qquad T - 1 < \sum_{i=1}^{n} f(x_i^*) \frac{b-a}{n} < T + 1$$

for any corresponding points x_i^*. For each $x \in [a, b]$ there is a $k \in \{1, \ldots, n\}$ so that $x \in [x_{k-1}, x_k]$. For this x choose $x_k^* = x$ and $x_i^* = x_i$ for $i \neq k$ in (6.2); this will allow us to find a bound for $f(x)$. To make this more explicit, for each $k \in \{1, \ldots, n\}$, let S_k be the sum

$$S_k = \sum_{i=1, i \neq k}^{n} f(x_i),$$

where we omit the kth term and choose x_i^* to be x_i for $i \neq k$. Then (6.2) yields

$$T - 1 < (S_k + f(x)) \frac{b-a}{n} < T + 1.$$

Equivalently,

$$(6.3) \qquad \frac{n}{b-a}(T-1) - S_k < f(x) < \frac{n}{b-a}(T+1) - S_k.$$

Let M be the maximum of the S_k, and let m be the minimum of the S_k as k ranges in $\{1, \ldots, n\}$. Thus, for all $x \in [a, b]$,

$$(6.4) \qquad \frac{n}{b-a}(T-1) - M < f(x) < \frac{n}{b-a}(T+1) - m,$$

showing that f is bounded on $[a, b]$. $\square$

Now that we have seen that a Riemann integrable function must be bounded, we will consider only functions that are bounded. In this case we can give another characterization of Riemann integrability. For a given partition $\mathcal{P} = \{x_0, \ldots, x_n\}$ of $[a, b]$, and a bounded function $f : [a, b] \to \mathbb{R}$, define the supremum and infimum of the values of the function on each interval of the partition by

$$M_i = \sup\{f(x) : x \in [x_{i-1}, x_i]\},$$
$$m_i = \inf\{f(x) : x \in [x_{i-1}, x_i]\}.$$

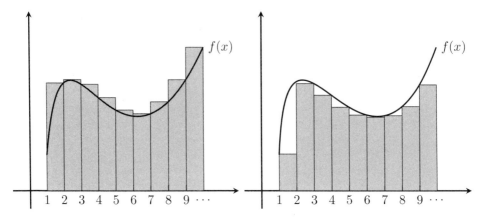

Figure 6.2. Upper and lower Riemann sums for $\Delta x = 1$.

Then over the interval $[x_{i-1}, x_i]$, the rectangle with base length $x_i - x_{i-1}$ and height M_i covers the graph of the function over this interval. Thus the area of the rectangle, namely $M_i(x_i - x_{i-1})$, can be thought of as an upper bound for whatever we think of as the area under the graph of f over the interval $[x_{i-1}, x_i]$. Similarly, the number $m_i(x_i - x_{i-1})$ is a lower bound for this area.

Definition 6.1.15. We define the **upper and lower Riemann sums** of f over $[a, b]$ (see Figure 6.2) by

$$U(f, \mathcal{P}) = \sum_{i=1}^{n} M_i \, (x_i - x_{i-1}),$$

$$L(f, \mathcal{P}) = \sum_{i=1}^{n} m_i \, (x_i - x_{i-1}). \qquad \diamond$$

These sums are sums of areas of rectangles and bound the number that we will eventually want to be the area of f over $[a, b]$. It is clear that $L(f, \mathcal{P}) \leq U(f, \mathcal{P})$.
R

The main content of the following lemma is that all Riemann sums are bounded above and below by the upper and lower Riemann sums.

Lemma 6.1.16. *Let $f : [a, b] \to \mathbb{R}$ be a function that is bounded above by M and below by m, i.e.,*

$$m \leq f(x) \leq M \text{ for all } x \in [a, b].$$

Then

$$m(b - a) \leq L(f, \mathcal{P}) \leq S(f, \mathcal{P}, \mathcal{P}^*) \leq U(f, \mathcal{P}) \leq M(b - a)$$

for every partition $\mathcal{P}$ of $[a, b]$ and corresponding points $\mathcal{P}^*$.

Proof. Let $\mathcal{P} = \{x_0, \ldots, x_n\}$. Then

$$m \leq m_i \leq f(x_i^*) \leq M_i \leq M$$

for all x_i^* in $[x_{i-1}, x_i]$ and $i \in \{0, \ldots, n\}$. So

$$S(f, \mathcal{P}, \mathcal{P}^*) = \sum_{i=1}^n f(x_i^*)(x_{i-1} - x_i) \leq \sum_{i=1}^n M_i(x_{i-1} - x_i)$$

$$\leq M \sum_{i=1}^n (x_{i-1} - x_i) = M(b - a).$$

The proof of the other inequality is similar. □

The following is now a consequence of the lemma, and its proof is left as an exercise.

Corollary 6.1.17. *Let $f : [a, b] \to \mathbb{R}$ be a Riemann integrable function such that $|f(x)| \leq M$ for all $x \in [a, b]$. Then*

$$-M(b-a) \leq \int_a^b f \leq M(b-a).$$

Theorem 6.1.20 will give a characterization of Riemann integrability in terms of upper and lower Riemann sums. In some developments of the theory it is used as the definition of what it means to be Riemann integrable. We start with some preliminary properties.

Lemma 6.1.18. *Let $f : [a, b] \to \mathbb{R}$ be a bounded function.*
 (a) *If $\mathcal{Q}$ is finer than $\mathcal{P}$, then $L(f, \mathcal{Q}) \geq L(f, \mathcal{P})$ and $U(f, \mathcal{Q}) \leq U(f, \mathcal{P})$.*
 (b) *For any partitions $\mathcal{P}$ and $\mathcal{Q}$ of $[a, b]$, $L(f, \mathcal{Q}) \leq U(f, \mathcal{P})$.*

Proof. For part (a) write

$$\mathcal{P} : x_0 = a < x_1 < \cdots < x_n = b \quad \text{and} \quad \mathcal{Q} : y_0 = a < y_1 < \cdots < y_k = b.$$

Since the partition $\mathcal{Q}$ is finer than $\mathcal{P}$, we know that $k \geq n$ and $y_1 \leq x_1$. For concreteness assume $n = 1$, $k = 2$, and $x_0 < y_1 < x_1$. The proof in this case will illustrate the main idea. Let

$$m_1 = \inf\{f(x) : x \in [x_0, x_1]\},$$
$$M_1 = \sup\{f(x) : x \in [x_0, x_1]\},$$
$$a_1 = \inf\{f(x) : x \in [y_0, y_1]\},$$
$$a_2 = \inf\{f(x) : x \in [y_1, y_2]\},$$
$$b_1 = \sup\{f(x) : x \in [y_0, y_1]\},$$
$$b_2 = \sup\{f(x) : x \in [y_1, y_2]\}.$$

Then $L(f, \mathcal{P}) = m_1(x_1 - x_0)$, $U(f, \mathcal{P}) = M_1(x_1 - x_0)$, and

$$L(f, \mathcal{Q}) = a_1(y_1 - y_0) + a_2(y_2 - y_1),$$
$$U(f, \mathcal{Q}) = b_1(y_1 - y_0) + b_2(y_2 - y_1).$$

Then
$$L(f, \mathcal{Q}) = a_1(y_1 - y_0) + a_2(y_2 - y_1)$$
$$\geq m_1(y_1 - y_0) + m_1(y_2 - y_1)$$
$$= m_1(y_1 - y_0 + y_2 - y_1) = m_1(y_2 - y_0)$$
$$= L(f, \mathcal{P}).$$

Similarly,
$$U(f, \mathcal{Q}) = b_1(y_1 - y_0) + b_2(y_2 - y_1)$$
$$\leq M_1(y_1 - y_0) + M_1(y_2 - y_1)$$
$$= M_1(y_2 - y_0)$$
$$= U(f, \mathcal{P}).$$

For part (b) we use part (a) and note that
$$L(f, \mathcal{Q}) \leq L(f, \mathcal{Q} \vee \mathcal{P}) \leq U(f, \mathcal{Q} \vee \mathcal{P}) \leq U(f, \mathcal{P}). \qquad \square$$

The following lemma tells us that the upper and lower Riemann sums can be well approximated by Riemann sums.

Lemma 6.1.19. *Let $f : [a, b] \to \mathbb{R}$ be a bounded function. Then for all partitions $\mathcal{P}$ of $[a, b]$ and all $\varepsilon > 0$, there exist sequences of corresponding points $\mathcal{P}', \mathcal{P}''$ so that*
$$|L(f, \mathcal{P}) - S(f, \mathcal{P}, \mathcal{P}')| < \varepsilon \quad \text{and} \quad |U(f, \mathcal{P}) - S(f, \mathcal{P}, \mathcal{P}'')| < \varepsilon.$$

Proof. Let $\mathcal{P} = \{x_0 = a, x_1, \ldots, x_n = b\}$. Write $M_i = \sup\{f(x) : x \in [x_{i-1}, x_i]\}$ and $m_i = \inf\{f(x) : x \in [x_{i-1}, x_i]\}$. Given $\varepsilon > 0$ there exists points $x_i', x_i'' \in [x_{i-1}, x_i]$, for $i = 1, \ldots, n$, so that
$$m_i \leq f(x_i') < m_i + \frac{\varepsilon}{b-a} \quad \text{and} \quad M_i - \frac{\varepsilon}{b-a} < f(x_i'') \leq M_i.$$
Then
$$\sum_{i=1}^n m_i(x_i - x_{i-1}) \leq \sum_{i=1}^n f(x_i')(x_i - x_{i-1}) < \sum_{i=1}^n [m_i + \frac{\varepsilon}{b-a}](x_i - x_{i-1}),$$
$$\sum_{i=1}^n [M_i - \frac{\varepsilon}{b-a}](x_i - x_{i-1}) < \sum_{i=1}^n f(x_i'')(x_i - x_{i-1}) \leq \sum_{i=1}^n M_i(x_i - x_{i-1}).$$
Since $\sum_{i=1}^n \frac{\varepsilon}{b-a}(x_i - x_{i-1}) = \frac{b-a}{b-a}\varepsilon = \varepsilon$, we have
$$L(f, \mathcal{P}) \leq S(f, \mathcal{P}, \mathcal{P}') < L(f, \mathcal{P}) + \varepsilon,$$
$$U(f, \mathcal{P}) - \varepsilon < S(f, \mathcal{P}, \mathcal{P}'') \leq U(f, \mathcal{P}),$$
completing the proof. $\square$

The following is a useful characterization for Riemann integration; note that in this case we do not need to have the value of the integral to check the condition for integrability.

6.1. The Riemann Integral

Theorem 6.1.20 (Darboux criterion). *Let $f : [a, b] \to \mathbb{R}$ be a bounded function. Then f is Riemann integrable if and only if for all $\varepsilon > 0$ there exists $\delta > 0$ so that*

$$U(f, \mathcal{P}) - L(f, \mathcal{P}) < \varepsilon \tag{6.5}$$

for any partition $\mathcal{P}$ such that $\|\mathcal{P}\| < \delta$.

Proof. Assume that for every $\varepsilon > 0$ there exists $\delta > 0$ so that (6.5) holds for any partition $\mathcal{P}$ with $\|\mathcal{P}\| < \delta$. First set

$$\int_a^b f = \inf\{U(f, \mathcal{P}) : \mathcal{P} \text{ is a partition of } [a, b]\}.$$

By Lemma 6.1.16 this infimum exists in $\mathbb{R}$. It follows that

$$L(f, \mathcal{P}) \leq \int_a^b f \leq U(f, \mathcal{P}). \tag{6.6}$$

From Lemma 6.1.16

$$L(f, \mathcal{P}) \leq S(f, \mathcal{P}, \mathcal{P}^*) \leq U(f, \mathcal{P}). \tag{6.7}$$

From (6.6) and (6.7)

$$-(U(f, \mathcal{P}) - L(f, \mathcal{P})) \leq S(f, \mathcal{P}, \mathcal{P}^*) - \int_a^b f \leq U(f, \mathcal{P}) - L(f, \mathcal{P})$$

for any partition $\mathcal{P}$ with $\|\mathcal{P}\| < \delta$ and any corresponding points $\mathcal{P}^*$. Using (6.5),

$$-\varepsilon < S(f, \mathcal{P}, \mathcal{P}^*) - \int_a^b f < \varepsilon,$$

which shows that f is Riemann integrable.

For the converse suppose f is Riemann integrable, and let $\varepsilon > 0$. Then there exists $\delta > 0$ so that

$$\left| \int_a^b f(x)\, dx - S(f, \mathcal{P}, \mathcal{P}^*) \right| < \frac{\varepsilon}{6}$$

for all partitions $\mathcal{P}$ such that $\|\mathcal{P}\| < \delta$ and for all corresponding $\mathcal{P}^*$.

Use Lemma 6.1.19 to find corresponding points $\mathcal{P}'$ and $\mathcal{P}''$ such that

$$|L(f, \mathcal{P}) - S(f, \mathcal{P}, \mathcal{P}')| < \frac{\varepsilon}{3} \quad \text{and} \quad |U(f, \mathcal{P}) - S(f, \mathcal{P}, \mathcal{P}'')| < \frac{\varepsilon}{3}.$$

Now

$$|S(f, \mathcal{P}, \mathcal{P}') - S(f, \mathcal{P}, \mathcal{P}'')| \leq |S(f, \mathcal{P}, \mathcal{P}') - \int_a^b f(x)\, dx|$$
$$+ |\int_a^b f(x)\, dx - S(f, \mathcal{P}, \mathcal{P}'')|$$
$$< \frac{\varepsilon}{6} + \frac{\varepsilon}{6} = \frac{\varepsilon}{3}.$$

Finally,
$$|U(f,\mathcal{P}) - L(f,\mathcal{P})| \le |U(f,\mathcal{P}) - S(f,\mathcal{P},\mathcal{P}'')| + |S(f,\mathcal{P},\mathcal{P}'') - S(f,\mathcal{P},\mathcal{P}')|$$
$$+ |S(f,\mathcal{P},\mathcal{P}') - L(f,\mathcal{P})|$$
$$< \frac{\varepsilon}{3} + \frac{\varepsilon}{3} + \frac{\varepsilon}{3} = \varepsilon.$$
□

The following is an important theorem in the theory of Riemann integrals as it gives a natural class of functions that are Riemann integrable.

Theorem 6.1.21. *If $f : [a,b] \to \mathbb{R}$ is a continuous function, then f is Riemann integrable.*

Proof. Since f is continuous on a compact interval, it is bounded; we will use Theorem 6.1.20 to show it is Riemann integrable. Also, f is uniformly continuous, so for any $\varepsilon > 0$ there exists $\delta > 0$ such that
$$|f(x) - f(y)| < \frac{\varepsilon}{b-a} \text{ for all } x, y \in [a,b] \text{ with } |x-y| < \delta.$$
Let $\mathcal{P} = \{x_i\}$ be a partition of $[a,b]$ with $||\mathcal{P}|| < \delta$. Let m_i and M_i be the infimum and supremum of f on the interval $[x_{i-1}, x_i]$, respectively. By the continuity of f, there exist $x_i^-, x_i^+ \in [x_{i-1}, x_i]$ such that
$$f(x_i^-) = m_i \quad \text{and} \quad f(x_i^+) = M_i \quad \text{for } i = 1, \ldots, n.$$
Therefore,
$$M_i - m_i = f(x_i^+) - f(x_i^-) < \frac{\varepsilon}{b-a}.$$
This implies that
$$U(f,\mathcal{P}) - L(f,\mathcal{P}) = \sum_{i=1}^n M_i (x_i - x_{i-1}) - \sum_{i=1}^n m_i (x_i - x_{i-1})$$
$$= \sum_{i=1}^n [M_i - m_i](x_i - x_{i-1})$$
$$< \frac{\varepsilon}{b-a} \cdot \left(\sum_{i=1}^n (x_i - x_{i-1}) \right)$$
$$= \frac{\varepsilon}{b-a} (b-a) = \varepsilon.$$
Theorem 6.1.20 implies that f is Riemann integrable. □

Question 6.1.22. Let $f : [a,b] \to \mathbb{R}$ be a monotone increasing function. Show it is Riemann integrable.

6.1.1. Properties of the Riemann Integral.
We start with a theorem that gives another version of the Darboux criterion (Theorem 6.1.20) and simplifies the condition to check for a function to be Riemann integrable.

Theorem 6.1.23. *Let $f : [a,b] \to \mathbb{R}$ be a bounded function. Then f is Riemann integrable if and only if for all $\varepsilon > 0$ there exists a partition $\mathcal{Q}$ such that*
$$U(f,\mathcal{Q}) - L(f,\mathcal{Q}) < \varepsilon.$$

6.1. The Riemann Integral

Proof. If f is Riemann integrable, then the partition $\mathcal{Q}$ exists by Theorem 6.1.20.

For the converse let $\varepsilon > 0$, and suppose there exists a partition $\mathcal{Q}$ such that
$$U(f, \mathcal{Q}) - L(f, \mathcal{Q}) < \varepsilon/2.$$
Write $\mathcal{Q} = \{x_0, \ldots, x_k\}$. We wish to use Theorem 6.1.20 to verify that f is Riemann integrable. Imagine a partition $\mathcal{P}$ that has small mesh relative to the mesh of $\mathcal{Q}$, say the mesh of $\mathcal{P}$ is less than $\frac{1}{2}\|\mathcal{Q}\|$. Then most subintervals of $\mathcal{P}$ will be wholly contained in some subinterval of $\mathcal{Q}$, in fact, there will be at most k subintervals of $\mathcal{P}$ that overlap two subintervals of $\mathcal{Q}$. This suggests the first upper bound for δ. As f is bounded, there is a number $B > 0$ such that for all $i = 1, \ldots, k$,
$$(\sup\{f(x) : x \in [x_{i-1}, x_i]\}) - (\inf\{f(x) : x \in [x_{i-1}, x_i]\}) < B.$$
Choose $\delta > 0$ such that
$$\delta < \frac{1}{2} \min\left\{\frac{\varepsilon}{Bk}, \|\mathcal{Q}\|\right\}.$$
Let $\mathcal{P}$ be an arbitrary partition such that $\|\mathcal{P}\| < \delta$. When calculating $U(f, \mathcal{P}) - L(f, \mathcal{P})$, first consider the subintervals that are wholly contained in a subinterval of $\mathcal{Q}$. It is clear that the sum over these subintervals is bounded by $U(f, \mathcal{Q}) - L(f, \mathcal{Q}) < \varepsilon/2$. Of the remaining subintervals there are at most k of them, and the sum over these is bounded by $kB\delta < \varepsilon/2$. Thus $U(f, \mathcal{P}) - L(f, \mathcal{P}) < \varepsilon$ when $\|\mathcal{P}\| < \delta$. $\square$

As we show below, this theorem implies that if f is integrable on $[a, b]$, then it is integrable on any subinterval of $[a, b]$.

Corollary 6.1.24. *Let $f : [a, b] \to \mathbb{R}$ be Riemann integrable, and let $c, d \in [a, b]$ with $c < d$. Then f is Riemann integrable on $[c, d]$.*

Proof. The function f is clearly bounded on $[c, d]$. Let $\varepsilon > 0$. As f is Riemann integrable on $[a, b]$, by Theorem 6.1.23 there is a partition $\mathcal{Q}$ such that $U(f, \mathcal{Q}) - L(f, \mathcal{Q}) < \varepsilon$. Let $\mathcal{Q}'$ consist of the partition $\mathcal{Q}$ with the addition of the extra points c and d, if $\mathcal{Q}$ does not already have them. Since $\mathcal{Q}'$ is a refinement of $\mathcal{Q}$, by Lemma 6.1.18 we have that $U(f, \mathcal{Q}') - L(f, \mathcal{Q}') < \varepsilon$. Now let $\mathcal{Q}''$ consist of the partition $\mathcal{Q}'$ restricted to $[c, d]$, namely, c and d will be the first and last points in the partition $\mathcal{Q}''$, respectively, and the remaining points will be points in $\mathcal{Q}'$ that are in (c, d). Then
$$U(f, \mathcal{Q}'') - L(f, \mathcal{Q}'') \leq U(f, \mathcal{Q}') - L(f, \mathcal{Q}') < \varepsilon.$$
This implies that f is integrable on $[c, d]$. $\square$

The following lemma is useful for calculations.

Lemma 6.1.25. *Let $f : [a, b] \to \mathbb{R}$ be a Riemann integrable function. Let (a_n) be a sequence of positive numbers decreasing to 0 (for example $a_n = 1/n$). For each $n \in \mathbb{N}$, let $\mathcal{P}_n$ be a partition of $[a, b]$ with $n + 1$ points such that $\|\mathcal{P}_n\| \leq a_n$, and let $\mathcal{P}_n^*$ be any sequence of corresponding points. Then*
$$\int_a^b f = \lim_{n \to \infty} S(f, \mathcal{P}_n, \mathcal{P}_n^*).$$

Proof. As f is Riemann integrable, there exists a number $\int_a^b f$ so that for all $\varepsilon > 0$, there exists $\delta > 0$ such that
$$|S(f, \mathcal{P}, \mathcal{P}^*) - \int_a^b f| < \varepsilon$$
for any partition $\mathcal{P}$ with $\|\mathcal{P}\| < \delta$ and any associated $\mathcal{P}^*$. Choose $N \in \mathbb{N}$ such that
$$a_N < \delta.$$
We know $\|\mathcal{P}_n\| \leq a_N < \delta$ for $n \geq N$. So if $n \geq N$,
$$|S(f, \mathcal{P}_n, \mathcal{P}_n^*) - \int_a^b f| < \varepsilon$$
for any $\mathcal{P}_n^*$. This completes the proof of the lemma. $\square$

The following is useful in some calculations, so we state it separately although it is a direct consequence of Lemma 6.1.25.

Lemma 6.1.26. *Let $f : [a, b] \to \mathbb{R}$ be a Riemann integrable function. For each $n \in \mathbb{N}$, let $\mathcal{P}_n$ be the uniform partition of $[a, b]$ with $n + 1$ points, and let $\mathcal{P}_n^*$ be any sequence of corresponding points. Then*
$$\int_a^b f = \lim_{n \to \infty} S(f, \mathcal{P}_n, \mathcal{P}_n^*).$$

Proof. Let $a_n = \frac{b-a}{n}$ in Lemma 6.1.25. $\square$

Question 6.1.27. Show that
$$\int_a^b c\, dx = (b - a)c.$$

Example 6.1.28. Let $f(x) = x^2$. As f is continuous, it is Riemann integrable; using Example 6.1.6 and Lemma 6.1.26, we find
$$\int_0^1 x^2\, dx = \lim_{n \to \infty} \frac{1}{6}(1 + \frac{1}{n})(2 + \frac{1}{n}) = \frac{1}{3}.$$

Theorem 6.1.29. *Let f, g be Riemann integrable on $[a, b]$. Then*

(1) *For each constant α the function αf is Riemann integrable and*
$$\int_a^b \alpha f = \alpha \int_a^b f.$$

(2) *The function $f + g$ is Riemann integrable and*
$$\int_a^b f + g = \int_a^b f + \int_a^b g.$$

(3) *If $c \in (a, b)$, then*
$$\int_a^b f = \int_a^c f + \int_c^b f.$$

6.1. The Riemann Integral

(4) $|f|$ is Riemann integrable on $[a,b]$ and
$$\left|\int_a^b f\right| \leq \int_a^b |f|.$$

Proof. For part (1) we first note that the result is clear when $\alpha = 0$. When $\alpha > 0$, then $U(\alpha f, \mathcal{P}) = \alpha U(f, \mathcal{P})$ and $L(\alpha f, \mathcal{P}) = \alpha L(f, \mathcal{P})$. Then by Theorem 6.1.20 with ε/α it follows that αf is integrable. When α is negative, the sup and the inf are switched so that $U(\alpha f, \mathcal{P}) = \alpha L(f, \mathcal{P})$ and $U(\alpha f, \mathcal{P}) = \alpha U(f, \mathcal{P})$. Then use the same theorem with $\varepsilon/-\alpha$. For the equality of the integrals, observe that for all partitions $\mathcal{P}$, the Riemann sums satisfy $S(\alpha f, \mathcal{P}, \mathcal{P}^*) = \alpha S(f, \mathcal{P}, \mathcal{P}^*)$. The proof is completed by Lemma 6.1.26.

For part (2), we use the definition. Let $\varepsilon > 0$. As f and g are Riemann integrable, there exists $\delta > 0$ so that

$$(6.8) \qquad \left|\int_a^b f - S(f, \mathcal{P}, \mathcal{P}^*)\right| < \frac{\varepsilon}{2} \quad \text{and} \quad \left|\int_a^b g - S(g, \mathcal{P}, \mathcal{P}^*)\right| < \frac{\varepsilon}{2}$$

whenever $\|\mathcal{P}\| < \delta$ and $\mathcal{P}^*$ is any sequence of corresponding points to $\mathcal{P}$. Next we observe that a direct consequence of the definition is that for all partitions $\mathcal{P}$, the Riemann sums obey

$$S(f+g, \mathcal{P}, \mathcal{P}^*) = S(f, \mathcal{P}, \mathcal{P}^*) + S(g, \mathcal{P}, \mathcal{P}^*).$$

Therefore,
$$\left|\int_a^b f + \int_a^b g - S(f+g, \mathcal{P}, \mathcal{P}^*)\right| = \left|\int_a^b f + \int_a^b g - S(f, \mathcal{P}, \mathcal{P}^*) - S(g, \mathcal{P}, \mathcal{P}^*)\right|$$
$$\leq \left|\int_a^b f - S(f, \mathcal{P}, \mathcal{P}^*)\right| + \left|\int_a^b g - S(g, \mathcal{P}, \mathcal{P}^*)\right|$$
$$< \frac{\varepsilon}{2} + \frac{\varepsilon}{2} = \varepsilon.$$

Thus $f+g$ is Riemann integrable over $[a,b]$ with integral $\int_a^b f + \int_a^b g$.

For part (3), we note that by Corollary 6.1.24 f is Riemann integrable on $[a,c]$ and on $[c,b]$. Let $\mathcal{P}_{1,n}$ be the uniform partition into n points of $[a,c]$, and let $\mathcal{P}_{2,n}$ be the uniform partition into n points of $[c,b]$. By Lemma 6.1.25,

$$\int_a^c f = \lim_{n \to \infty} S(f, \mathcal{P}_{1,n}, \mathcal{P}^*_{1,n}) \quad \text{and} \quad \int_c^b f = \lim_{n \to \infty} S(f, \mathcal{P}_{2,n}, \mathcal{P}^*_{2,n}).$$

Putting the two partitions together to obtain a partition of $[a,b]$, Lemma 6.1.25 completes the proof.

Part (4) is left as an exercise. $\square$

Remark 6.1.30. When $b = a$ all the intervals in $[a,a]$ have zero length so it makes sense to define $\int_a^a f = 0$. When $b < a$, we define

$$\int_a^b f = -\int_b^a f.$$

Then it can be shown (Exercise 6.1.10) that for all $a, b, c \in \mathbb{R}$, if f is Riemann integrable on the largest interval containing the points a, b, c,

$$\int_a^b f = \int_a^c f + \int_c^b f.$$

The following corollary will be used in the next section.

Corollary 6.1.31. *Let $f : (a, b) \to \mathbb{R}$ be a continuous and bounded function. Let $g : [a, b] \to \mathbb{R}$ be any function such that $g(x) = f(x)$ for all $x \in (a, b)$. Then g is integrable on $[a, b]$, and its integral is independent of the value of g at the endpoints.*

Proof. As f is bounded, there exists a constant M so that $|g(x)| < M$ for all $x \in [a, b]$. We will use Theorem 6.1.23 to prove that g is integrable. Let $\varepsilon > 0$. Choose $n \in \mathbb{N}$ so that $M(4/n) < \varepsilon/2$. For each $n \in \mathbb{N}$, the function f is Riemann integrable on the interval $[a + \frac{1}{n}, b - \frac{1}{n}]$. Then by Theorem 6.1.23 there exists a partition $\mathcal{P}_n$ of this interval so that

$$U(f, \mathcal{P}_n) - L(f, \mathcal{P}_n) < \frac{\varepsilon}{2}.$$

Let $\mathcal{P}'_n$ be the partition of $[a, b]$ consisting of $\mathcal{P}_n$ plus the points a and b. Then $U(g, \mathcal{P}'_n)$ differs from $U(f, \mathcal{P}_n) = U(g, \mathcal{P}_n)$ in the intervals $[a, a+1/n]$ and $[b-1/n, b]$ (and similarly for $L(g, \mathcal{P}'_n)$ and $L(f, \mathcal{P}_n)$). Since g is bounded by M on both intervals and each is of length $2/n$, we obtain

$$U(g, \mathcal{P}'_n) - L(g, \mathcal{P}'_n) < \frac{2M}{n} + U(g, \mathcal{P}_n) - L(g, \mathcal{P}_n) + \frac{2M}{n}$$
$$= \frac{2M}{n} + U(f, \mathcal{P}_n) - L(f, \mathcal{P}_n) + \frac{2M}{n}$$
$$< \varepsilon.$$

Therefore, g is Riemann integrable on $[a, b]$. The fact that the integral is independent of the value of g at the endpoints follows from Lemma 6.1.26. □

Remark 6.1.32. Corollary 6.1.31 does not hold if f is not bounded on (a, b). For example, if $f(x) = 1/x$ for $x \in (0, 1)$, then every extension g of f to the interval $[0, 1]$ is not Riemann integrable (see Exercise 6.1.17).

The following corollary is an immediate consequence but we state it as it is a useful result.

Corollary 6.1.33. *Let $f : [a, b] \to \mathbb{R}$ be a function, and suppose that f is continuous and bounded on (a, b). Then f is Riemann integrable on $[a, b]$.*

Proof. The function f is an extension of $f : (a, b) \to \mathbb{R}$ to $[a, b]$. So Corollary 6.1.31 completes the proof. □

Example 6.1.34. We find error estimates when using various Riemann sums to approximate the integral of a function. Let $f : [a, b] \to \mathbb{R}$ be a continuous function that we will assume has a second derivative on an open set containing the interval $[a, b]$ (or we could just assume the one-sided derivatives at the endpoints exist).

6.1. The Riemann Integral

We will consider uniform partitions $\mathcal{P}_n = \{x_i\}$ of $[a,b]$ into n subintervals. So the partition points are given by $x_i = a + i \cdot \frac{b-a}{n}$, for $i = 0, \ldots, n$, and

$$x_i - x_{i-1} = \frac{b-a}{n}.$$

Also, set

$$m_i = \frac{x_{i-1} + x_i}{2}, \text{ the midpoint of the interval } [x_{i-1}, x_i].$$

We will consider the corresponding points to be the right endpoints, which we denote by $\mathcal{R}_n$, or the midpoints, which we denote by $\mathcal{M}_n$. Then the corresponding Riemann sums are

(6.9) $$S(f, \mathcal{P}_n, \mathcal{R}_n) = \sum_{i=1}^{n} f(x_i)(x_i - x_{i-1}),$$

(6.10) $$S(f, \mathcal{P}_n, \mathcal{M}_n) = \sum_{i=1}^{n} f(m_i)(x_i - x_{i-1}).$$

We first use Taylor's theorem to find an approximation to f by a constant polynomial about the point x_i to obtain, for some z_i between x and x_i,

(6.11) $$f(x) = f(x_i) + f'(z_i)(x - x_i).$$

Then we estimate the difference between the integral and the Riemann sum using the right endpoints. In the estimates below we use (6.11) and that we have a bound on the derivative: $|f'(x)| \leq M_1$ for all $x \in [a,b]$. Also, since $f(x_i)$ is constant for the integral, we have

$$f(x_i)(x_i - x_{i-1}) = \int_{x_{i-1}}^{x_i} f(x_i)\, dx.$$

Then

$$\left| \int_a^b f(x)\, dx - S(f, \mathcal{P}_n, \mathcal{R}_n) \right| = \left| \sum_{i=1}^{n} \int_{x_{i-1}}^{x_i} f(x)\, dx - \sum_{i=1}^{n} f(x_i)(x_i - x_{i-1}) \right|$$

$$= \left| \sum_{i=1}^{n} \int_{x_{i-1}}^{x_i} (f(x) - f(x_i))\, dx \right|$$

$$= \left| \sum_{i=1}^{n} \int_{x_{i-1}}^{x_i} f'(z_i)(x - x_i)\, dx \right|.$$

Here we note that

$$\sum_{i=1}^{n} \int_{x_{i-1}}^{x_i} (x - x_i)\, dx = \sum_{i=1}^{n} \left[\frac{(x_i - x_i)^2}{2} - \frac{(x_{i-1} - x_i)^2}{2} \right] = -\frac{(b-a)^2}{2n^2}.$$

Therefore,

$$\left| \sum_{i=1}^{n} \int_{x_{i-1}}^{x_i} f'(z_i)(x - x_i)\, dx \right| \leq \sum_{i=1}^{n} M_1 \cdot \frac{(b-a)^2}{2n^2} \leq M_1 \cdot \frac{(b-a)^2}{2n}.$$

This gives us the following error estimate for the Riemann sum when using the right endpoints:

(6.12) $$\left| \int_a^b f(x)\,dx - S(f, \mathcal{P}_n, \mathcal{R}_n) \right| \le M_1 \cdot \frac{(b-a)^2}{2n}.$$

We now obtain a second error estimate when using midpoints in the Riemann sums. For this we will use Taylor's theorem to find a linear approximation to f by a polynomial about the point m_i to obtain, for some w_i between x and m_i,

$$f(x) = f(m_i) + f'(m_i)(x - m_i) + \frac{f''(w_i)}{2!}(x - m_i)^2.$$

We outline the calculations, with the details left as an exercise. We mention though an important simplification that comes from the nature of the midpoints: one can compute that

$$\int_{x_{i-1}}^{x_i} (x - m_i)\,dx = 0.$$

We also have a bound on the second derivative: $|f''(x)| \le M_2$ for all $x \in [a, b]$. Then we have the following error estimate when using the midpoints:

(6.13) $$\left| \int_a^b f(x)\,dx - S(f, \mathcal{P}_n, \mathcal{M}_n) \right| = \left| \sum_{i=1}^n \int_{x_{i-1}}^{x_i} f(x)\,dx - \sum_{i=1}^n f(m_i) \frac{b-a}{n} \right|$$

$$= \left| \sum_{i=1}^n \int_{x_{i-1}}^{x_i} (f(x) - f(m_i))\,dx \right|$$

$$\le M_2 \cdot \frac{(b-a)^3}{24n^2}.$$

This shows that in the limit, the error estimate when using the midpoints is better than when using the right endpoints.

Exercises: The Riemann Integral

6.1.1 Let $f(x) = 5x + 3$. Find $\int_1^2 f$.

6.1.2 Let $f(x) = 4x^2 - 2$. Find $\int_2^4 f$.

6.1.3 Let $f : [a, b] \to \mathbb{R}$ be a bounded function. Let $U = \inf U(f, \mathcal{P})$ over all partitions $\mathcal{P}$ and $L = \sup L(f, \mathcal{P})$ over all partitions $\mathcal{P}$. Prove that f is Riemann integrable if and only if $L = U$.

6.1.4 Let $f : [a, b] \to \mathbb{R}$ be Riemann integrable. Prove that $\int f = \inf U(f, \mathcal{P})$ over all partitions $\mathcal{P}$ and $\int f = \sup L(f, \mathcal{P})$ over all partitions $\mathcal{P}$.

6.1.5 Let $f : [a, b] \to \mathbb{R}$ be Riemann integrable. Suppose that $g : [a, b] \to \mathbb{R}$ agrees with f at all but a finite number of points in the interval $[a, b]$. Show that g is Riemann integrable on $[a, b]$ and has the same integral as f.

6.1.6 Prove Corollary 6.1.17.

6.1.7 Let $f : [a, b] \to \mathbb{R}$ be a bounded function. Suppose that f is continuous on $[a, b]$ except at a finite number of points. Show that f is Riemann integrable.

6.1.8 Let $f : [a, b] \to \mathbb{R}$ be a continuous function. Suppose that $f(x) \geq 0$ for all $x \in [a, b]$ and $\int_a^b f = 0$. Show that $f(x) = 0$ for all $x \in [a, b]$.

6.1.9 Let $f : [a, b] \to \mathbb{R}$ be a bounded function. Are the upper and lower sums of f monotone sequences with respect to n?

6.1.10 Prove the second part of Remark 6.1.30.

6.1.11 Prove part (4) of Theorem 6.1.29.

6.1.12 Let $a < c < b$. Suppose f is Riemann integrable on $[a, c]$ and on $[c, b]$. Prove it is Riemann integrable on $[a, b]$.

6.1.13 Prove that the modified Dirichlet function is Riemann integrable.

6.1.14 Find the number of points necessary to estimate $\int_0^{\pi/2} \sin x^2 \, dx$ to two decimal places.

6.1.15 Estimate $\int_0^1 e^{-x^2} \, dx$ to one decimal place.

6.1.16 Complete the details in the proof of the estimate in (6.13).

6.1.17 Let $f(x) = 1/\sqrt{x}$ for $x \in (0, 1)$. Prove that if g is any extension of f to the interval $[0, 1]$ (i.e., $g(x) = f(x)$ for $x \in (0, 1)$), then g is not Riemann integrable on $[0, 1]$.

6.1.18 Show that one does not get a better estimate for the error when approximating the integral with right-hand endpoints if one uses a higher-degree Taylor approximation.

6.1.19 Estimate the error when approximating integrals by Riemann sums when using the left endpoints of the subintervals.

6.1.20 Estimate the error when approximating integrals by Riemann sums when using the trapezoid rule for the subintervals (here the area over the ith subinterval is $\frac{1}{2}(f(x_{i-1}) + f(x_i))(x_i - x_{i-1})$).

6.2. The Fundamental Theorem of Calculus

The Fundamental Theorem of Calculus has two parts and establishes a striking connection between differentiation and integration; it says that we can think of the two processes as inverses of each other.

Theorem 6.2.1 (Fundamental Theorem of Calculus—first form). *Let $f : [a, b] \to \mathbb{R}$ be a Riemann integrable function. Let*

$$F(x) = \int_a^x f \text{ for all } x \in [a, b].$$

Then F is continuous on $[a, b]$. Furthermore, if f is continuous at $x \in (a, b)$, then F is differentiable at x and

$$F'(x) = f(x).$$

Proof. First we show that F is Lipschitz. As f is Riemann integrable, we know it is bounded, so there exists a constant M such that $|f(x)| \leq M$ for all $x \in [a,b]$. Then, for all $x, y \in [a,b]$,

$$(6.14) \qquad |F(x) - F(y)| = |\int_y^x f| \leq M|x-y|.$$

This shows that F is Lipschitz, hence it is continuous. Suppose now that f is continuous at x. Then

$$F'(x) = \lim_{h \to 0} \frac{F(x+h) - F(x)}{h} = \lim_{h \to 0} \frac{\int_a^{x+h} f - \int_a^x f}{h}$$
$$= \lim_{h \to 0} \frac{1}{h} \int_x^{x+h} f$$
$$= f(x).$$

We justify the last equality. By continuity of f at x, for $\varepsilon > 0$ there exists $\delta > 0$ such that

$$|f(x) - f(t)| < \varepsilon \text{ whenever } |x-t| < \delta, \text{ and } t \in [a,b].$$

First consider the case when $h > 0$. We can write

$$f(x) = \frac{1}{h}\int_x^{x+h} f(x)\, dt \text{ since } f(x) \text{ is a constant with respect to } t.$$

So if $h < \delta$, as $t \in (x, x+h)$, then $|x - t| < \delta$. Therefore,

$$|(\frac{1}{h}\int_x^{x+h} f(t)\, dt) - f(x)| = |\frac{1}{h}\int_x^{x+h} (f(t) - f(x))\, dt|$$
$$\leq \frac{1}{h}\int_x^{x+h} |f(t) - f(x)|\, dt$$
$$< \frac{1}{h}\int_x^{x+h} \varepsilon\, dt$$
$$= \frac{1}{h}(x + h - x)\varepsilon = \varepsilon.$$

When $h < 0$, the same inequality is obtained using that $\int_x^{x+h} f = -\int_{x+h}^x f$. Thus

$$\lim_{h \to 0} \frac{1}{h}\int_x^{x+h} f(t)\, dt = f(x). \qquad \square$$

Remark 6.2.2. An interesting consequence of Theorem 6.2.1 is that a continuous function has an *antiderivative*. We say that a function $F : [a,b] \to \mathbb{R}$ is an **antiderivative** of $f : [a,b] \to \mathbb{R}$ if F is continuous on $[a,b]$, differentiable on (a,b), and $F'(x) = f(x)$ for all $x \in (a,b)$. By Theorem 6.2.1 an antiderivative of a continuous function $f : [a,b] \to \mathbb{R}$ is given by $F(x) = \int_a^x f$ for $x \in [a,b]$.

Theorem 6.2.3 (Fundamental Theorem of Calculus—second form). *Let $f : [a,b] \to \mathbb{R}$ be a Riemann integrable function. Suppose there exists a continuous function $G : [a,b] \to \mathbb{R}$ such that $G : (a,b) \to \mathbb{R}$ is differentiable and*

$$G'(x) = f(x) \text{ for all } x \in (a,b).$$

6.2. The Fundamental Theorem of Calculus

Then
$$\int_a^b f = G(b) - G(a).$$

Proof. For $n \in \mathbb{N}$, let $\mathcal{P} = \{x_0, \ldots, x_{n+1}\}$ be the uniform partition of $[a,b]$ with $n+1$ points. By the mean value theorem, for each $i \in \{1, \ldots, n\}$, there exists $x_i^* \in (x_{i-1}, x_i)$ such that
$$G'(x_i^*) = \frac{G(x_i) - G(x_{i-1})}{x_i - x_{i-1}}.$$

Let $\mathcal{P}^* = \{x_1^*, \ldots, x_n^*\}$. By Lemma 6.1.25,
$$\int_a^b f = \lim_{n \to \infty} S(f, \mathcal{P}_n, \mathcal{P}_n^*).$$

We calculate
$$S(f, \mathcal{P}_n, \mathcal{P}_n^*) = \sum_{i=1}^n f(x_i^*)(x_i - x_{i-1})$$
$$= \sum_{i=1}^n G'(x_i^*)(x_i - x_{i-1})$$
$$= \sum_{i=1}^n G(x_i) - G(x_{i-1})$$
$$= G(b) - G(a).$$

Therefore,
$$\int_a^b f = \lim_{n \to \infty} S(f, \mathcal{P}_n, \mathcal{P}_n^*) = \lim_{n \to \infty} G(b) - G(a) = G(b) - G(a). \quad \square$$

Remark 6.2.4. A consequence of Corollary 6.1.31 is that if a continuous function $f : [a,b] \to \mathbb{R}$ is differentiable on (a,b), with a continuous and bounded derivative on (a,b), then the integral $\int_a^b f'$ makes sense, since we can take any extension of f' to $[a,b]$ and this extension will be Riemann integrable. Then it follows from Theorem 6.2.3 that
$$\int_a^b f' = f(b) - f(a).$$

We observe that f' may not have a continuous extension to $[a,b]$, but any extension of f' to $[a,b]$ is Riemann integrable (provided f' is continuous and bounded on (a,b)), and this is all we need. This covers many, if not most, calculus applications. For example, this is the case when f is differentiable on $[a,b]$ (where differentiability at the endpoints is understood one-sidedly), since then f' is continuous on $[a,b]$.

Theorem 6.2.5 (Integration by parts). *Let $f, g : [a,b] \to \mathbb{R}$ be continuous functions that are differentiable on (a,b) and such that their derivatives are continuous and bounded. Then*
$$\int_a^b f(x)g'(x)dx = f(b)g(b) - f(a)g(a) - \int_a^b f'(x)g(x)\,dx.$$

Proof. The function $(f(x)g(x))'$ is continuous and bounded on (a,b). By the product rule, for all $x \in (a,b)$,

(6.15) $$(f(x)g(x))' = f(x)g'(x) + f'(x)g(x).$$

Then by Theorem 6.2.1 (see Remark 6.2.4), after integrating both sides of (6.15), we obtain

$$f(b)g(b) - f(a)g(a) = \int_a^b f(x)g'(x)\,dx + \int_a^b f'(x)g(x)\,dx. \qquad \square$$

Theorem 6.2.6 (Integration by substitution). *Let $g : [a,b] \to \mathbb{R}$ be a continuous function that is differentiable on (a,b) with a continuous and bounded derivative. Let f be a continuous function on the closed interval with endpoints $g(a)$ and $g(b)$. Then*

$$\int_a^b f(g(x))g'(x)\,dx = \int_{g(a)}^{g(b)} f.$$

Proof. To first see the function $f(g(x))g'(x)$ as the result of applying the chain rule, we look for a function whose derivative is f. So for a fixed c in the interval with endpoints $g(a)$ and $g(b)$, define, for t in that interval,

$$F(t) = \int_c^t f.$$

By Theorem 6.2.1, $F'(t) = f(t)$, so we can write

$$(F(g(x)))' = f(g(x))g'(x) \text{ for } x \in (a,b).$$

For each $x \in [a,b]$, $g(x)$ is in the closed interval with endpoints $g(a)$ and $g(b)$. The function $f(g(x))g'(x)$ is continuous on $[a,b]$, so it is Riemann integrable on the interval $[a,b]$. By Theorem 6.2.3

$$\int_a^b f(g(x))g'(x)\,dx = \int_a^b (F(g(x)))'\,dx$$
$$= F(g(b)) - F(g(a))$$
$$= \int_c^{g(b)} f - \int_c^{g(a)} f$$
$$= \int_c^{g(b)} f + \int_{g(a)}^c f$$
$$= \int_{g(a)}^{g(b)} f. \qquad \square$$

6.2.1. The Logarithmic Function. Here we define and develop the basic properties of the **natural logarithmic function**, denoted $\ln x$. Define, for $x \in (0, \infty)$,

$$\ln x = \int_1^x \frac{1}{t}\,dt.$$

We recall that when $0 < x < 1$, we interpret $\int_x^1 \frac{1}{t}\,dt$ as $-\int_1^x \frac{1}{t}\,dt$. Also, the integral over the interval $[1,1]$ is 0, so $\ln 1 = 0$.

6.2. The Fundamental Theorem of Calculus

Proposition 6.2.7. *The natural logarithmic function satisfies the following properties.*

(1) $\ln x$ *is strictly increasing on* $(0, \infty)$.

(2) $\ln x$ *is differentiable on* $(0, \infty)$, *and*
$$D_x(\ln x) = \frac{1}{x}.$$

(3) *For all* $a, b \in (0, \infty)$,
$$\ln(ab) = \ln a + \ln b.$$

(4) *For all* $a, b \in (0, \infty)$,
$$\ln\left(\frac{a}{b}\right) = \ln a - \ln b.$$

(5) *For all* $a \in (0, \infty)$ *and* $n \in \mathbb{Z}$,
$$\ln(a^n) = n \ln a.$$

(6) $\lim_{x \to \infty} \ln x = \infty$ *and* $\lim_{x \to 0} \ln x = -\infty$.

Proof. For part (1), let $y > x$, and assume $x > 1$ (the proof for $x < 1$ is similar). Then
$$\ln y = \int_1^y \frac{1}{t}\, dt = \int_1^x \frac{1}{t}\, dt + \int_x^y \frac{1}{t}\, dt > \int_1^x \frac{1}{t}\, dt = \ln x,$$
since $\int_x^y \frac{1}{t}\, dt > 0$. (This can also be seen as a consequence of (2).)

For (2), the fact that $\ln x$ is differentiable for all $x \in (0, \infty)$ follows from Theorem 6.2.1. Also, by this theorem,
$$D_x(\ln x) = D_x \int_1^x \frac{1}{t}\, dt = \frac{1}{x}.$$

For (3), let us first assume that $1 < a < ab$. So
$$\ln(ab) = \int_1^{ab} \frac{1}{t}\, dt = \int_1^a \frac{1}{t}\, dt + \int_a^{ab} \frac{1}{t}\, dt.$$
Now we apply substitution (Theorem 6.2.6) with $f(t) = \frac{1}{t}$, $g(t) = \frac{t}{a}$, and $g'(t) = \frac{1}{a}$ to obtain that
$$\int_a^{ab} \frac{1}{t}\, dt = \int_a^{ab} \frac{a}{t}\frac{1}{a}\, dt = \int_a^{ab} f(g(t))g'(t)\, dt$$
$$= \int_{g(a)}^{g(ab)} f(t)\, dt$$
$$= \int_1^b \frac{1}{t}\, dt = \ln b.$$

Thus $\ln(ab) = \ln a + \ln b$. The other cases are similar, and they are left as an exercise and use identities that follow from substitution, such as $\int_{ab}^a 1/t\, dt = -\ln b$.

For part (4) we note that applying (3) we have
$$\ln\left(\frac{a}{b}\right) = \ln\left(a \cdot \frac{1}{b}\right) = \ln a + \ln\left(\frac{1}{b}\right).$$

To evaluate $\ln(1/b)$, we first assume that $1/b > 1$ and apply substitution (Theorem 6.2.6) with $f(t) = \frac{1}{t}, g(t) = \frac{1}{t}, g'(t) = -\frac{1}{t^2}$ to obtain

$$\int_1^{1/b} \frac{1}{t}\, dt = \int_1^{1/b} \frac{1}{(1/t)} \frac{1}{t^2}\, dt = -\int_1^{1/b} f(g(t))g'(t)\, dt$$
$$= -\int_{g(1)}^{g(1/b)} f(t)\, dt$$
$$= -\int_1^b \frac{1}{t}\, dt = -\ln b.$$

Thus $\ln(a/b) = \ln a - \ln b$.

For part (5) first assume that $n \in \mathbb{N}$. The case $n = 1$ is clear. For the inductive step we apply (3) to see that

$$\ln(a^{n+1}) = \ln(a^n \cdot a) = \ln(a^n) + \ln(a) = n\ln a + \ln a = (n+1)\ln a.$$

Now if $-n \in \mathbb{N}$, using (4) and the part we have shown,

$$\ln a^n = \ln(1/a^{-n}) = \ln 1 - \ln a^{-n} = -(-n)\ln a = n\ln a.$$

For part (6) we first find a number a such that $\ln a > 1$. (There are several ways to do this. For example, we know that $\ln 2 > 0$, so choose $k \in \mathbb{N}$ such that $k > 1/\ln 2$ and set $a = 2^k$. Then $\ln a = k\ln 2 > 1$.) Then

$$\lim_{n \to \infty} \ln a^n = \lim_{n \to \infty} n\ln a = \infty.$$

It follows that for every $A > 0$ there exists $M \in \mathbb{N}$ such that $n\ln a > A$ whenever $n \geq M$. Since the logarithm is an increasing function, whenever $x > a^M$, we have that $\ln x > M\ln a > A$. Therefore, $\lim_{x \to \infty} \ln x = \infty$. For the last part we note that $\ln x = -\ln(1/x)$, so $\lim_{x \to 0} \ln x = -\infty$. $\square$

Exercises: The Fundamental Theorem of Calculus

6.2.1 Let $f : [a,b] \to \mathbb{R}$ be a continuous function. Show that there exists $c \in (a,b)$ such that
$$f(c) = \frac{1}{b-a} \int_a^b f.$$

6.2.2 Let $f : [a,b] \to \mathbb{R}$ be a continuous function. For $x \in [a,b]$, let $F(x) = \int_x^b f$. Prove that for each $x \in (a,b)$, $F(x)$ is differentiable and
$$D_x \int_x^b f(t)\, dt = -f(x).$$

6.2.3 Find
$$D_x \int_{\sin x^2}^{2+\cos^3 x} \sqrt{t^3 + 1}\, dt.$$

6.2.4 Prove that $\ln : (0, \infty) \to \mathbb{R}$ is a bijection.

6.2.5 Complete the proof of part (3) in Proposition 6.2.7.

6.2.6 *Taylor's theorem with integral remainder*: Let $f : (a, b) \to \mathbb{R}$ be $n + 1$ times differentiable, for $n \in \mathbb{N}_0$. Let $c \in (a, b)$. Show that for every $x \in (a, b)$, we can write $f(x) = P_n(x) + R_n(x)$, where $P_n(x)$ is the nth degree Taylor polynomial of f centered at c, and $R_n(x)$, called the **integral form of the remainder**, is given by

$$R_n(x) = \frac{1}{n!} \int_c^x f^{(n+1)}(t)(x-t)^n \, dt.$$

(*Hint*: Use induction and the Fundamental Theorem of Calculus.)

6.3. Improper Riemann Integrals

We have seen that if a function f is Riemann integrable over an interval $[a, b]$, then the function f must be bounded over $[a, b]$. This means that a function such as $\frac{1}{x^2}$ on $(0, 1]$ cannot be extended to a Riemann integrable function on $[0, 1]$. (If there existed a function $f : [0, 1] \to \mathbb{R}$ that is Riemann integrable with $f(x) = \frac{1}{x^2}$ on $(0, 1]$, then $\frac{1}{x^2}$ on $(0, 1]$ would be bounded, a contradiction.) Also, sometimes we may be interested in integrating a function such as $\frac{1}{\sqrt{x}}$ over an infinite interval like $[1, \infty)$. We are interested in extending the notion of Riemann integration to these classes of functions.

Definition 6.3.1. Let f be a function defined on an interval $(a, b]$ and such that f is Riemann integrable on $[a+\varepsilon, b]$ for all $\varepsilon > 0$. We define the **improper Riemann integral** of f over $[a, b]$ by

$$\int_a^b f = \lim_{\varepsilon \to 0^+} \int_{a+\varepsilon}^b f. \qquad \diamond$$

In the definition of the improper Riemann integral here we are considering the one-sided limit as ε approaches 0 from the right. If the function f is already Riemann integrable on $[a, b]$, then this definition agrees with the usual definition of the Riemann integral (Exercise 6.3.1). We also note that we do not introduce a new notation; the fact that it is improper should be recognized from the context.

Definition 6.3.2. Let f be defined on $[a, b)$ and such that f is Riemann integrable on $[a, b-\varepsilon]$ for all $\varepsilon > 0$. Define the **improper Riemann integral** of f over $[a, b]$ by

$$\int_a^b f = \lim_{\varepsilon \to 0^+} \int_a^{b-\varepsilon} f. \qquad \diamond$$

Example 6.3.3. The functions $1/x^2$ and $1/\sqrt{x}$ are unbounded on $(0, 1]$, so they do not have an extension to a Riemann integrable function on $[0, 1]$. We find the improper Riemann integrals of $1/x^2$ and $1/\sqrt{x}$ over $[0, 1]$. We first observe that for all $\varepsilon > 0$, the functions $1/x^2$ and $1/\sqrt{x}$ are continuous on $[\varepsilon, 1]$, so they are integrable on this interval. Then we calculate

$$\int_0^1 \frac{1}{x^2} \, dx = \lim_{\varepsilon \to 0^+} \int_\varepsilon^1 \frac{1}{x^2} \, dx.$$

We note that $D_x(-\frac{1}{x}) = \frac{1}{x^2}$, so by Theorem 6.2.3,
$$\lim_{\varepsilon \to 0^+} \int_\varepsilon^1 \frac{1}{x^2}\,dx = \lim_{\varepsilon \to 0^+} -\frac{1}{1} + \frac{1}{\varepsilon} = +\infty.$$
In this case we say that the improper Riemann integral of $1/x^2$ over $[0,1]$ diverges.

For the second function, again using Theorem 6.2.3 and that $D_x(2\sqrt{x}) = \frac{1}{\sqrt{x}}$,
$$\int_0^1 \frac{1}{\sqrt{x}}\,dx = \lim_{\varepsilon \to 0^+} \int_\varepsilon^1 \frac{1}{\sqrt{x}}\,dx = \lim_{\varepsilon \to 0^+} 2\sqrt{1} - 2\sqrt{\varepsilon} = 2.$$
We say that this function has an improper Riemann integral.

The second type of improper integrals we consider is when the domain is an infinite interval.

Definition 6.3.4. Let f be defined on $[a, \infty)$, and suppose it is Riemann integrable on $[a, b]$ for all $b > a$. Then we define
$$\int_a^\infty f = \lim_{t \to \infty} \int_a^t f.$$
When the limit exists and is finite, we say the integral **converges** or that the function has an improper Riemann integral. Similarly one defines $\int_{-\infty}^b f$. ◊

Definition 6.3.5. Let f be defined on $[a, c)$ and $(c, b]$, where $a < c < b$, and suppose it is Riemann integrable on $[a, c - \varepsilon]$ and $[c + \varepsilon, b]$ for all $\varepsilon > 0$. Then we define
$$\int_a^b f = \lim_{\varepsilon \to 0^+} \int_a^{c-\varepsilon} f. + \lim_{\varepsilon \to 0^+} \int_{c+\varepsilon}^b f.$$
When the limit exists and is finite, we say the integral **converges** or that the function has an improper Riemann integral. ◊

Example 6.3.6. We find the improper Riemann integrals of $1/x^2$ and $1/\sqrt{x}$ over the interval $[1, \infty)$. We first observe that both the functions $1/x^2$ and $1/\sqrt{x}$ are continuous on $[1, \infty)$, so they are Riemann integrable on the interval $[1, t]$ for all $t > 1$. Then we calculate
$$\int_1^\infty \frac{1}{x^2}\,dx = \lim_{t \to \infty} \int_1^t \frac{1}{x^2}\,dx$$
$$= \lim_{t \to \infty} -\frac{1}{t} + 1$$
$$= 1.$$
We say this integral converges to 1. For the second function, again using Theorem 6.2.3,
$$\int_1^\infty \frac{1}{\sqrt{x}}\,dx = \lim_{t \to \infty} \int_1^t \frac{1}{\sqrt{x}}\,dx$$
$$= \lim_{t \to \infty} 2\sqrt{t} - 2\sqrt{1}$$
$$= \infty.$$
In this case we say this integral diverges.

6.3. Improper Riemann Integrals

Now we consider an integral that combines the two previous cases, for example, an integral such as $\int_0^\infty \frac{1}{x^2}\, dx$.

Definition 6.3.7. Suppose f is defined on (a, ∞) and is Riemann integrable on $[a+\varepsilon, b]$ for all $\varepsilon > 0$ and $b > a$. Then we define the **improper Riemann integral** of f over $[a, \infty)$ by

$$\int_a^\infty f = \lim_{\varepsilon \to 0^+} \int_{a+\varepsilon}^{a+1} f + \lim_{t \to \infty} \int_{a+1}^{t} f.$$

We say the integral **converges** if both limits exist and are finite. ◇

Exercise 6.3.2 shows a couple of ways of evaluating this.

Finally, we define the improper integral over the whole real line.

Definition 6.3.8.
$$\int_{-\infty}^\infty f = \int_{-\infty}^0 f + \int_0^\infty f,$$
for a function f that is integrable on $[a, b]$ for all $a < 0$ and $b > 0$. ◇

Exercises: Improper Riemann Integrals

6.3.1 Let $f : [a, b] \to \mathbb{R}$ be a Riemann integrable function. Show
$$\int_a^b f = \lim_{\varepsilon \to 0^+} \int_{a+\varepsilon}^b f = \lim_{\varepsilon \to 0^+} \int_a^{b-\varepsilon} f.$$

6.3.2 Suppose that $\int_a^\infty f$ converges. Show that it can also be evaluated by
$$\lim_{\varepsilon \to 0^+} \lim_{t \to \infty} \int_{a-\varepsilon}^t f$$
and by
$$\lim_{t \to \infty} \lim_{\varepsilon \to 0^+} \int_{a-\varepsilon}^t f.$$

6.3.3 Evaluate $\int_{-1}^1 \frac{1}{t^2}\, dt$.

6.3.4 Evaluate $\int_0^1 \frac{1}{x^p}\, dx$ for $p > 0$.

6.3.5 Evaluate $\int_1^\infty \frac{1}{x^p}\, dx$ for $p > 0$.

6.3.6 Evaluate $\int_0^\infty \frac{\sin x}{x}\, dx$. Show that $\int_0^\infty \left|\frac{\sin x}{x}\right|\, dx$ does not converge.

Chapter 7

Series

7.1. Series of Real Numbers

Series are infinite sums. Addition, however, is defined only for finitely many terms, so we need to specify what we mean by an infinite sum. This is not a trivial question, and historically there have been several ways to approach it. The way we understand a series now is as a sequence of finite sums, and we understand the sum of the series as the limit of the sequence of finite sums.

Definition 7.1.1. A **series** (of real numbers) is an expression of the form

(7.1) $$\sum_{n=1}^{\infty} a_n,$$

where $a_n \in \mathbb{R}$ for $n \in \mathbb{N}$. Its **sequence of partial sums** is the sequence

$$\left(\sum_{i=1}^{n} a_i \right)_{n \in \mathbb{N}}.$$

The expression in (7.1) also stands for the limit of this sequence of partial sums. ◊

Given a series $\sum_{n=1}^{\infty} a_n$, we often denote its sequence of partial sums by (S_n), where $S_n = \sum_{i=1}^{n} a_i$ for $n \in \mathbb{N}$. Note that the sequence (S_n) is built from the sequence (a_n) but is different from it.

Definition 7.1.2. We say that the series $\sum_{n=1}^{\infty} a_n$ **converges** to a real number S if its sequence of partial sums (S_n) converges to S, i.e., $\lim_{n \to \infty} S_n = S$. In this case we write

$$\sum_{n=1}^{\infty} a_n = S.$$

We may also say that S is the **sum** of the series. ◊

Definition 7.1.3. If the sequence (S_n) does not converge, we say that the series **diverges**. In the case when $\lim_{n \to \infty} S_n = \infty$, we say that the series diverges to infinity and write $\sum_{n=1}^{\infty} a_n = \infty$. Similarly, we write $\sum_{n=1}^{\infty} a_n = -\infty$ when $\lim_{n \to \infty} S_n = -\infty$. ◊

Diverging to $\pm\infty$ is a special case of divergence, but the series may diverge by simply having the limit of the sequence of partial sums not exist.

Example 7.1.4. For our first example we consider the series

(7.2) $$\sum_{n=1}^{\infty} (-1)^{n+1}.$$

Its sequence of partial sums is

$$S_n = \sum_{i=1}^{n} (-1)^{i+1}.$$

To determine whether the sequence (S_n) converges, we first consider the case when n is even. Then we can group two terms at a time to obtain

$$S_n = (1-1) + (1-1) + \cdots = 0 + 0 + \cdots = 0.$$

Now if n is odd, grouping the terms two at a time after the first term, we obtain

$$S_n = 1 + (-1+1) + (-1+1) + \cdots = 1 + 0 + \cdots = 1.$$

Therefore, the sequence (S_n) does not converge, which means that the series diverges.

Remark 7.1.5. The series in Example 7.1.4 represents the *infinite sum*

(7.3) $$1 - 1 + 1 - 1 + \cdots.$$

The discussion of the partial sums shows that the way terms are grouped matters in the case of infinite sums, while we know that for finite sums the grouping does not make a difference.

Example 7.1.6. Now consider the infinite sum of a constant number such as 1:

$$1 + 1 + 1 + \cdots.$$

Here the series is $\sum_{n=1}^{\infty} 1$ and its sequence of partial sums is given by

$$S_n = \sum_{i=1}^{n} 1 = n.$$

So $\lim_{n \to \infty} S_n = \infty$, and the series diverges to ∞.

Example 7.1.7. We consider the following series and show it converges.

$$\sum_{n=1}^{\infty} \frac{1}{n(n+1)}.$$

We want to study its sequence of partial sums, which is given by

$$S_n = \sum_{i=1}^{n} \frac{1}{i(i+1)}.$$

In this case we note that

$$\frac{1}{n(n+1)} = \frac{1}{n} - \frac{1}{n+1}.$$

7.1. Series of Real Numbers

Then we can write

$$S_n = \frac{1}{1 \cdot 2} + \frac{2}{2 \cdot 3} + \frac{1}{3 \cdot 4} + \cdots + \frac{1}{n \cdot (n+1)}$$

$$= \frac{1}{1} - \frac{1}{2} + \frac{1}{2} - \frac{1}{3} + \frac{1}{3} - \frac{1}{4} + \cdots + \frac{1}{n} - \frac{1}{n+1}$$

$$= 1 - \frac{1}{n+1}.$$

When we see this sort of cancellation in a sum, we will say it is a *telescoping sum*; it is very particular to the algebra of the problem. Then $\lim_{n \to \infty} S_n = 1$. So the series converges to 1.

Example 7.1.8. Consider the series

$$\sum_{n=1}^{\infty} \left(\frac{1}{2}\right)^{n-1} = 1 + \frac{1}{2} + \frac{1}{2^2} + \cdots + \frac{1}{2^i} + \cdots.$$

The partial sums are given by

$$S_n = 1 + \frac{1}{2} + \frac{1}{2^2} + \cdots + \frac{1}{2^{n-1}}.$$

By Exercise 1.1.12 we have

$$S_n = 2 - \frac{1}{2^{n-1}}.$$

It follows that

$$\lim_{n \to \infty} S_n = 2.$$

Therefore, this series converges to 2, and we write

$$\sum_{n=1}^{\infty} \frac{1}{2^{n-1}} = 2.$$

We may also write this series as $\sum_{n=0}^{\infty} \frac{1}{2^n}$. This is a special case of the geometric series, covered in Lemma 7.1.10.

The following proposition states some basic properties of series.

Proposition 7.1.9. *Suppose $\sum_{n=1}^{\infty} a_n$ and $\sum_{n=1}^{\infty} b_n$ are convergent series.*

(1) *For each $c \in \mathbb{R}$, the series $\sum_{n=1}^{\infty} c a_n$ converges and*

$$\sum_{n=1}^{\infty} c a_n = c \sum_{n=1}^{\infty} a_n.$$

(2) *The series $\sum_{n=1}^{\infty} (a_n + b_n)$ converges and*

$$\sum_{n=1}^{\infty} (a_n + b_n) = \sum_{n=1}^{\infty} a_n + \sum_{n=1}^{\infty} b_n.$$

(3) *For each $k \in \mathbb{N}, k \geq 2$, the series $\sum_{n=k}^{\infty} a_n$ converges and*

$$\sum_{n=k}^{\infty} a_n = \sum_{n=1}^{\infty} a_n - \sum_{n=1}^{k-1} a_n.$$

Proof. We prove part (2); the other parts are similar and are left to the reader. Let $S_n = \sum_{i=1}^n a_n$, $T_n = \sum_{i=1}^n b_n$, and $Q_n = \sum_{i=1}^n (a_n + b_n)$. By assumption the limits $\lim_{n \to \infty} S_n$ and $\lim_{n \to \infty} T_n$ exist; denote them by S and T, respectively. Clearly, $Q_n = S_n + T_n$. Then $\lim_{n \to \infty} Q_n$ exists and equals $S + T$. □

The following series is very useful and is an example where we can easily calculate the sum when it converges. We have seen this mainly in the context of limits of finite sums in Subsection 2.2.5, but we cover all the details again for completeness.

Lemma 7.1.10 (Geometric series). *Let $r \in \mathbb{R}$. Then the series*

$$\sum_{n=1}^{\infty} r^{n-1} = \begin{cases} \frac{1}{1-r} & \text{when } |r| < 1, \\ \text{diverges to } \infty & \text{when } r \geq 1, \\ \text{diverges} & \text{when } r \leq -1. \end{cases}$$

Proof. In Exercise 1.1.12 the reader was asked to prove the equality in (7.4) by induction. Here we give a different argument. We show that

(7.4) $$\sum_{i=1}^n r^{i-1} = \frac{1 - r^n}{1 - r}.$$

We calculate the product

$$(1 - r) \sum_{i=1}^n r^{i-1} = \sum_{i=1}^n r^{i-1} - \sum_{i=1}^n r^i$$
$$= \sum_{i=1}^n r^{i-1} - \sum_{i=2}^{n+1} r^{i-1}$$
$$= 1 - r^n.$$

This completes the proof of (7.4). Thus the partial sums are given by

$$S_n = \frac{1 - r^n}{1 - r}.$$

When $|r| < 1$, $\lim_{n \to \infty} r^n = 0$, so $\lim_{n \to \infty} S_n = 1/(1-r)$. When $r > 1$, $\lim_{n \to \infty} r^n = \infty$, so $\lim_{n \to \infty} S_n = \infty$. The case when $r = 1$ has already been considered in Example 7.1.6. Finally, when $r \leq -1$, the sequence (S_n) does not converge as r^n oscillates between being positive and negative, so the series diverges. □

Remark 7.1.11. The series $\sum_{n=1}^{\infty} r^{n-1}$ can also be written as $\sum_{n=0}^{\infty} r^n$; it will be useful to rewrite series in different ways. Also, from Proposition 7.1.9 we obtain another useful formula for the geometric series:

$$\sum_{n=1}^{\infty} r^n = \frac{1}{1-r} - 1 = \frac{r}{1-r}, \quad \text{when } |r| < 1.$$

7.1. Series of Real Numbers

7.1.1. Convergence Tests. We explore a necessary condition for convergence, which gives rise to the first divergence test. Proposition 7.1.12 is the simplest test for showing that a series does not converge.

Proposition 7.1.12 (First divergence test). *Suppose that $\lim_{n \to \infty} a_n$ does not exist or, when it exists, that*
$$\lim_{n \to \infty} a_n \neq 0.$$
Then the series $\sum_{n=1}^{\infty} a_n$ does not converge.

Proof. We prove the contrapositive, i.e., we prove that if $\sum_{n=1}^{\infty} a_n$ converges, then $\lim_{n \to \infty} a_n$ exists and $\lim_{n \to \infty} a_n = 0$. So assume that the series converges. Then its sequence of partial sums (S_n) converges to a number S. As $\{S_{n-1}\}_{n \geq 2}$ is a subsequence, it also converges to S. Since $a_n = S_n - S_{n-1}$, we have
$$\begin{aligned}\lim_{n \to \infty} a_n &= \lim_{n \to \infty} (S_n - S_{n-1}) \\ &= \lim_{n \to \infty} S_n - \lim_{n \to \infty} S_{n-1} \\ &= S - S \\ &= 0.\end{aligned}$$
$\square$

We note that the converse of Proposition 7.1.12 does not hold. For example, we shall see in Example 7.1.21 that the series $\sum_{n=1}^{\infty} 1/n$ does not converge even though $a_n = 1/n \to 0$. The first divergence test does not apply for this series; so if all we know for a series is that $a_n \to 0$, we cannot conclude that it converges.

Example 7.1.13. The series
$$\sum_{n=1}^{\infty} \frac{n^2 + 1}{5n^2 + n}$$
diverges. This follows from the first divergence test as
$$\lim_{n \to \infty} \frac{n^2 + 1}{5n^2 + n} = \frac{1}{5} \neq 0.$$

Question 7.1.14. Define what it means for an infinite product $\prod_{n=1}^{\infty} a_n$ to converge. Formulate and prove a necessary condition for convergence of infinite products similar to the first divergence test.

Proposition 7.1.15 (Comparison test). *Let a_n and b_n be nonnegative sequences such that*
$$a_n \leq b_n \text{ for all } n \in \mathbb{N}.$$
(1) *If $\sum_{n=1}^{\infty} b_n$ converges, then $\sum_{n=1}^{\infty} a_n$ converges.*
(2) *If $\sum_{n=1}^{\infty} a_n$ diverges, then $\sum_{n=1}^{\infty} b_n$ diverges.*

Proof. Suppose the series $\sum_{n=1}^{\infty} b_n$ converges to S. We have that $\sum_{i=1}^{n} b_i \leq S$, for all $n \in \mathbb{N}$, since the terms of the series are nonnegative. Then

(7.5) $$\sum_{i=1}^{n} a_i \leq \sum_{i=1}^{n} b_i \leq S.$$

We note that the sequence $(\sum_{i=1}^{n} a_i)$ is monotone in n (it consists of sums of nonnegative terms), and (7.5) shows that it is bounded. The monotone sequence theorem implies that the sequence of partial sums, $\sum_{i=1}^{n} a_i$, converges, so its series converges.

Now suppose $\sum_{n=1}^{\infty} a_n$ diverges. As the terms are nonnegative, it must diverge to infinity. Since $\sum_{i=1}^{n} a_i \leq \sum_{i=1}^{n} b_i$, it follows that $\sum_{i=1}^{n} b_i$ diverges to infinity. □

Example 7.1.16 uses the comparison test.

Example 7.1.16. Consider the two series
$$\sum_{n=1}^{\infty} \frac{1}{n3^n} \quad \text{and} \quad \sum_{n=1}^{\infty} \frac{n^2}{3^n}.$$

For the first series note that
$$\frac{1}{n3^n} \leq \frac{1}{3^n} \text{ for all } n \in \mathbb{N}.$$

Since $\sum_{n=1}^{\infty} \frac{1}{3^n}$ is a geometric series that converges, the comparison test implies that $\sum_{n=1}^{\infty} \frac{1}{n3^n}$ converges. For the second series we first note that $n^2 < 2^n$ for all $n \geq 4$ (this can be seen by induction; see Exercise 0.5.7). Then
$$\frac{n^2}{3^n} \leq \frac{2^n}{3^n} = \left(\frac{2}{3}\right)^n \text{ for all } n \geq 4.$$

Again, as $\sum_{n=4}^{\infty} \left(\frac{2}{3}\right)^n$ is a converging geometric series, we obtain that $\sum_{n=4}^{\infty} \frac{n^2}{3^n}$ converges by the comparison test, so $\sum_{n=1}^{\infty} \frac{n^2}{3^n}$ converges.

Question 7.1.17. Let (a_n) be a sequence of nonnegative terms. Suppose that the series $\sum_{n=1}^{\infty} a_n$ converges. Does the series $\sum_{n=1}^{\infty} a_{2^n}$ converge? Suppose further that the sequence (a_n) is monotone decreasing. Show that $\sum_{n=1}^{\infty} 2^n a_{2^n}$ converges.

The following is another version of a comparison test that is useful in cases when Proposition 7.1.15 does not readily apply.

Proposition 7.1.18 (Limit comparison test)**.** *Let a_n and b_n be nonnegative sequences with $b_n > 0$ and such that the following limit exists:*
$$L = \lim_{n \to \infty} \frac{a_n}{b_n}.$$
If $0 < L < \infty$, then $\sum_{n=1}^{\infty} b_n$ converges if and only if $\sum_{n=1}^{\infty} a_n$ converges.

Proof. Let $\varepsilon = \frac{L}{2}$. There exists an integer N so that $|L - \frac{a_n}{b_n}| < L/2$ for all $n \geq N$. Then
$$L - \frac{L}{2} < \frac{a_n}{b_n} < L + \frac{L}{2},$$
$$\frac{L}{2} < \frac{a_n}{b_n} < \frac{3L}{2}.$$

If $\sum_{n=1}^{\infty} b_n$ converges since
$$a_n < \frac{3L}{2} \cdot b_n$$
by the comparison test, it follows that $\sum_{n=1}^{\infty} a_n$ converges. Similarly, if $\sum_{n=1}^{\infty} a_n$ converges, we use that
$$b_n < \frac{2}{L} \cdot a_n$$

to deduce that $\sum_{n=1}^{\infty} b_n$ converges. □

Example 7.1.19. We show that the series
$$\sum_{n=1}^{\infty} \frac{n^2}{3^n - 2n^2}$$
converges. Set $a_n = \frac{n^2}{3^n - 2n^2}$. We compare it with $b_n = \frac{n^2}{3^n}$. In this case we have
$$\lim_{n \to \infty} \frac{a_n}{b_n} = \lim_{n \to \infty} \frac{n^2}{3^n - 2n^2} \cdot \frac{3^n}{n^2} = 1.$$
Now we show that $\sum_{n=1}^{\infty} b_n$ converges. For this we compare it to a geometric series. We note that
$$\frac{n^2}{3^n} \leq \frac{2^n}{3^n} \text{ for all } n \geq 4.$$
Again, as $\sum_{n=1}^{\infty} (\frac{2}{3})^n$ is a converging geometric series, we obtain that our series converges by the comparison test.

The next test provides us with an important class of convergent and divergent series.

Proposition 7.1.20 (The integral test). *Let $\sum_{n=1}^{\infty} a_n$ be a series of positive terms. Let $f : [1, \infty) \to (0, \infty)$ be a continuous function such that*

(1) f *is decreasing;*

(2) $f(n) = a_n$ *for all $n \in \mathbb{N}$.*

Then the series $\sum_{n=1}^{\infty} a_n$ converges if and only if the sequence $(\int_1^n f)_{n=1}^{\infty}$ converges.

Proof. We first note that the numbers a_i can be interpreted as the area of the rectangle with height a_i and base $[i, i+1]$, and as f is decreasing, this area is an upper bound for the area under f over $[i, i+1]$. Thus
$$a_i \geq \int_i^{i+1} f.$$
Next we interpret a_{i+1} as the area of the rectangle with height a_{i+1} and base $[i, i+1]$. Thus
$$\int_i^{i+1} f \geq a_{i+1};$$
see Figure 7.1.

From these two inequalities we obtain
$$\sum_{i=2}^{n} a_i \leq \int_1^n f \leq \sum_{i=1}^{n-1} a_i.$$
It is clear that the partial sums (S_n) converge if and only if $\{\int_1^n f\}$ converges. □

We recall that as far as the convergence or divergence behavior of a series is concerned, it does not matter if the sum of the series starts at $n = 1$ or at any other integer, so sometimes the integral may start at an integer other than 1.

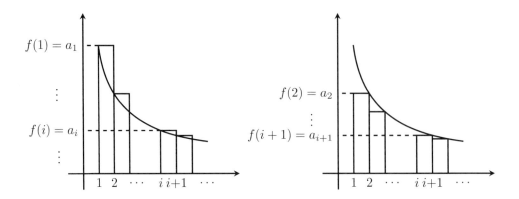

Figure 7.1. Integral test illustration.

Example 7.1.21.

(a) The series $\sum_{n=2}^{\infty} \frac{1}{n \ln^2 n}$ converges. We see this from the integral test by choosing $f(x) = \frac{1}{x \ln^2 x}$. It is clear that f is continuous and decreasing as $x \ln^2 x$ is increasing. Also, $f(n) = \frac{1}{n \ln^2 n}$. By the integral test we need to evaluate

$$\int_2^n \frac{1}{x \ln^2 x}\, dx.$$

To apply Theorem 6.2.6, we observe that if $h(x) = 1/x^2$ and $g(x) = \ln x$, then $h(g(x))g'(x) = \frac{1}{x \ln^2 x}$. Then

$$\int_2^n \frac{1}{x \ln^2 x}\, dx = \int_2^{\ln n} \frac{1}{x^2}\, dx = -\frac{1}{\ln n} + \frac{1}{2}.$$

As $\{\int_2^n 1/x \ln^2 x\}$ clearly converges, it follows that the series converges.

(b) The series $\sum_{n=1}^{\infty} 1/n$ diverges. We can see this from the integral test by choosing $f(x) = 1/x$. But we have a more direct method that does not use the integral test. Observe that

$$\frac{1}{3} + \frac{1}{4} > \frac{1}{2}$$

$$\frac{1}{5} + \frac{1}{6} + \frac{1}{7} + \frac{1}{8} > \frac{1}{2}$$

$$\vdots$$

$$\frac{1}{2^i + 1} + \cdots + \frac{1}{2^i + 2^i} > \frac{1}{2}.$$

Based on this, we make the claim

(7.6) $\qquad S_{2^{i+1}} > \frac{1}{2}(i+1)$ for all $i \in \mathbb{N}$.

7.1. Series of Real Numbers

We prove (7.6) by induction. The base case $i = 1$ is clear. Assume the inequality holds for i. Now

$$S_{2^{i+1}} = S_{2^i} + \frac{1}{2^i + 1} + \frac{1}{2^i + 2} + \cdots + \frac{1}{2^i + 2^i}$$
$$> \frac{1}{2} i + \frac{1}{2} = \frac{1}{2}(i+1).$$

It follows that the sequence of partial sums diverges, so $\sum_{n=1}^{\infty} 1/n$ diverges. (This proof goes back to Nicole Oresme in the 14th century; there are now many proofs of the divergence of this series.) As we have mentioned, this example shows that the converse of Proposition 7.1.12 does not hold.

Definition 7.1.22. The series $\sum_{n=1}^{\infty} 1/n$ is called the **harmonic series**. (The name comes from the harmonic series of sounds in music.) ◇

Exercises: Series of Real Numbers

7.1.1 Complete the proof of Proposition 7.1.9.

7.1.2 Determine the sum of the following series.
   (a) $\sum_{n=1}^{\infty} \frac{1}{n(n+5)}$
   (b) $\sum_{n=1}^{\infty} \frac{1}{n^2 - 4}$

7.1.3 Let $p \in \mathbb{Q}$. Show that for $p > 1$, the following series converges, and that it diverges for $p < 1$.
$$\sum_{n=1}^{\infty} \frac{1}{n^p}$$

7.1.4 Show that the following series converge and find their sum.
   (a)
   $$-\frac{5^2}{7} + \frac{5^3}{7^2} - \frac{5^4}{7^3} + \cdots$$
   (b)
   $$\frac{2^3}{3^3} + \frac{2^4}{3^4} + \frac{2^5}{3^5} + \cdots$$

7.1.5 Show that the following series converge and find their sum. (*Hint*: Use the geometric series.)
   (a) $\sum_{n=1}^{\infty} \frac{n}{2^n}$
   (b) $\sum_{n=1}^{\infty} \frac{n^2}{2^n}$

7.1.6 Determine whether the following series converge or diverge.
   (a) $\sum_{n=2}^{\infty} \frac{5}{n-1}$
   (b) $\sum_{n=2}^{\infty} \frac{1}{3n^2 + 2}$
   (c) $\sum_{n=1}^{\infty} \frac{5n^2 - 3n + 2}{n^2 - n}$
   (d) $\sum_{n=1}^{\infty} \frac{1}{n^{1/3} + 1}$
   (e) $\sum_{n=2}^{\infty} \frac{1}{n^{4/3} + 5n}$
   (f) $\sum_{n=2}^{\infty} \frac{1}{n \ln n}$

(g) $\sum_{n=1}^{\infty} \frac{n^2+5n+2}{5n^3+n}$

(h) $\sum_{n=1}^{\infty} \frac{n^3}{4^n}$

(i) $\sum_{n=1}^{\infty} \frac{n^3}{4^n+2n-1}$

(j) $\sum_{n=1}^{\infty} \frac{1}{n \ln^2 n}$

(k) $\sum_{n=1}^{\infty} \frac{n^2+2n}{n^5+n}$

7.1.7 Give examples to show that if, in the limit comparison test, $L = 0$ or $L = \infty$, the test is inconclusive.

7.1.8 Let (a_n) be a sequence such that $a_n \geq 0, a_{n+1} \leq a_n$, and $\lim_{n\to\infty} a_n = 0$. Prove that $\sum_{n=1}^{\infty} a_n$ converges if and only if $\sum_{n=1}^{\infty} 2^n a_{2^n}$ converges. (One direction is done in Question 7.1.17. This test is named after Cauchy.)

7.1.9 Use Exercise 7.1.8 to show $\sum_n \frac{1}{n^2}$ converges (without using the integral test).

7.1.10 Let $\sum_{n=1}^{\infty} a_n$ be a series that converges to S and satisfies the conditions of the integral test, and let f be the corresponding function. Let $S_n = \sum_{i=1}^{n} a_i$. Prove that $S - S_n \leq \int_n^{\infty} f(x)dx$. Use this to estimate $\sum_{n=1}^{\infty} 1/n^2$ to two decimal places.

7.1.11 Using your definition in Question 7.1.14, determine whether the following products converge or diverge:

$$\prod_{n=2}^{\infty}\left(1 - \frac{1}{n}\right) \quad \text{and} \quad \prod_{n=2}^{\infty}\left(1 - \frac{1}{n^2}\right).$$

(*Hint*: Use logarithms and the limit comparison test.)

7.2. Alternating Series and Absolute Convergence

In Section 7.1 we have mainly considered series of nonnegative terms. While the definition allowed the terms a_n to be any real number, the basic comparison test and the integral test assumed the terms of the series to be nonnegative. If all the terms of the series are negative, one could just consider the additive inverse of these terms and thus obtain a series of nonnegative terms. One example where we had terms that were both negative as well as positive was the case of the geometric series when r was a negative number. This example is a special case of an alternating series, which we now define.

Definition 7.2.1. An **alternating series** is a series of the form

$$\sum_{n=1}^{\infty}(-1)^{n+1}a_n = a_1 - a_2 + \cdots, \quad \text{where } a_n > 0. \qquad \diamond$$

Example 7.2.2. We show that the following alternating series converges.

$$\sum_{n=1}^{\infty}(-1)^{n+1}\frac{1}{n} = 1 - \frac{1}{2} + \frac{1}{3} - \frac{1}{4} + \cdots$$

7.2. Alternating Series and Absolute Convergence

First we consider the even partial sums

$$S_{2k} = (1 - \frac{1}{2}) + \cdots + (\frac{1}{2k-1} - \frac{1}{2k}) = \sum_{i=1}^{k}(\frac{1}{i} - \frac{1}{i+1}).$$

We claim that the sequence (S_{2k}) is bounded and increasing. To see that it is bounded we write, for example, S_6 as

$$S_{2 \cdot 3} = 1 + (-\frac{1}{2} + \frac{1}{3}) + (-\frac{1}{4} + \frac{1}{5}) - \frac{1}{6}.$$

As each term in parenthesis is negative, we see that $S_{2 \cdot 3}$ is bounded above by 1. The same idea works for any S_{2k}. To see that it is increasing write, for example,

$$S_{2 \cdot 2} = (1 - \frac{1}{2}) + (\frac{1}{3} - \frac{1}{4}),$$
$$S_{2 \cdot 3} = (1 - \frac{1}{2}) + (\frac{1}{3} - \frac{1}{4}) + (\frac{1}{5} - \frac{1}{6}).$$

As each term in parenthesis is positive, and the same argument works for S_{2k}, we see that (S_{2k}) is increasing. Then by the monotone sequence theorem (Theorem 2.2.1), (S_{2k}) converges, say, to a number $S \in \mathbb{R}$.

To find the limit of the odd partial sums S_{2k+1}, we note

$$S_{2k+1} = S_{2k} + \frac{1}{2k+1}.$$

From this it follows that (S_{2k+1}) also converges to S as $\frac{1}{2k+1}$ converges to 0. By Exercise 2.2.14 it follows that the sequence (S_n) converges to S. Therefore, the series converges to S.

Similar ideas can be used to prove the following theorem, though we outline a different proof using the strong nested intervals theorem instead of the monotone sequence theorem.

Theorem 7.2.3 (Alternating series test). *Let (a_n) be a sequence such that*

$$a_n > 0 \text{ and } a_n > a_{n+1} \text{ for all } n \in \mathbb{N}.$$

The series

$$\sum_{n=1}^{\infty}(-1)^n a_n$$

converges if and only if

$$\lim_{n \to \infty} a_n = 0.$$

Proof. Suppose $\lim_{n \to \infty} a_n = 0$. We will construct a decreasing sequence of nested closed intervals $I_n = [x_n, y_n]$ whose intersection is a single point that will be the sum of the series. Let $y_1 = a_1$ and $x_1 = a_1 - a_2$. Since $0 < a_2 < a_1$, we have that $x_1 < y_1$. Next let $y_2 = x_1 + a_3$ and $x_2 = y_2 - a_4$. Since $0 < a_3 < a_2$, we have $x_1 < y_2 < y_1$, and since $0 < a_4 < a_3$, we have $x_1 < x_2 < y_2$. Continue in this way to set

$$y_n = S_{2n-1}, x_n = S_{2n}, I_n = [x_n, y_n].$$

The intervals I_n form a nested sequence of closed intervals, and since $|I_n| = a_{2n+1}$, they intersect to a single point that we denote by S. Also, $S_{2n} \in I_n$ so $|S - S_{2n}| \leq a_{2n+1}$. It follows that S_{2n} converges to S. As $S_{2n+1} = S_{2n} + a_{2n+1}$, an argument similar to the one in Example 7.2.2 completes the proof. $\square$

Example 7.2.4. Consider the series $\sum_{n=1}^{\infty} (-1)^{n+1} \frac{1}{n}$. Here $a_n = \frac{1}{n}$ and clearly $a_n > a_{n+1}$ for all $n \in \mathbb{N}$. Therefore the alternating series test applies and shows that the series converges since $a_n \to 0$. Note that we have already seen that $\sum_{n=1}^{\infty} \frac{1}{n}$ does not converge.

Definition 7.2.5. A series $\sum_{n=1}^{\infty} a_n$ is said to **converge absolutely** if

$$\sum_{n=1}^{\infty} |a_n|$$

converges. $\diamond$

If the terms a_n are nonnegative, then $|a_n| = a_n$, and then convergence and absolute convergence are the same. So this is a new concept only when infinitely many of the terms a_n are negative and infinitely many are positive. Lemma 7.2.6 shows that absolute convergence implies convergence, and Example 7.2.4 shows that the converse of the lemma is not true.

Lemma 7.2.6. *If the series $\sum_{n=1}^{\infty} a_n$ converges absolutely, then it converges.*

Proof. We know that

$$-|a_n| \leq a_n \leq |a_n| \text{ for all } n \in \mathbb{N}.$$

Therefore,

$$0 \leq a_n + |a_n| \leq 2|a_n|.$$

Since $\sum_{n=1}^{\infty} 2|a_n|$ converges, the comparison test implies that $\sum_{n=1}^{\infty} (a_n + |a_n|)$ converges. By Proposition 7.1.9, $\sum_{n=1}^{\infty} a_n$ converges. $\square$

Definition 7.2.7. A series $\sum_{n=1}^{\infty} a_n$ is said to **converge conditionally** if $\sum_{n=1}^{\infty} a_n$ converges but $\sum_{n=1}^{\infty} |a_n|$ does not converge, i.e., the series converges but does not converge absolutely. $\diamond$

Example 7.2.8. We have seen in Example 7.2.4 that the series

(7.7) $$\sum_{n=1}^{\infty} (-1)^{n+1} \frac{1}{n}$$

converges, and that when considered in absolute value, i.e., $\sum_{n=1}^{\infty} |(-1)^{n+1} 1/n|$, we obtain the harmonic series, which we know does not converge. Therefore, the series in (7.7) converges conditionally.

The following test is very useful for absolute convergence.

7.2. Alternating Series and Absolute Convergence

Theorem 7.2.9 (Ratio test). *Let $\sum_{n=1}^{\infty} a_n$ be a series.*

(1) *If*
$$\limsup_{n \to \infty} \left| \frac{a_{n+1}}{a_n} \right| < 1,$$
then the series $\sum_{n=1}^{\infty} a_n$ converges absolutely.

(2) *If*
$$\liminf_{n \to \infty} \left| \frac{a_{n+1}}{a_n} \right| > 1,$$
then the series $\sum_{n=1}^{\infty} a_n$ diverges.

(3) *Otherwise, the test is inconclusive.*

Proof. For part (1), choose ρ so that $\limsup_{n \to \infty} \left| \frac{a_{n+1}}{a_n} \right| < \rho < 1$. Then there exists $N \in \mathbb{N}$ such that if $n \geq N$,
$$\left| \frac{a_{n+1}}{a_n} \right| < \rho.$$
So
$$|a_{n+1}| < \rho |a_n| \text{ for all } n \geq N.$$
Then $|a_{n+2}| < \rho |a_{n+1}| < \rho^2 |a_n|$, and one shows by induction that
$$|a_{N+i}| < \rho^i |a_N| \text{ for all } i \geq 1.$$
Since $\rho < 1$, by the comparison test with the geometric series, the series $\sum_{i=1}^{\infty} |a_{N+i}|$ converges. It follows that $\sum_{n=1}^{\infty} |a_n|$ converges.

Now for part (2), we can choose $\rho > 1$ so that there exists $N \in \mathbb{N}$ such that for all $n \geq N$,
$$\rho < \left| \frac{a_{n+1}}{a_n} \right|.$$
So
$$|a_N| \rho^i < |a_{N+i}| \text{ for all } i \geq 1.$$
As $\rho > 1$, by the comparison test, the series $\sum_{n=1}^{\infty} |a_n|$ diverges.

Finally, we see that the test is inconclusive if either
$$\limsup_{n \to \infty} \left| \frac{a_{n+1}}{a_n} \right| \geq 1 \quad \text{or} \quad \liminf_{n \to \infty} \left| \frac{a_{n+1}}{a_n} \right| \leq 1.$$
If we let $a_n = 1/n$, then
$$\lim_{n \to \infty} \left| \frac{a_{n+1}}{a_n} \right| = \lim_{n \to \infty} \frac{\frac{1}{n+1}}{\frac{1}{n}} = 1,$$
and we note that the series $\sum_{n=1}^{\infty} a_n$ diverges. Now if we let $a_n = 1/n^2$, we also have
$$\lim_{n \to \infty} \left| \frac{a_{n+1}}{a_n} \right| = \lim_{n \to \infty} \frac{\frac{1}{(n+1)^2}}{\frac{1}{n^2}} = 1,$$
and in this case the series $\sum_{n=1}^{\infty} a_n$ converges. $\square$

Example 7.2.10. The series
$$\sum_{n=1}^{\infty}(-1)^n\frac{n^5}{2^n}$$
converges absolutely by the ratio test as
$$\rho=\lim_{n\to\infty}\left|\frac{(n+1)^5 2^n}{n^5 2^{n+1}}\right|=\frac{1}{2}.$$

The following test is also useful in many cases, in particular when the ratio test does not apply. The reader will note the proof is similar to the proof for the ratio test, though it is possible to obtain the ratio test from the root test using some inequalities.

Theorem 7.2.11 (Root test). *Let $\sum_{n=1}^{\infty} a_n$ be a series.*

(1) *If*
$$\limsup_{n\to\infty}|a_n|^{1/n}<1,$$
then the series $\sum_{n=1}^{\infty} a_n$ converges absolutely.

(2) *If*
$$\limsup_{n\to\infty}|a_n|^{1/n}>1,$$
then the series $\sum_{n=1}^{\infty} a_n$ diverges.

(3) *Otherwise, the test is inconclusive.*

Proof. For part (1), choose ρ so that $\limsup_{n\to\infty}|a_n|^{1/n}<\rho<1$. Then there exists $N\in\mathbb{N}$ such that if $n\geq N$,
$$|a_n|^{1/n}<\rho.$$
So
$$|a_n|<\rho^n \text{ for all } n\geq N.$$
Since $\rho<1$, by the comparison test with the geometric series, the series converges.

For part (2), we know there exist infinitely many n so that
$$|a_n|^{1/n}>1 \quad\text{or}\quad |a_n|>1.$$
So $\lim|a_n|\neq 0$ and the series $\sum_{n=1}^{\infty}|a_n|$ diverges. □

The following establishes the Cauchy property for series.

Theorem 7.2.12. *The series of real numbers $\sum_{n=1}^{\infty} a_n$ converges if and only if for every $\varepsilon>0$ there exists $N\in\mathbb{N}$ such that*

(7.8)
$$\left|\sum_{n=k}^{\ell} a_n\right|<\varepsilon \text{ whenever } k,\ell\geq N.$$

Proof. Let
$$S_n=\sum_{i=1}^{n} a_i.$$

7.2. Alternating Series and Absolute Convergence

Let $p, k \in \mathbb{N}$. We observe that
$$S_{k+p} - S_{k-1} = \sum_{i=k}^{k+p} a_i.$$
If the series converges, then (S_n) is a Cauchy sequence, so given $\varepsilon > 0$ there exists $N \in \mathbb{N}$ so that $|S_{k+p} - S_{k-1}| < \varepsilon$ for all $k \geq N$ and all $p \in \mathbb{N}_0$. So letting $k \in \mathbb{N}$ and $\ell = k + p$, we have
$$\left| \sum_{n=k}^{\ell} a_n \right| < \varepsilon.$$
For the converse we note that, by a similar argument, (7.8) implies that (S_n) is a Cauchy sequence, so the series converges. □

A statement analogous to Theorem 7.2.12 but for the case of absolute convergence is in Exercise 7.2.3.

Now we see an interesting consequence of absolute convergence regarding changing the order of the terms in a series.

Definition 7.2.13. A **rearrangement** of a series $\sum_{n=1}^{\infty} a_n$ has the form $\sum_{n=1}^{\infty} a_{\alpha(n)}$, where $\alpha : \mathbb{N} \to \mathbb{N}$ is a bijection. ◇

Theorem 7.2.14. *If series $\sum_{n=1}^{\infty} a_n$ converges absolutely to S, then for every bijection $\alpha : \mathbb{N} \to \mathbb{N}$, the rearrangement $\sum_{n=1}^{\infty} a_{\alpha(n)}$ converges to S.*

Proof. Write
$$S_n = \sum_{i=1}^{n} a_i \quad \text{and} \quad S_n^{\alpha} = \sum_{i=1}^{n} a_{\alpha(i)}.$$
Let $\varepsilon > 0$. Since the series converges to S, there exists $N_1 \in \mathbb{N}$ so that if $n \geq N_1$, then
$$|S - S_n| < \frac{\varepsilon}{2} \text{ for all } n \geq N_1.$$
As the series converges absolutely, by Exercise 7.2.3, there exists $N_2 \in \mathbb{N}$, which we choose greater than N_1, such that
$$\sum_{i=n}^{n+\ell} |a_i| < \frac{\varepsilon}{2} \text{ for all } n \geq N_2 \text{ and all } \ell \in \mathbb{N}.$$

Now choose $N_3 \in \mathbb{N}$ so that $a_1, \ldots, a_{N_2}$ are all in $\{a_{\alpha(1)}, \ldots, a_{\alpha(N_3)}\}$. This means that for $n \geq N_3$, $S_n - S_n^{\alpha}$ consists wholly of terms a_i with $i > N_2$. Therefore, if $n \geq N_3$,
$$|S_n - S_n^{\alpha}| \leq \sum_{i=n}^{\infty} |a_i| \leq \frac{\varepsilon}{2}.$$

Then, for $n \geq N_3$,
$$|S - S_n^{\alpha}| = |S - (S_n^{\alpha} - S_n + S_n)| \leq |S - S_n| + |S_n^{\alpha} - S_n| < \frac{\varepsilon}{2} + \frac{\varepsilon}{2} = \varepsilon,$$
showing that S_n^{α} converges to S. □

A consequence of Theorem 7.2.14 is that for a convergent series of nonnegative terms, the order of the sums is not relevant.

Corollary 7.2.15. *Suppose the series $\sum_{n=1}^{\infty} a_n$ converges absolutely to S. Let P be a subset of $\mathbb{N}$. (We are mainly interested in the case where P and $P^c = \mathbb{N} - P$ are infinite.) Then the series $\sum_{n \in P} a_n$ converges absolutely and to the same value independently of how P is written as a sequence. Also, $S = \sum_{n \in P} a_n + \sum_{n \in P^c} a_n$.*

Regarding a converse of Theorem 7.2.14, we note that for a conditionally convergent series, for every real number A, there is a rearrangement of the series that converges to A (Exercise 7.2.9). We illustrate this in the case of an alternating series.

Example 7.2.16. Consider the series $\sum_{n=1}^{\infty}(-1)^{n+1}\frac{1}{n}$; we know this series converges conditionally. We will show that given any $A \in \mathbb{R}$, there is a rearrangement of the series that converges to A. To simplify the argument we just consider the case when A is positive. Write

$$S_n = \sum_{i=1}^{n} \frac{1}{2i-1} \quad \text{and} \quad T_n = \sum_{i=1}^{n} \frac{1}{2i}.$$

We know that each of the partial sum sequences (S_n) and (T_n) diverges to ∞. Find the first n, call it n_1, so that $S_{n_1} > A$. It has to exist as (S_n) is strictly increasing and diverging. Next find the smallest $n > n_1$, call it n_2, so that $S_{n_1} - T_{n_2} < A$. Again, it has to exist as (T_n) is strictly increasing. Next find the smallest $n_3 > n_2$ so that $S_{n_3} - T_{n_2} > A$. Next choose n_4 the smallest integer greater than n_3 so that $S_{n_3} - T_{n_4} < A$. Continuing in this way we generate an increasing sequence (n_i). Note that as $S_{n_{i+1}-1} - T_{n_i} \leq A$. By the definition of n_i, we must have

$$0 < (S_{n_{i+1}} - T_{n_i}) - A < \frac{1}{n_{i+1}}.$$

Therefore, we have a rearrangement of the partial sums that converges to A. A similar argument can find rearrangements converging to $+\infty$ or $-\infty$. Also, essentially the same argument works for any alternating series $\sum_{n=1}^{\infty}(-1)^{n+1}a_n$ with a_n decreasing to 0 but such that $\sum_{n=1}^{\infty} a_n$ does not converge.

Exercises: Alternating Series and Absolute Convergence

7.2.1 Determine whether the following series are absolutely convergent, conditionally convergent, or divergent.
   (a) $\sum_{n=1}^{\infty}(-1)^{n+1}\frac{3}{4n-2}$
   (b) $\sum_{n=1}^{\infty}(-1)^n n$
   (c) $\sum_{n=2}^{\infty}(-1)^{n+1}\frac{n}{3n^2-4}$
   (d) $\sum_{n=1}^{\infty}\frac{(-1)^{n+1}}{n^2+2n}$
   (e) $\sum_{n=2}^{\infty}(-1)^{n+1}\frac{n}{\ln n^2}$
   (f) $\sum_{n=2}^{\infty}(-1)^{n+1}\frac{n}{3n^3-4}$
   (g) $\sum_{n=1}^{\infty}\frac{3^n}{2^n+3^n}$
   (h) $\sum_{n=1}^{\infty}\frac{n^4}{3^n}$
   (i) $\sum_{n=1}^{\infty}\frac{2^n}{n!}$

7.2. Alternating Series and Absolute Convergence

(j) $\sum_{n=1}^{\infty} \frac{n!}{10^n}$

(k) $\sum_{n=1}^{\infty} \frac{3^n}{n^n}$

(l) $\sum_{n=1}^{\infty} (-1)^{n+1} \sin(\frac{1}{n})$

7.2.2 Give an example of an alternating series $\sum_{n=1}^{\infty}(-1)^{n+1}a_n$, where $a_n > 0$ and $\lim_{n\to\infty} a_n = 0$ but such that the series diverges. (*Hint*: The a_n cannot be decreasing.)

7.2.3 Let $\sum_{n=1}^{\infty} a_n$ be a series. Prove that $\sum_{n=1}^{\infty} a_n$ converges absolutely if and only if for every $\varepsilon > 0$ there exists $N \in \mathbb{N}$ so that if $n \geq N$, then $\sum_{i=N}^{N+\ell} |a_i| < \varepsilon$ for all $\ell \in \mathbb{N}$.

7.2.4 Let $\sum_{n=1}^{\infty} a_n$ be a series. Prove that if it converges absolutely, then for each $\ell \in \mathbb{N}$, the series $\sum_{n=\ell}^{\infty} |a_n|$ converges, and $\lim_{\ell \to \infty} \sum_{n=\ell}^{\infty} |a_n| = 0$.

7.2.5 Does the series $\sum_{n=1}^{\infty} \sin(\frac{1}{n^2})$ converge? How about the series $\sum_{n=1}^{\infty} \sin^2(\frac{1}{n^2})$?

7.2.6 Let $\sum_{n=1}^{\infty}(-1)^n a_n$ be an alternating series satisfying the hypotheses of Theorem 7.2.3, and let S and S_n be the sum and the partial sum of the series, respectively. Prove that $|S - S_n| \leq a_{n+1}$.

7.2.7 Let $\sum_{n=1}^{\infty} a_n$ be a series that converges absolutely, and let (b_n) be a sequence whose limit exists and is finite. Prove that $\sum_{n=1}^{\infty} a_n b_n$ converges.

7.2.8 Prove Corollary 7.2.15.

* 7.2.9 Let $\sum_{n=1}^{\infty} a_n$ be a series. Prove that if it converges conditionally, then for each $A \in \mathbb{R}$ there is a rearrangement of the series that converges to A.

Chapter 8

Sequences and Series of Functions

8.1. Pointwise Convergence

Instead of studying sequences of numbers, we will now consider sequences of functions.

Definition 8.1.1. Let A be a set in $\mathbb{R}$. If for each $n \in \mathbb{N}$ there is a function $f_n : A \to \mathbb{R}$, we call (f_n) a **sequence of functions**. ◇

The functions (f_n) all share the same domain A, and often A is assumed to be an open or closed interval.

Definition 8.1.2. We say that the sequence (f_n) **converges pointwise**, or sometimes simply that it **converges**, to the function $f : A \to \mathbb{R}$ if for each $x \in A$ the sequence of numbers $(f_n(x))$ converges to $f(x)$. In this case we write

$$\lim_{n \to \infty} f_n(x) = f(x) \text{ for } x \in A \quad \text{or} \quad \lim_{n \to \infty} f_n(x) = f(x) \text{ on } A.$$ ◇

Pointwise convergence of a sequence of functions reduces to the usual convergence of numbers at each point in the domain of the functions. So (f_n) converges pointwise to f on A if and only if for each $x \in A$, for every $\varepsilon > 0$ there exists $N \in \mathbb{N}$ (in general depending on x and of course also on ε) such that

$$\text{if } n \geq N, \text{ then } |f(x) - f_n(x)| < \varepsilon.$$

We study some standard examples.

Example 8.1.3. Define $f_n : [0, 1] \to \mathbb{R}$ and $g : [0, 1] \to \mathbb{R}$ by

$$f_n(x) = x^n \text{ and}$$

$$g(x) = \begin{cases} 0 & \text{if } 0 \leq x < 1, \\ 1 & \text{if } x = 1. \end{cases}$$

We show that
$$\lim_{n\to\infty} f_n(x) = g(x) \text{ on } [0,1].$$
By Exercise 2.1.7, x^n converges to 0 for each x such that $0 \leq x < 1$. Finally for $x = 1$, $f_n(1) = 1$ and since $g(1) = 1$, $f_n(1)$ converges to $g(1)$. We have shown that $\lim_{n\to\infty} f_n(x) = g(x)$ on $[0,1]$.

Remark 8.1.4. Example 8.1.3 illustrates one of the problems with pointwise convergence. It shows that continuity is not preserved under pointwise convergence as all the functions f_n are continuous but their pointwise limit g is not.

Example 8.1.5. Let (r_n) be an enumeration of the rational numbers in $[0,1]$. Define $f_n : [0,1] \to \mathbb{R}$ by
$$f_n(x) = \begin{cases} 1 & \text{if } x \in \{r_1, \ldots, r_n\}, \\ 0 & \text{otherwise.} \end{cases}$$
Next let $f : [0,1] \to \mathbb{R}$ be Dirichlet's function D. We now show that (f_n) converges to D pointwise on $[0,1]$.

Let $x \in [0,1]$, and let $\varepsilon > 0$. If $x \in [0,1] \setminus \mathbb{Q}$ choose $N = 1$. Then for any $n \geq N$, $|f_n(x) - D(x)| = |0 - 0| = 0 < \varepsilon$. If $x \in [0,1] \cap \mathbb{Q}$, there exists $k \in \mathbb{N}$ such that $x = r_k$. Choose $N = k$. Hence, if $n \geq N$,
$$|f_n(x) - D(x)| = |1 - 1| = 0 < \varepsilon.$$
This completes the proof that the sequence f_n converges pointwise to D on $[0,1]$. Note that each f_n is Riemann integrable, while D is not.

Exercises: Pointwise Convergence

8.1.1 Let $f_n : [0,1] \to \mathbb{R}$ be defined by $f_n(x) = x^2 + \frac{x}{n}$. Show that (f_n) converges pointwise to x^2 on $[0,1]$.

8.1.2 Let $f_n : [0,1] \to \mathbb{R}$ be defined by
$$f_n(x) = \begin{cases} n & \text{if } 0 < x \leq \frac{1}{n}, \\ 0 & \text{if } \frac{1}{n} < x \leq 1 \text{ or } x = 0. \end{cases}$$
Show that (f_n) converges pointwise to 0 on $[0,1]$. Does $\int_0^1 f_n$ converge to $\int_0^1 0$?

8.1.3 Let $f_n : [0,1] \to \mathbb{R}$ be defined by $f_n(x) = x^{1/n}$. Find $\lim_{n\to\infty} f_n(x)$, and prove your answer.

8.1.4 Let $f_n : [-1,1] \to \mathbb{R}$ be defined by $f_n(x) = \frac{x}{1+x^{2n}}$. Find $\lim_{n\to\infty} f_n(x)$, and prove your answer.

8.1.5 Let $f_n : [0,\infty) \to \mathbb{R}$ be defined by $f_n(x) = \frac{1}{1+x^n}$. Find $\lim_{n\to\infty} f_n(x)$, and prove your answer. Show that each f_n is continuous. Is the limit continuous?

8.2. Uniform Convergence

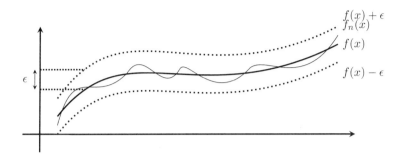

Figure 8.1. Uniform convergence.

8.2. Uniform Convergence

Uniform convergence is a global concept, defined for a sequence of functions, that in general depends on the domain of the function in a way similar to the notion of uniform continuity.

Definition 8.2.1. Let A be a set in $\mathbb{R}$, and let $f_n : A \to \mathbb{R}$ be a sequence of functions. We say that the sequence (f_n) **converges uniformly on** A to a function $f : A \to \mathbb{R}$ and write

$$\lim_{n \to \infty} f_n(x) = f(x) \text{ uniformly on } A$$

if

for every $\varepsilon > 0$ there exists $N \in \mathbb{N}$ such that
if $x \in A$ and $n \geq N$, then $|f_n(x) - f(x)| < \varepsilon$.

To emphasize that N does not depend on x, we often write this as

for every $\varepsilon > 0$ there exists $N \in \mathbb{N}$ such that
$|f_n(x) - f(x)| < \varepsilon$ whenever $n \geq N$ and for all $x \in A$. ◇

It follows directly from the definition that if (f_n) converges uniformly to f on A, then (f_n) converges pointwise to f on A. The difference between pointwise and uniform convergence is that in the case of uniform convergence, the number N can be chosen independently of x. This means that the approximation $|f_n(x) - f(x)| < \varepsilon$ holds uniformly for all x in the domain of the function, provided n is picked greater than or equal to N. This is illustrated in Figure 8.1, where we see that the "window" of approximation around f holds for all x in the domain of the functions.

It also follows from the definition that if f_n converges uniformly to f on A, and $B \subseteq A$, then (f_n) converges uniformly to f on B.

Question 8.2.2. Suppose $f_n : A \to \mathbb{R}$ converges uniformly to $f : A \to \mathbb{R}$ and that $g_n : A \to \mathbb{R}$ converges uniformly to $g : A \to \mathbb{R}$.

(a) Show that $f_n + g_n$ converges uniformly to $f + g$.
(b) Assume one sequence is uniformly bounded: $|f_n(x)| < M$ for all n and $x \in A$. Show that $f_n g_n$ converges uniformly to fg.

Question 8.2.3. Suppose $f_n : A \to \mathbb{R}$ converges uniformly to $f : A \to \mathbb{R}$. Prove that if each function f_n is bounded on A, then f is bounded on A.

We start with a simple example.

Example 8.2.4. Let $f_n(x) = x^n$. We show that f_n converges uniformly to 0 on $[0, 1/2]$. Let $\varepsilon > 0$. Choose N so that $2^N > 1/\varepsilon$. We know this can be done as $2^n > n$ for all $n \in \mathbb{N}$. (Alternatively, we could use the logarithm and say that by the Archimedean principle there exists $N > \ln(1/\varepsilon)$—assuming $\varepsilon < 1$ so that $\ln(1/\varepsilon) > 0$.) Hence, if $n \geq N$,

$$(8.1) \qquad |x^n - 0| \leq \frac{1}{2^n} \leq \frac{1}{2^N} < \varepsilon$$

for all $x \in [0, 1/2]$. We emphasize here that the inequalities in (8.1) hold for all x in $[0, 1/2]$ (we also say that they hold uniformly on $[0, 1/2]$). In Exercise 8.2.1 the reader is asked to show that (f_n) converges uniformly to 0 on $[0, a]$ for any $a < 1$, and in Example 8.2.6 we show it does not converge uniformly on $[0, 1]$.

Remark 8.2.5. We state what it means for a sequence of functions not to converge uniformly. Let (f_n) be a sequence of functions that converges pointwise to f on A. The sequence (f_n) does not converge uniformly on A if and only if

there exists $\varepsilon > 0$ such that for all $N \in \mathbb{N}$,

$|f_n(x) - f(x)| \geq \varepsilon$ for some $n \geq N$ and for some $x \in A$.

Example 8.2.6. We show that $f_n(x) = x^n$ does not converge uniformly on $[0, 1]$. Set $\varepsilon = 1/2$. For each $N \in \mathbb{N}$ choose $i \in \mathbb{N}$ so that $2^{i-1} > N$, and set

$$n = 2^{i-1} \quad \text{and} \quad x = \left(1 - \frac{1}{2^i}\right).$$

Clearly, $x \in (0, 1)$. Then, by Bernoulli's inequality,

$$|x^n - 0| = x^n = \left(1 - \frac{1}{2^i}\right)^n \geq 1 - \frac{2^{i-1}}{2^i} = \frac{1}{2}.$$

This shows that (f_n) does not converge uniformly on $[0, 1]$. Figure 8.2 illustrates that one can find arbitrarily large integers, namely $n = 2^{i-1}$, and corresponding points, namely $1 - 1/2^i$, where the value of the function f_n at the point is not $1/2$-close to the limit 0.

We need a technical lemma that will have several applications for showing uniform convergence. For this we introduce the notion of uniform Cauchy.

Definition 8.2.7. Let A be a set in $\mathbb{R}$, and let $f_n : A \to \mathbb{R}$ be a sequence of functions. We say the sequence is **uniformly Cauchy on A** if

for every $\varepsilon > 0$ there exists $N \in \mathbb{N}$ such that

$|f_n(x) - f_m(x)| < \varepsilon$ whenever $n, m \geq N$ and for all $x \in A$. ◊

Lemma 8.2.8. *Let A be a subset of $\mathbb{R}$, and let $f_n : A \to \mathbb{R}$ be a sequence of functions. Then the sequence (f_n) is uniformly Cauchy on A if and only if it converges uniformly to a function $f : A \to \mathbb{R}$.*

8.2. Uniform Convergence

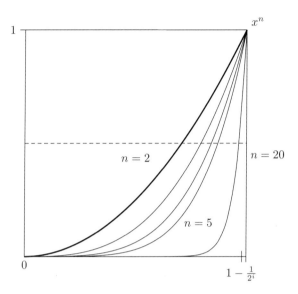

Figure 8.2. The sequence of functions x^n is not uniformly convergent.

Proof. Suppose the sequence (f_n) is uniformly Cauchy on A. For each $x \in A$, the sequence $(f_n(x))$ is Cauchy, so it converges to a point in $\mathbb{R}$ that we may denote by $f(x)$. This defines a function $f : A \to \mathbb{R}$.

Let $\varepsilon > 0$. There exists $N \in \mathbb{N}$ such that

$$(8.2) \qquad |f_n(x) - f_m(x)| < \frac{\varepsilon}{2} \text{ for all } x \in A \text{ and whenever } m, n \geq N.$$

Since the sequence $(f_n(x))$ converges pointwise to $f(x)$, for each $x \in A$ there exists $m_x \in \mathbb{N}$, which can be chosen so that $m_x > N$, and satisfies

$$|f(x) - f_{m_x}(x)| < \frac{\varepsilon}{2}.$$

Then,

$$\begin{aligned} |f_n(x) - f(x)| &= |f_n(x) - f_{m_x}(x) + f_{m_x}(x) - f(x)| \\ &\leq |f_n(x) - f_{m_x}(x)| + |f_{m_x}(x) - f(x)| \\ &< \frac{\varepsilon}{2} + \frac{\varepsilon}{2} = \varepsilon \end{aligned}$$

for all $x \in A$. Therefore, the sequence converges uniformly. (An alternative proof follows by taking the limit as $m \to \infty$, for fixed n, in (8.2). Then one has $|f_n(x) - f(x)| \leq \varepsilon/2$ for all $x \in A$, which implies uniform convergence. The first time one does this it may seem like magic; the calculations above can be seen as a justification that the method works.)

For the converse suppose that the sequence converges uniformly. Let $\varepsilon > 0$. There exists $N \in \mathbb{N}$ such that

$$|f_n(x) - f(x)| < \frac{\varepsilon}{2} \text{ for all } x \in A \text{ and } n \geq N.$$

Then, if $n, m \geq N$,
$$|f_n(x) - f_m(x)| \leq |f_n(x) - f(x)| + |f(x) - f_m(x)| < \varepsilon,$$
for all $x \in A$. Therefore, (f_n) is uniformly Cauchy on A. □

8.2.1. Uniform Convergence and Continuity. The limit of a sequence of continuous functions need not be continuous as the following example shows. However, by Theorem 8.2.10, if the convergence is uniform, the limit will be continuous.

Example 8.2.9. We recall Example 8.1.3 where we defined the functions $f_n(x) = x^n$ for $x \in [0, 1]$ and saw that the sequence (f_n) converges pointwise to a function g on $[0, 1]$ that is not continuous at $x = 1$, while each f_n is continuous on $[0, 1]$.

Theorem 8.2.10. Let $f_n : A \to \mathbb{R}$ be a sequence of continuous functions. If (f_n) converges uniformly to f on A, then f is continuous.

Proof. Fix $x_0 \in A$. We wish to show that f is continuous at x_0. Let $\varepsilon > 0$. There exists $N \in \mathbb{N}$ so that if $n \geq N$, then
$$|f_n(x) - f(x)| < \frac{\varepsilon}{3} \text{ for all } x \in A.$$
As f_N is continuous at x_0 there exists $\delta > 0$ so that
$$|f_N(y) - f_N(x_0)| < \frac{\varepsilon}{3} \text{ whenever } |y - x_0| < \delta.$$
Then
$$|f(y) - f(x_0)| = |f(y) - f_N(y) + f_N(y) - f_N(x_0) + f_N(x_0) - f(x_0)|$$
$$\leq |f(y) - f_N(y)| + |f_N(y) - f_N(x_0)| + |f_N(x_0) - f(x_0)|$$
$$\leq \frac{\varepsilon}{3} + \frac{\varepsilon}{3} + \frac{\varepsilon}{3} = \varepsilon$$
whenever $|x_0 - y| < \delta$. Therefore, f is continuous at each point of A. □

8.2.2. Uniform Convergence and Integration. There are many applications where one would like to exchange the order of the limit of a sequence of functions with their integral. In general, this does not work, as the following example shows. One answer, that has applications to power series, is that one can exchange the order when there is uniform convergence (Theorem 8.2.13).

Example 8.2.11. We will define a sequence of functions t_n on $[0, 1]$ whose pointwise limit is the function 0. One can do this, however, and still keep the integral of each t_n to be a nonzero constant, say the constant 1. We do this by constructing over the interval $[0, 1/n]$ a very tall triangle so that its area is 1. As n increases the length of the interval that is the base of the triangle will converge to 0, but the area of the triangles will remain the constant 1. So define
$$t_n(x) = \begin{cases} 4n^2 x & \text{when } 0 \leq x < 1/2n, \\ (4n^2 - 2)x + 1/n & \text{when } 1/2n \leq x < 1/n, \\ 0 & \text{when } x > 1/n. \end{cases}$$
We have chosen t_n so that the area of the triangle over $[0, 1/n]$ is 1, see Figure 8.3. (For a simpler example one could have set the value of the function over the

8.2. Uniform Convergence

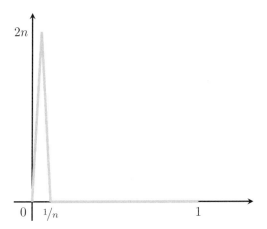

Figure 8.3. The function t_n.

interval $[0, 1/n]$ to be n, but this would not be continuous while t_n is.) As t_n is 0 in the remainder of the interval, we have

$$\int_0^1 t_n = 1.$$

It is clear that for each $x \in (0, 1]$, one can find $N \in \mathbb{N}$ so that $1/N < x$, so for $n \geq N$, we have $t_n(x) = 0$. (Also, $t_n(0) = 0$.) Thus $\lim_{n \to \infty} t_n(x) = 0$ for all $x \in [0, 1]$, so $\int_0^1 \lim_{n \to \infty} t_n(x) = 0$. At the same time we have that $\lim_{n \to \infty} \int_0^1 t_n(x) = \lim_{n \to \infty} 1 = 1$. Therefore, in this case

$$\lim_{n \to \infty} \int_0^1 t_n(x) \, dx \neq \int_0^1 \lim_{n \to \infty} f_n(x) \, dx.$$

Example 8.2.12. This example is of a different nature, but it is also interesting. We have shown that Dirichlet's function is not Riemann integrable. Now we show that each function f_n of Example 8.1.5 is Riemann integrable. Fix $n \in \mathbb{N}$, and let $\varepsilon > 0$. Choose $\delta = \varepsilon/n$. Let $\|\mathcal{P}\| < \delta$. We note that f_n is positive only at the points $\{r_1, \ldots, r_n\}$. So the sum $S(f_n, \mathcal{P}, \mathcal{P}^*)$ has at most n nonzero summands, and each summand is of the form $x_i - x_{i+1}$. By the property of the partition, each of these summands is less than $\|\mathcal{P}\|$. Therefore

$$S(f_n, \mathcal{P}, \mathcal{P}^*) \leq n \cdot \|\mathcal{P}\| < n \, \delta = \varepsilon.$$

Therefore, each f_n is Riemann integrable. So we have a sequence of Riemann integrable functions that converges pointwise to a function that is not Riemann integrable.

Theorem 8.2.13. *Let $f_n : [a, b] \to \mathbb{R}$ be a sequence of Riemann integrable functions on $[a, b]$. If (f_n) converges uniformly to f on $[a, b]$, then f is Riemann integrable on $[a, b]$ and*

(8.3)
$$\lim_{n \to \infty} \int_a^b f_n = \int_a^b f.$$

Proof. We first use the Darboux criterion (Theorem 6.1.20) to show f is Riemann integrable. From Question 8.2.3 we know that f is bounded. We may assume $b > a$. Let $\varepsilon > 0$. Since the sequence converges uniformly, there exists $N \in \mathbb{N}$ such that

(8.4) $\qquad |f_n(x) - f(x)| < \dfrac{\varepsilon}{4(b-a)}$ for all $n \geq N$ and $x \in [a, b]$.

Then, in particular, we have that for all $x \in [a, b]$,

(8.5) $\qquad -\dfrac{\varepsilon}{4(b-a)} + f_N(x) < f(x) < f_N(x) + \dfrac{\varepsilon}{4(b-a)}.$

Since f_N is Riemann integrable, there exists $\delta > 0$ such that for all partitions $\mathcal{P}$ of $[a, b]$,

(8.6) $\qquad$ if $||\mathcal{P}|| < \delta,$ then $U(f_N, \mathcal{P}) - L(f_N, \mathcal{P}) < \dfrac{\varepsilon}{2}.$

From (8.4) and from the definition of the upper and lower Riemann sums, we obtain

(8.7) $\qquad -\dfrac{\varepsilon}{4} + U(f_N, \mathcal{P}) < U(f, \mathcal{P}) < U(f_N, \mathcal{P}) + \dfrac{\varepsilon}{4},$

(8.8) $\qquad -\dfrac{\varepsilon}{4} + L(f_N, \mathcal{P}) < L(f, \mathcal{P}) < L(f_N, \mathcal{P}) + \dfrac{\varepsilon}{4}.$

Then, and using (8.6),

(8.9) $\qquad U(f, \mathcal{P}) - L(f, \mathcal{P}) < U(f_N, \mathcal{P}) - L(f_N, \mathcal{P}) + \dfrac{\varepsilon}{4} + \dfrac{\varepsilon}{4}$

$$< \dfrac{\varepsilon}{2} + \dfrac{\varepsilon}{2} = \varepsilon.$$

Therefore, f is Riemann integrable on $[a, b]$.

(If we knew that the functions f_n are continuous on $[a, b]$, then by Theorem 8.2.10 f would be continuous, thus Riemann integrable, and it would simplify this part of the argument.)

Finally, to prove (8.3) we start from (8.4). Then

(8.10) $\qquad \left| \displaystyle\int_a^b f_n(x)\, dx - \int_a^b f(x)\, dx \right| = \left| \int_a^b (f_n - f)\, dx \right|$

$$\leq \int_a^b |f_n(x) - f(x)|\, dx$$

$$< \dfrac{\varepsilon}{4(b-a)} (b-a) < \varepsilon.$$

Therefore, $\int_a^b f_n$ converges to $\int_a^b f$. $\qquad \square$

8.2.3. Uniform Convergence and Differentiation. We state a condition under which the derivative of the limit of a sequence of functions equals the limit of the derivatives of the sequence. While the hypotheses of the theorem are rather general, Corollary 8.2.17 below gives a simpler version of the theorem that is used more often in applications. We start with an example showing that in general one cannot exchange the limit with the derivative.

8.2. Uniform Convergence

Example 8.2.14. Let $f_n(x) = \frac{x^{n+1}}{n+1}$ for $x \in [-1, 1]$. Then
$$|f_n(x)| \leq \frac{1}{n+1} \text{ for } |x| \leq 1.$$
It follows that (f_n) converges to 0 on $[0, 1]$, in fact, one can see that the convergence is uniform on $[0, 1]$. Also, $f'_n(x) = x^n$ for $x \in [0, 1]$, where at the endpoints the derivative is understood one-sidedly. However, $f'(\pm 1) = \pm 1$. So
$$(\lim_{n \to \infty} f_n(x))' \neq \lim_{n \to \infty} f'_n(x) \text{ for } x = \pm 1.$$
For an example where the convergence of the derivatives fails at interior points of the interval see Exercise 8.2.11.

Theorem 8.2.15. *Let $f_n : [a, b] \to \mathbb{R}$ be a sequence of continuous functions such that each $f_n : (a, b) \to \mathbb{R}$ is differentiable. Suppose that there exists a point $x_0 \in [a, b]$ such that $(f_n(x_0))$ converges to a number that we denote $f(x_0)$ and that the sequence (f'_n) converges uniformly to a function g on (a, b). Then*

(1) *(f_n) converges uniformly to a function f on $[a, b]$; and*

(2) *f is differentiable on (a, b) and $f'(x) = g(x)$ for all $x \in (a, b)$, thus*
$$f'(x) = (\lim_{n \to \infty} f_n(x))' = \lim_{n \to \infty} f'_n(x) \text{ for all } x \in (a, b).$$

Proof. First we show that (f_n) is a uniformly Cauchy sequence on $[a, b]$; this implies that (f_n) converges uniformly on $[a, b]$, completing part (1). Let $\varepsilon > 0$. To find an upper bound for $|f_n(x) - f_m(x)|$, when x ranges over $[a, b]$, we write

(8.11)
$$|f_n(x) - f_m(x)| = |f_n(x) + f_n(x_0) - f_n(x_0) + f_m(x_0) - f_m(x_0) - f_m(x)|$$
$$\leq |f_n(x_0) - f_m(x_0)| + |(f_n(x) - f_m(x)) - (f_n(x_0) - f_m(x_0))|.$$

Since $(f_n(x_0))$ converges, it is a Cauchy sequence; therefore, there exists $N_1 \in \mathbb{N}$ such that
$$|f_n(x_0) - f_m(x_0)| < \frac{\varepsilon}{2} \text{ for all } n, m \geq N_1.$$
This gives an upper bound for the first summand in (8.11). For the second term in the sum, observe that, for each $n, m \in \mathbb{N}$, the function $f_n - f_m$ is continuous on $[a, b]$ and differentiable on (a, b). Therefore by the mean value theorem, thinking of $(f_n - f_m)$ as a single function, for each $x \in [a, b]$ there exists $z \in (a, b)$ such that

(8.12) $\quad (f_n(x) - f_m(x)) - (f_n(x_0) - f_m(x_0)) = (f'_n(z) - f'_m(z))(x - x_0).$

Since (f'_n) converges uniformly, it is uniformly Cauchy and thus there exists $N_2 \in \mathbb{N}$ such that

(8.13) $\quad |f'_n(w) - f'_m(w)| < \frac{\varepsilon}{2(b-a)}$ for all $n, m \geq N_2$ and all $w \in (a, b)$.

Replacing this in (8.12),

(8.14) $\quad |(f_n(x) - f_m(x)) - (f_n(x_0) - f_m(x_0))| < \frac{\varepsilon}{2(b-a)} |x - x_0| \leq \frac{\varepsilon}{2}.$

Now using (8.12) and (8.14) in (8.11),
$$|f_n(x) - f_m(x)| < \frac{\varepsilon}{2} + \frac{\varepsilon}{2} = \varepsilon$$

for all $m, n \geq \max\{N_1, N_2\}$ and all $x \in [a,b]$. Therefore (f_n) converges uniformly on $[a,b]$.

We now prove the second part but only under the additional assumption that each f_n is *continuously differentiable*.

Definition 8.2.16. A function f is **continuously differentiable** if it is differentiable and f' is continuous. ◇

Let f_n be continuously differentiable on (a,b). So f_n' and g are continuous on (a,b), and thus Riemann integrable on any closed interval contained in (a,b). (These assumptions will be satisfied in all applications of this theorem in Section 8.4.)

Choose a point $a_0 \in (a,b)$. By the Fundamental Theorem of Calculus, for each $x \in (a,b)$,
$$\int_{a_0}^{x} f_n'(t)\, dt = f_n(x) - f_n(a_0).$$

Using Theorem 8.2.13, we have
$$\int_{a_0}^{x} g(t)\, dt = \int_{a_0}^{x} \lim_{n \to \infty} f_n'(t)\, dt$$
$$= \lim_{n \to \infty} \int_{a_0}^{x} f_n'(t)\, dt$$
$$= \lim_{n \to \infty} f_n(x) - f_n(a_0)$$
$$= f(x) - f(a_0).$$

Again, by the Fundamental Theorem of Calculus, differentiating both sides of the equation,
$$g(x) = f'(x),$$
completing the proof. □

The following corollary follows from Theorem 8.2.15 and it is stated in a form that is easier to apply.

Corollary 8.2.17. *Let $f_n : [a,b] \to \mathbb{R}$ be a sequence of continuous functions such that $f_n : (a,b) \to \mathbb{R}$ are continuously differentiable functions. Suppose that $(f_n(x))$ converges to a function $f(x)$ and that the sequence (f_n') is uniformly Cauchy on (a,b). Then*
$$f \text{ is differentiable and } f'(x) = \lim_{n \to \infty} f_n'(x) \text{ for } x \in (a,b).$$

Proof. Since (f_n') is uniformly Cauchy, it converges uniformly to some function g on (a,b). Then we can apply Theorem 8.2.15 directly to finish the proof. □

Remark 8.2.18. Sometimes we may need to apply Corollary 8.2.17 when the functions f_n are defined on an open interval. Given

$f_n : (a,b) \to \mathbb{R}$ continuously differentiable and converging to f on (a,b),

and such that (f_n') is uniformly Cauchy on (a,b),

8.2. Uniform Convergence

we can apply Corollary 8.2.17 to the functions f_n on the intervals $[a', b']$ for $a < a' < b' < b$ to obtain the same conclusion on the interval (a', b'). That is, we obtain that

f is differentiable at x and $f'(x) = \lim_{n \to \infty} f'_n(x)$ for $x \in (a, b)$.

However, it is possible to modify the proof of Theorem 8.2.15 to obtain uniform convergence on (a, b) (in this case of course $x_0 \in (a, b)$ and the mean value theorem is applied on the interval $[x_0, x]$ or $[x, x_0]$).

8.2.4. An Example: The Cantor Function. We define a very interesting function, the Cantor function, that is naturally related to the Cantor set. Let $(G_n)_{n \geq 0}$ be the open sets in $[0, 1]$ that are used in the definition of the Cantor set. Each G_n is a union of 2^n disjoint open intervals. For each $n \in \mathbb{N}_0$ we define a function f_n that is constant on each of the subintervals of G_n. We start with $n = 0$; let f_0 be $1/2$ on $G_0 = (1/3, 2/3)$, and in the two remaining subintervals that are complements of G_0 is $[0, 1]$, the function f_0 is chosen to be linear and such that the resulting function is continuous and increasing. So, since $y = \frac{3}{2}x$ is the line from the point $(0, 0)$ to the point $(1/3, 1/2)$, and $y = \frac{3}{2}x - 1/2$ is the line from the point $(2/3, 1/2)$ to $(1, 1)$, we define f_0 by

$$f_0(x) = \begin{cases} \frac{3}{2}x & \text{when } 0 \leq x \leq 1/3, \\ \frac{1}{2} & \text{when } 1/3 < x < 2/3, \\ \frac{3}{2}x - \frac{1}{2} & \text{when } 1/3 \leq x \leq 2/3. \end{cases}$$

Next we define f_1, to agree with f_0 on G_0, to be $1/4$ on the first subinterval of G_1, and $3/4$ on the second and last subinterval of G_1, and on the remaining subintervals it is defined to be linear so that the resulting function is increasing and continuous. Figure 8.4 shows the graph of f_0 and f_1.

Continue in this way and define f_n so that it agrees with f_{n-1} on $G_0, \ldots, G_{n-1}$ and on the subintervals of G_n the function f_n is the constant value equal to the midpoint of the corresponding dyadic subintervals. For example, f_2 agrees with f_1

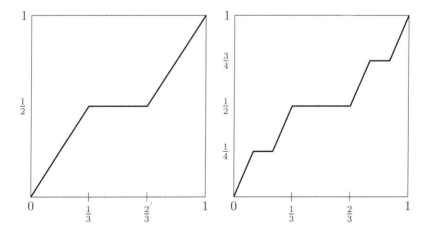

Figure 8.4. The functions f_0 and f_1 in the construction of the Cantor function.

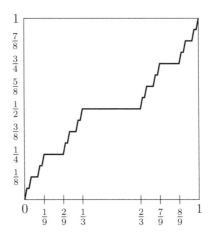

Figure 8.5. Illustration of the Cantor function.

on $G_0 = (1/3, 2/3)$ and on $G_1 = (1/9, 2/9) \cup (7/9, 8/9)$. There are now four dyadic intervals left that one can see on the vertical axis of the graph of f_1 in Figure 8.4: $[0, 1/4], [1/4, 1/2], [1/2, 3/4], [3/4, 1]$. Then looking at the subintervals of G_2, on $(1/27, 2/27)$ the function f_2 takes the value $1/8$, on the subinterval $(7/27, 8/27)$ it takes the value $3/8$, etc.

It follows that the sequence (f_n) is uniformly Cauchy (Exercise 8.2.7). (The idea is that as f_{n+1} and f_n differ only on F_{n+1}, a rough estimate can be given by $|f_{n+1}(x) - f_n(x)| \leq 1/2^{n+1}$.) Thus the sequence converges to a continuous function C on $[0, 1]$.

Definition 8.2.19. The function C that the sequence (f_n) converges to is called the **Cantor function** (see Figure 8.5). ◊

We observe that if x is a point outside the Cantor set, then $x \in G_n$ for some n. Then $f_m(x) = f_n(x)$ for all $m \geq n$. Also, for x in any of the open intervals of G_n, f_n is constant so $f'_m(x) = 0$ for all $m \geq n$ and $x \in G_n$. It follows that for each $x \in G_n$, for some $n \in \mathbb{N}_0$, the Cantor function is differentiable and $C'(x) = 0$. So we have a continuous function that is monotone increasing and whose derivative is 0 at every point outside a set of measure 0.

8.2.5. Dominated and Monotone Convergence Theorems. Here we state, without proof, the dominated convergence theorem. As we have discussed, the question of exchanging two limits—in particular, the question of whether the limit of an integral is the integral of the limit—has been an important question in analysis. Theorem 8.2.13 provides an answer to this, but it has the requirement that the functions converge uniformly, and this condition is not satisfied in many cases where one would like to apply the theorem. It was not until Lebesgue's integration theory of 1902 that answers under more general conditions were found. One such answer is the dominated convergence theorem of Lebesgue, which originally was proved using the theory of Lebesgue integration, but it can be stated for the case of Riemann integrals. In the Riemann case, it is necessary to assume that the limit

8.2. Uniform Convergence

function is Riemann integrable; in the case of Lebesgue integrals, the theorem is more general. We state it without proof as it is a useful theorem; the bibliography has some references.

Theorem 8.2.20 (Dominated convergence theorem). *Let $f_n : [a,b] \to \mathbb{R}$ be a sequence of Riemann integrable functions on $[a,b]$ converging to a function f on $[a,b]$. If f is Riemann integrable on $[a,b]$ and there exists a Riemann integrable function $g : [a,b] \to \mathbb{R}$ such that*

$$|f_n(x)| \leq g(x) \text{ for all } n \in \mathbb{N} \text{ and } x \in [a,b],$$

then

$$\lim_{n \to \infty} \int_a^b f_n = \int_a^b f. \tag{8.15}$$

The following is also a very useful theorem, which can be obtained as a consequence of Theorem 8.2.20.

Theorem 8.2.21 (Monotone convergence theorem). *Let $f_n : [a,b] \to \mathbb{R}$ be a sequence of Riemann integrable functions such that for all $x \in [a,b]$,*

$$0 \leq f_1(x) \leq f_2(x) \leq \cdots \leq f_n(x) \leq \cdots$$

and $\lim_{n \to \infty} f_n(x) = f(x)$ exists and is Riemann integrable. Then

$$\lim_{n \to \infty} \int_a^b f_n = \int_a^b f.$$

Proof. From the hypothesis we have that $|f_n(x)| \leq f(x)$ for all $x \in [a,b]$. As f is Riemann integrable, Theorem 8.2.20 completes the proof. □

Exercises: Uniform Convergence

8.2.1 Let $f_n : [0,1] \to \mathbb{R}$ be defined by $f_n(x) = x^n$. Let $a \in (0,1)$. Show that (f_n) converges uniformly to the function 0 on $[0,a]$.

8.2.2 Prove that the converse of Theorem 8.2.10 does not hold.

8.2.3 Let $f_n : A \to \mathbb{R}$, for $A \subseteq \mathbb{R}$, be uniformly continuous. Show that if (f_n) converges uniformly to the function f, then f is uniformly continuous.

8.2.4 Let $A \subseteq \mathbb{R}$, and let $f_n, f : A \to \mathbb{R}$ be bounded functions. Prove that (f_n) converges uniformly to f if and only if for all $\varepsilon > 0$ there is $N \in \mathbb{N}$ such that

$$n \geq N \text{ implies } \sup\{|f_n(x) - f(x)| : x \in A\} < \varepsilon.$$

8.2.5 Let $x \in [0, \infty)$, and define

$$f_n(x) = \frac{nx}{nx+1}.$$

Find the pointwise limit $f(x) = \lim_{n \to \infty} f_n(x)$ for $x \in [0, \infty)$ and prove your answer. Prove that the convergence is uniform on $[1, \infty)$ and is not uniform on $[0, \infty)$. Is it uniform on $(0, \infty)$? Prove or disprove.

8.2.6 Let $x \in [0, \infty)$, and define
$$f_n(x) = \frac{nx^2}{nx+1}.$$
Find the pointwise limit $f(x) = \lim_{n \to \infty} f_n(x)$ for $x \in [0, \infty)$, and prove your answer. Prove that the convergence is uniform on $[1, \infty)$ and is not uniform on $[0, \infty)$. Is it uniform on $(0, \infty)$? Prove or disprove.

8.2.7 Prove that the sequence (f_n) in Example 8.2.4 is uniformly Cauchy.

* 8.2.8 Give another proof of Theorem 8.2.13 using the definition of Riemann integration instead of the Cauchy criterion.

8.2.9 Prove that the Cantor function C satisfies $C'(x) = 0$ for all x outside a set of measure zero. Is the Cantor function Riemann integrable?

8.2.10 Let $f(x) = x + C(x)$, where $C(x)$ is the Cantor function and $x \in [0, 1]$. Prove that f is continuous and strictly increasing. Classify the points where f is differentiable.

8.2.11 Let $f_n(x) = \frac{\sin(nx)}{\sqrt{n}}$ for $x \in [0, \pi/2]$ and $n \in \mathbb{N}$. Prove that (f_n) converges uniformly to 0 on $[0, \pi/2]$, but (f'_n) does not converge on $[0, \pi/2]$.

8.2.12 Prove (without using Theorem 8.2.20) that the functions t_n of Example 8.2.11 are not bounded by an integrable function; i.e., there is no Riemann integrable function g such that $|t_n(x)| \leq g(x)$ for all $x \in [0, 1]$.

8.2.13 *Dini's theorem*: Let $f_n : [a, b] \to \mathbb{R}$ be a sequence of continuous functions such that $f_{n+1}(x) \leq f_n(x)$ for all $x \in [a, b]$ and $n \in \mathbb{N}$. Prove that if $f(x) = \lim_{n \to \infty} f_n(x)$ is continuous on $[a, b]$, then the convergence is uniform on $[a, b]$.

8.3. Series of Functions

Definition 8.3.1. Let A be a subset of $\mathbb{R}$, and let $f_n : A \to \mathbb{R}$ be a sequence of functions. A **series of functions** has the form

(8.16) $$\sum_{n=1}^{\infty} f_n(x).$$

(Although we start our series at $n = 1$, the same ideas apply to any other starting value.) ◊

For each $x \in A$, (8.16) defines a series of numbers. Convergence of (8.16) depends on convergence of the sequence of partial sums

$$S_n(x) = \sum_{i=1}^{n} f_i(x).$$

Definition 8.3.2. We say that a series $\sum_{n=1}^{\infty} f_n(x)$ **converges** or **converges pointwise** to a function $f : A \to \mathbb{R}$ if for each $x \in A$ the sequence of partial sums $(S_n(x))$ converges to $f(x)$. Similarly, we say that the series $\sum_{n=1}^{\infty} f_n(x)$ **converges uniformly** to a function $f : A \to \mathbb{R}$ if the sequence of partial sums (S_n) converges uniformly to f on A. ◊

8.3. Series of Functions

The following is a very useful test used to show uniform convergence of a series.

Theorem 8.3.3 (Weierstrass M-test). *Let (M_n) be a sequence of nonnegative numbers such that*
$$\sum_{n=1}^{\infty} M_n < \infty.$$
Suppose $f_n : A \to \mathbb{R}$ is a sequence of functions such that

(8.17) $\qquad |f_n(x)| \leq M_n$ *for all $x \in A$.*

Then $\sum_{n=1}^{\infty} f_n(x)$ converges uniformly and absolutely on A.

Proof. The sequence $(\sum_{i=1}^{n} M_i)_n$ is Cauchy as it converges, so for each $\varepsilon > 0$ there exists $N \in \mathbb{N}$ so that
$$\sum_{i=n}^{n+\ell} M_i < \varepsilon \text{ for all } n > N, \text{ and } \ell > 1.$$

By (8.17),
$$\sum_{i=n}^{n+\ell} |f_i(x)| \leq \sum_{i=n}^{n+\ell} M_i < \varepsilon \text{ for all } x \in A.$$

Since
$$\left| \sum_{i=1}^{n+\ell} |f_i(x)| - \sum_{i=1}^{n} |f_i(x)| \right| = \sum_{i=n+1}^{n+\ell} |f_i(x)| < \varepsilon$$

whenever $x \in A, n > N$ and $\ell \geq 1$, it follows that the sequence $(\sum_{i=1}^{n} |f_i(x)|)_n$ is uniformly Cauchy. By Lemma 8.2.8 it converges uniformly and absolutely. $\square$

The following theorems give useful conditions for integrating and differentiating a series of functions.

Theorem 8.3.4. *For each $n \in \mathbb{N}$, let $f_n : [a,b] \to \mathbb{R}$ be a continuous function. If the series $\sum_{n=1}^{\infty} f_n(x)$ converges uniformly on $[a,b]$, then*
$$\int_a^t \sum_{n=1}^{\infty} f_n = \sum_{n=1}^{\infty} \int_a^t f_n \text{ for all } t \in [a,b].$$

Proof. The sequence of partial sums $S_k(x) = \sum_{n=1}^{k} f_n(x)$ is continuous so each S_k is Riemann integrable. By Theorem 8.2.13 the limit of the S_k is integrable, and we have that
$$\int_a^t \sum_{n=1}^{\infty} f_n = \int_a^t \lim_{k \to \infty} \sum_{n=1}^{k} f_n = \lim_{k \to \infty} \int_a^t \sum_{n=1}^{k} f_n$$
$$= \lim_{k \to \infty} \sum_{n=1}^{k} \int_a^t f_n = \sum_{n=1}^{\infty} \int_a^t f_n. \qquad \square$$

The following theorem holds under more general conditions, but we state it in a simpler form that will be sufficient for later applications.

Theorem 8.3.5. *For each $n \in \mathbb{N}$, let $f_n : (a,b) \to \mathbb{R}$ be continuously differentiable on (a,b). If the series $\sum_{n=1}^{\infty} f_n(x)$ converges on (a,b) and $\sum_{n=1}^{\infty} f'_n(x)$ converges uniformly on (a,b), then*

$$\left(\sum_{n=1}^{\infty} f_n(x)\right)' = \sum_{n=1}^{\infty} f'_n(x) \text{ for all } x \in (a,b).$$

Proof. Again consider the sequence of partial sums and apply Corollary 8.2.17 and Remark 8.2.18. □

8.3.1. Continuous Nowhere Differentiable Functions. The existence of functions that are continuous on the real line but differentiable at no point caused great surprise. The first published example is due to Weierstrass (1872), but Bolzano had an example a few years before, though it was not published. Since then many other examples have been proposed, and we will present an example of Takagi, also discussed by van der Waerden.

We need a definition that will be used in the construction.

Definition 8.3.6. Let $f : \mathbb{R} \to \mathbb{R}$ be a function. If there is a number $p \in \mathbb{R} \setminus \{0\}$ such that
$$f(x+p) = f(x) \text{ for all } x \in \mathbb{R},$$
we say that f is a **periodic function** and p is called a **period** of f. If f is periodic and if there exists a smallest $p > 0$ such that $f(x+p) = f(x)$ for all $x \in \mathbb{R}$, then p is called the **fundamental period** of f. Sometimes the fundamental period is called simply the period, and it will be clear from the context if we refer to it as *the* period of f. Note that if f has fundamental period p, then np is also a period for all $n \in \mathbb{Z}$.

We start with a function s on the real line that has been called the *saw-tooth function*. On the unit interval it is a kind of inverted absolute value. Define

$$s(x) = \begin{cases} x & \text{when } 0 \leq x \leq 1/2, \\ 1-x & \text{when } 1/2 \leq x \leq 1. \end{cases}$$

Now we extend s to the whole real line so that its pattern gets repeated on each interval of the form $[n, n+1]$; i.e., we extend it to be a periodic function of period 1. So define a function s_0 in the following way. Given a real number x, there is a unique $n_x \in \mathbb{Z}$ such that $x - n_x \in [0,1)$. Then set

$$s_0(x) = s(x - n_x).$$

It follows that
$$s_0(x+1) = s_0(x) \text{ for all } x \in \mathbb{R}.$$
Another way to think of $s_0(x)$ is as the closest distance from x to $\mathbb{Z}$.

Clearly, $s_0(x)$ is 0 when x is an integer, it attains its maximum value, $\frac{1}{2}$, at the points of the form $x = 1/2 + n$, is continuous on $\mathbb{R}$, and has slope $+1$ or -1 at any point $x \neq n/2$, where $n \in \mathbb{Z}$. Of course, s_0 is not differentiable at these points, though it has one-sided derivatives. Let us call the points where it is not differentiable *corner* points.

8.3. Series of Functions

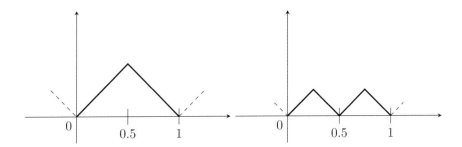

Figure 8.6. The functions s_0 and s_1 in the construction of the Takagi function.

The function $s_0(2x)$ is similar to $s_0(x)$ but it now has (fundamental) period $1/2$ instead of 1, and its corner points are of the form $n/4, n \in \mathbb{Z}$. We will consider the functions
$$s_n(x) = \frac{s(2^n x)}{2^n} \text{ for } n \in \mathbb{N}.$$
First note that s_n is clearly continuous as it is the composition and product of continuous functions. We next observe that s_n has a maximum value of $1/2^{n+1}$, has period $1/2^n$, and is not differentiable at its corner points, which have the form $k/2^{n+1}, k \in \mathbb{Z}$. Also, s_n has slope $+1$ or -1 at noncorner points. There is one more property that will be needed. Since s_n has (fundamental) period $1/2^n$, it also has period $1/2^{n-1}$, as $s(x + 1/2^{n-1}) = s(x + 1/2^n + 1/2^n) = s(x)$. So in general

(8.18) $$s_n(x + \frac{1}{2^i}) = s_n(x) \text{ for } i = 0, 1, 2, \ldots, n.$$

Figure 8.6 shows part of the functions s_0 and s_1.

Theorem 8.3.7. *The function*
$$T(x) = \sum_{n=0}^{\infty} s_n(x)$$
is continuous on $\mathbb{R}$ but differentiable at no point of $\mathbb{R}$.

Proof. For continuity we first observe that for all $x \in \mathbb{R}$,
$$|s_n(x)| \leq \frac{1}{2^{n+1}}.$$
Since $\sum_n \frac{1}{2^{n+1}}$ converges, by the Weierstrass M-test, $\sum_{n=0}^{\infty} s_n(x)$ converges uniformly, and as each s_n is continuous, the series is continuous.

Now let $x \in \mathbb{R}$. We will show T is not differentiable at x. Since every s_n has period 1, T is periodic of period 1, so from now on we can assume $x \in [0, 1)$. We recall that the dyadic rationals in $[0, 1]$ have the form
$$\frac{k}{2^n} \text{ for } n \in \mathbb{N}, k = 0, 1, \ldots, 2^n,$$
and are obtained by subsequently subdividing the interval in half. Then, given x, for each $m \in \mathbb{N}$, there exists a unique $\ell_m \in \mathbb{N}$ so that
$$\frac{\ell_m}{2^m} \leq x < \frac{\ell_m + 1}{2^m}.$$

Set
$$p_m = \frac{\ell_m}{2^m} \quad \text{and} \quad q_m = \frac{\ell_m + 1}{2^m}.$$

From the definition of p_m and q_m, it follows that they converge to x and
$$0 < p_m \le x < q_m < 1 \text{ for all } m \in \mathbb{N}.$$

By Question 5.1.16, to show that T is not differentiable at x, it suffices to show that the limit
$$\lim_{m \to \infty} \frac{T(q_m) - T(p_m)}{q_m - p_m}$$
does not exist. We first examine
$$s_n(q_m) - s_n(p_m).$$

Note that $q_m - p_m = 1/2^m$, so by (8.18), when $n \ge m$,
$$s_n(q_m) - s_n(p_m) = 0.$$

This implies that we can simplify the infinite sum in the following way:
$$\sum_{n=0}^{\infty}(s_n(q_m) - s_n(p_m)) = \sum_{n=0}^{m-1}(s_n(q_m) - s_n(p_m)).$$

Next we observe that when $n < m$, p_m and q_m are on the same side of adjacent corner points of s_n, so
$$\frac{s_n(q_m) - s_n(p_m)}{q_m - p_m} = +1 \text{ or } -1.$$

Therefore,
$$\lim_{m \to \infty} \frac{T(q_m) - T(p_m)}{q_m - p_m} = \lim_{m \to \infty} \frac{\sum_{n=0}^{\infty} s_n(q_m) - \sum_{n=0}^{\infty} s_n(p_m)}{q_m - p_m}$$
$$= \lim_{m \to \infty} \frac{\sum_{n=0}^{\infty}(s_n(q_m) - s_n(p_m))}{q_m - p_m}$$
$$= \lim_{m \to \infty} \frac{\sum_{n=0}^{m-1}(s_n(q_m) - s_n(p_m))}{q_m - p_m}$$
$$= \lim_{m \to \infty} \sum_{n=0}^{m-1} \frac{s_n(q_m) - s_n(p_m)}{q_m - p_m}$$
$$= \lim_{m \to \infty} \sum_{n=0}^{m-1} (+1 \text{ or } -1),$$

which implies the limit does not exist. $\square$

Exercises: Series of Functions

8.3.1 Let $p \geq 2$. Prove that the series
$$\sum_{n=1}^{\infty} \frac{\sin nx}{n^p}$$
defines a differentiable function on $\mathbb{R}$.

8.3.2 Prove that the series
$$f(x) = \sum_{n=1}^{\infty} \frac{\cos(2^n x)}{2^n}$$
converges uniformly on $\mathbb{R}$ and defines a continuous function. Show $f(x)$ is not differentiable at $x = 0$ and at any $x = p/2^k$ for $p, k \in \mathbb{N}$. (The function is known to be not differentiable at any x.)

8.3.3 Prove that the series
$$f(x) = \sum_{n=1}^{\infty} \frac{x^n}{n^2}$$
converges uniformly on $[0, 1]$ and defines a continuous function.

8.3.4 Give an example of a series $\sum_{n=1}^{\infty} f_n(x)$ that converges uniformly on some interval $[a, b]$, but such that the series $\sum_{n=1}^{\infty} f'_n(x)$ diverges on this interval.

8.3.5 For the Takagi function, can you give more information regarding differentiability at the dyadic points?

8.3.6 Suppose $\sum_{n=1}^{\infty} f_n(x)$ and $\sum_{n=1}^{\infty} g_n(x)$ converge uniformly on an interval $[a, b]$. Does $\sum_{n=1}^{\infty} (f_n(x) + g_n(x))$ converge uniformly on $[a, b]$?

8.4. Power Series

Definition 8.4.1. A **power series** is a series of the form

(8.19) $$\sum_{n=0}^{\infty} a_n(x - c)^n,$$

where $a_n, c \in \mathbb{R}$ and x is in some interval. $\diamond$

We will largely be interested in the case when $c = 0$ as the general case reduces to this by setting $y = x - c$. In general, we start our power series at $n = 0$ although of course it may happen that $a_0 = 0$.

The following lemma tells us an interesting property regarding the convergence of power series.

Lemma 8.4.2. *If the series*

(8.20) $$\sum_{n=0}^{\infty} a_n x^n$$

converges for $x = b$, where $b \neq 0$, then it converges uniformly and absolutely on $[-c, c]$ for each $c > 0$ such that $c < |b|$.

Proof. If the series in (8.20) converges for $x = b$, then by the first divergence test (Proposition 7.1.12) there exists a constant $C > 0$ so that
$$|a_n b^n| < C \text{ for all } n \in \mathbb{N}.$$
Let $c \in \mathbb{R}$ be such that
$$0 < c < |b|.$$
If $x \in [-c, c]$, then
$$|a_n x^n| = |a_n||x^n| \leq |a_n||c^n| = |a_n b^n| \frac{|c|^n}{|b|^n} < C \left|\frac{c}{b}\right|^n.$$
As
$$\left|\frac{c}{b}\right| < 1,$$
if we let $M_n = \left|\frac{c}{b}\right|^n$, then the series $\sum_{n=0}^{\infty} M_n$ converges, so by the Weierstrass M-test the series
$$\sum_{n=0}^{\infty} a_n x^n$$
converges uniformly and absolutely $[-c, c]$. □

So convergence always occurs on an interval.

Definition 8.4.3. The **radius of convergence** of the power series
$$(8.21) \qquad \sum_{n=0}^{\infty} a_n x^n$$
is defined to be the number $R \in [0, \infty]$ such that the power series converges absolutely for all x with $|x| < R$ and diverges for all x such that $|x| > R$. If $R = \infty$, we understand it to converge for all $x \in \mathbb{R}$, and if $R = 0$ it means that the series converges absolutely only for $x = 0$. (In general, nothing is known for x such that $|x| = R$.) ◊

Remark 8.4.4. It follows from Lemma 8.4.2 that if a power series has radius of convergence R, then it converges uniformly and absolutely on any closed interval inside the interval $(-R, R)$. The series may or may not converge at the endpoints of the interval $(-R, R)$.

The following theorem gives a way to find the radius of convergence and will be used in Theorem 8.4.9.

Theorem 8.4.5. *Let $\sum_{n=0}^{\infty} a_n x^n$ be a power series. Let*
$$\rho = \limsup_{n \to \infty} |a_n|^{1/n}.$$
The radius of convergence is given by
$$(8.22) \qquad R = \frac{1}{\rho}.$$
If $\rho = 0$, we write $R = \infty$, and if $\rho = \infty$, we write $R = 0$.

8.4. Power Series

Proof. We need to show that if R is finite, then the series converges absolutely for all x such that $|x| < R$ and diverges for x such that $|x| > R$, and if $R = \infty$, then the series converges absolutely for all $x \in \mathbb{R}$.

For a fixed $x \in \mathbb{R}$, by the root test (Theorem 7.2.11) the series $\sum_{n=0}^{\infty} a_n x^n$ converges absolutely if
$$\limsup_{n \to \infty} |a_n x^n|^{1/n} < 1.$$

Equivalently, the series converges for x such that

(8.23) $$\limsup_{n \to \infty} |a_n|^{1/n} |x| < 1 \quad \text{or}$$

(8.24) $$\rho |x| < 1.$$

If $\rho = 0$, the series converges absolutely for all x. If $0 < \rho < \infty$, it converges absolutely for $|x| < 1/\rho$, and the series diverges for x such that $|x| > R$. If $\rho = \infty$, it only converges for $x = 0$. □

The following theorem gives another way to compute the radius of convergence that, while not as general as Theorem 8.4.5, is easier to apply in some cases.

Theorem 8.4.6. *Let $\sum_{n=0}^{\infty} a_n x^n$ be a power series. If the limit*

(8.25) $$\lim_{n \to \infty} \left| \frac{a_n}{a_{n+1}} \right|$$

exists or is ∞, then this limit is the radius of convergence of the power series.

Proof. Set

(8.26) $$R = \lim_{n \to \infty} \left| \frac{a_n}{a_{n+1}} \right|.$$

For a fixed $x \in \mathbb{R}$, by the ratio test (Theorem 7.2.9) the series converges absolutely if
$$\lim_{n \to \infty} \frac{|a_{n+1} x^{n+1}|}{|a_n x^n|} < 1.$$

Equivalently, the series converges for x such that

(8.27) $$\lim_{n \to \infty} \frac{|a_{n+1}|}{|a_n|} |x| < 1.$$

If $0 < R < \infty$, then
$$\lim_{n \to \infty} \frac{|a_{n+1}|}{|a_n|} = \frac{1}{R},$$
so this is equivalent to saying that the series converges absolutely for all x such that
$$|x| < R.$$

By a similar argument the series diverges for x such that $|x| > R$. If $R = \infty$, then $\lim_{n \to \infty} \frac{|a_{n+1}|}{|a_n|} = 0$, and therefore the inequality in (8.27) is satisfied for all $x \in \mathbb{R}$. The case $R = 0$ is similar. □

Example 8.4.7. Consider the series
$$\sum_{n=0}^{\infty} \frac{x^n}{n!}.$$
Here $a_n = 1/n!$, for all $n \geq 0$, and
$$R = \lim_{n \to \infty} \frac{\frac{1}{n!}}{\frac{1}{(n+1)!}} = \lim_{n \to \infty} \frac{(n+1)!}{n!} = \infty.$$
Therefore, the series converges absolutely for all $x \in \mathbb{R}$.

Example 8.4.8. Consider the series
$$\sum_{n=0}^{\infty} x^n.$$
In this case $a_n = 1$ for all $n \geq 0$. Hence $R = 1$. The series converges absolutely for x such that $|x| < 1$. The reader may have recognized this as the geometric series. It diverges for $|x| \geq 1$.

We have the following property that will be useful when considering the derivative and integral of a power series.

Theorem 8.4.9. *Let $\sum_{n=0}^{\infty} a_n x^n$ be a power series. Let R be its radius of convergence. Then for every $0 \leq a < R$, the series converges uniformly on $(-a, a)$ and can be integrated and differentiated term-by-term on this interval. Furthermore, the series*
$$\sum_{n=1}^{\infty} n a_n x^{n-1} \quad \text{and} \quad \sum_{n=0}^{\infty} \frac{a_n}{n+1} x^{n+1}$$
also have radius of convergence R.

Proof. The Weierstrass M-test gives uniform convergence on any interval $(-a, a)$ where $a < R$. Then Theorems 8.3.4 and 8.3.5 show that the derivative and the integral of the series are obtained by differentiating and integrating the partial sums, which gives term-by-term differentiation and integration.

Finally, we note that $\sum_{n=1}^{\infty} n a_n x^{n-1}$ has the same radius of convergence as $x \sum_{n=1}^{\infty} n a_n x^{n-1} = \sum_{n=1}^{\infty} n a_n x^n$. For this series we calculate
$$\limsup (n a_n)^{1/n} = \limsup (n)^{1/n} (a_n)^{1/n} = \limsup (a_n)^{1/n},$$
which is the radius of convergence of $\sum_{n=0}^{\infty} a_n x^n$. A similar argument works for $\sum_{n=0}^{\infty} \frac{a_n}{n+1} x^{n+1} = x \sum_{n=0}^{\infty} \frac{a_n}{n+1} x^n$. □

A direct consequence of Theorem 8.4.9 is that a power series can be differentiated inside its radius of convergence and its derivative has the same radius of convergence. We state this explicitly as it will be useful, though it is essentially a rephrasing of Theorem 8.4.9. This technique can be used to obtain power series for many functions.

8.4. Power Series

Corollary 8.4.10. *Let $\sum_{n=0}^{\infty} a_n x^n$ be a power series with radius of convergence R. Define the function f on $(-R, R)$ by*

$$f(x) = \sum_{n=0}^{\infty} a_n x^n.$$

Then f is infinitely differentiable and Riemann integrable, and for all $x \in (-R, R)$,

(8.28) $$f'(x) = \sum_{n=1}^{\infty} n a_n x^{n-1} \quad \text{and}$$

(8.29) $$\int_0^x f(t)\, dt = \sum_{n=0}^{\infty} \frac{a_n}{n+1} x^{n+1}.$$

Proof. By Theorem 8.4.9 we obtain the derivative of f by differentiating term-by-term, so this establishes (8.28). Again by Theorem 8.4.9, the series of f' has radius of convergence R, so it can be differentiated again. Induction shows f is infinitely differentiable on $(-R, R)$.

Now assume $x > 0$. We know that f is integrable on $[0, x]$. Integrating the nth term of the series of $f(t)$, we obtain

$$\int_0^x a_n t^n\, dt = \frac{1}{n+1} a_n x^{n+1} - \frac{1}{n+1} a_n (0)^{n+1} = \frac{1}{n+1} a_n x^{n+1},$$

which gives (8.29). $\square$

Example 8.4.11. In Example 8.4.7 we saw that the series

$$\sum_{n=0}^{\infty} \frac{x^n}{n!}$$

has infinite radius of convergence; let $f(x)$ denote this series. Here $a_n = 1/n!$. By Corollary 8.4.10,

$$f'(x) = \sum_{n=0}^{\infty} \left(\frac{x^n}{n!}\right)' = \sum_{n=1}^{\infty} n \cdot \frac{x^{n-1}}{n!} = \sum_{n=1}^{\infty} \frac{x^{n-1}}{(n-1)!} = \sum_{n=0}^{\infty} \frac{x^n}{n!}.$$

Therefore, $f'(x) = f(x)$ for all $x \in \mathbb{R}$. We will study this function in more detail in Section 8.5.

Example 8.4.12. We have seen that the geometric series

$$\sum_{n=0}^{\infty} x^n$$

has radius of convergence $R = 1$ and that it converges to $1/(1-x)$ for all $x \in (-1, 1)$. So

$$\frac{1}{1-x} = \sum_{n=0}^{\infty} x^n.$$

Replacing x with $-x$, we obtain, for $|x| < 1$,

$$\frac{1}{1+x} = \sum_{n=0}^{\infty} (-1)^n x^n = 1 - x + x^2 - x^3 + \cdots.$$

Since
$$\ln x = \int_0^x \frac{1}{t}\, dt,$$
using substitution (in Theorem 6.2.6 let $f(t) = 1/t$ and $g(t) = t+1$), we obtain
$$\ln(1+x) = \int_1^{x+1} \frac{1}{t}\, dt = \int_0^x \frac{1}{t+1}\, dt.$$
Therefore, for $|x| < 1$,

(8.30) $$\ln(1+x) = \sum_{n=0}^\infty \int_0^x (-1)^n t^n\, dt = \sum_{n=0}^\infty (-1)^n \frac{x^{n+1}}{n+1}$$
$$= x - \frac{x^2}{2} + \frac{x^3}{3} - \frac{x^4}{4} + \cdots.$$

Exercises: Power Series

8.4.1 Find the radius of convergence of the following series, and check convergence at the endpoints when the radius is finite.

(a)
$$\sum_{n=0}^\infty n x^n$$

(b)
$$\sum_{n=0}^\infty n! x^n$$

(c)
$$\sum_{n=1}^\infty \frac{1}{n} x^n$$

(d)
$$\sum_{n=1}^\infty \frac{n^2}{3^n} x^n$$

(e)
$$\sum_{n=1}^\infty (-1)^n \frac{2^n x^n}{n!}$$

8.4.2 Find the radius of convergence of the following power series, and check convergence at the endpoints when the radius is finite.

(a)
$$\sum_{n=1}^\infty \frac{(-1)^n}{n \cdot 5^n} x^{2n}$$

(b)
$$\sum_{n=1}^\infty \frac{1}{3^{n^2}} (x-3)^{3n}$$

8.4.3 Prove that (8.30) still holds if we let $x = 1$, giving a series expression for $\ln 2$.

8.4.4 Let $\sum_{n=0}^{\infty} a_n x^n$ be a power series with R its radius of convergence. Prove that

$$\liminf_{n \to \infty} \left| \frac{a_n}{a_{n+1}} \right| \leq R \leq \limsup_{n \to \infty} \left| \frac{a_n}{a_{n+1}} \right|. \tag{8.31}$$

8.5. Taylor Series

An important class of power series are the *Taylor series* of a function. Suppose f is a function defined on an interval (a, b), where a could be $-\infty$ and b could be ∞. Let $c \in (a, b)$, and assume f is infinitely differentiable at c (i.e., the derivatives $f^{(n)}(c)$ exist for all $n \in \mathbb{N}$), and recall that $f^{(0)} = f$.

Definition 8.5.1. The **Taylor series about** c, or **centered at** c, of the function f is the power series

$$\sum_{n=0}^{\infty} \frac{f^{(n)}(c)}{n!} (x - c)^n. \qquad \diamond$$

The radius of convergence of the series depends on the coefficients of $(x - c)^n$, and the series may converge only for $x = c$, or for all x in $\mathbb{R}$, or for x in some interval around c. The reader may also have noticed a connection with the Taylor polynomial of a function. A question one can ask at this moment is whether a function is equal to its Taylor series on the interval where the series converges. As we shall see, this need not always be the case, even when the function is infinitely differentiable. In many cases, though, such as in the case of exp, sin, and cos, the function is equal to its Taylor series. In Lemma 8.5.4 we state a condition that guarantees when equality occurs.

Before stating the lemma, we see that the representation of a function as a power series is unique.

Lemma 8.5.2. *Let $c \in \mathbb{R}$, $R > 0$, and let $f : (c - R, c + R) \to \mathbb{R}$ be a function. Suppose f can be written as a convergent power series*

$$f(x) = \sum_{n=0}^{\infty} a_n (x - c)^n \text{ for all } x \in (c - R, c + R).$$

Then f is infinitely differentiable on $(c - R, c + R)$ and

$$a_n = \frac{f^{(n)}(c)}{n!}, \ n \in \mathbb{N}_0.$$

Proof. Since by assumption the series converges on $(c - R, c + R)$, by Corollary 8.4.10 the function f is infinitely differentiable, and its derivative is given term-by-term, so we can compute

$$f(x) = a_0 + a_1(x-c) + a_2(x-c)^2 + a_2(x-c)^3 + \cdots$$
$$f'(x) = a_1 + 2a_2(x-c) + 3a_2(x-c)^2 + \cdots$$
$$f''(x) = 2!a_2 + 3!a_2(x-c) + \cdots$$
$$\vdots$$
$$f^{(n)}(x) = n!a_n + (n+1)!a_{n+1}(x-c) + \frac{(n+2)!}{2}a_{n+1}(x-c)^2 \cdots.$$

Substituting $x = c$ in each of the equations gives $f^{(n)}(c) = n!a_n$ for $n \geq 0$, completing the proof. $\square$

Remark 8.5.3. Lemma 8.5.2 shows that a power series is equal to its Taylor series in its interval of convergence.

Lemma 8.5.4. *Let $c \in \mathbb{R}$, $R > 0$, and let $f : (c-R, c+R) \to \mathbb{R}$ be a function. Suppose f is infinitely differentiable on $(c-R, c+R)$. Then $f(x)$ equals its Taylor series around c,*

$$\sum_{n=0}^{\infty} \frac{f^{(n)}(c)}{n!}(x-c)^n$$

on $(c-R, c+R)$ if and only if $\lim_{n\to\infty} R_n(x) = 0$ for $x \in (c-R, c+R)$, where R_n is the nth remainder of its Taylor polynomial.

Proof. By Taylor's theorem, we know that for each $n \in \mathbb{N}$,

$$f(x) = P_n(x) + R_n(x),$$

where $P_n(x)$ is its n degree Taylor polynomial and $R_n(x)$ is its remainder:

$$P_n(x) = \sum_{i=0}^{n} \frac{f^{(i)}(c)}{i}(x-c)^i \quad \text{and} \quad R_n(x) = \frac{f^{(n+1)}(\zeta)}{(n+1)!}(x-c)^{n+1}$$

for some ζ between x and c. If $\lim_{n\to\infty} R_n(x) = 0$, it follows that P_n converges to the Taylor series. Conversely, if $f(x)$ equals its Taylor series, then

$$\lim_{n\to\infty} f(x) - P_n(x) = 0,$$

which implies that $\lim_{n\to\infty} R_n(x) = 0$. $\square$

Corollary 8.5.5. *Let $c \in \mathbb{R}$ and $R > 0$, and let $f : (c-R, c+R) \to \mathbb{R}$ be a function. Suppose f is infinitely differentiable on $(c-R, c+R)$. If there exists $M > 0$ such that*

$$|f^{(n)}(x)| \leq M^n \quad \text{for all } x \in (c-R, c+R), \text{ and } n \in \mathbb{N}_0$$

then f is equal to its Taylor series on $(c-R, c+R)$:

$$f(x) = \sum_{n=0}^{\infty} \frac{f^{(n)}(c)}{n!}(x-c)^n.$$

8.5. Taylor Series

Proof. We have
$$|R_n(x)| = \left|\frac{f^{(n+1)}(\zeta)}{(n+1)!}(x-c)^{n+1}\right| \le \frac{M^{n+1}}{(n+1)!}R^{n+1}$$
for some ζ between x and c. Since $\lim_{n\to\infty}(MR)^{n+1}/(n+1)! = 0$, then $\lim_{n\to\infty} R_n(x) = 0$, and Lemma 8.5.4 completes the proof. $\square$

Corollary 8.5.6. *Let $f : \mathbb{R} \to \mathbb{R}$ be a differentiable function satisfying the equation*
$$f'(x) = f(x) \quad \text{for all } x \in \mathbb{R}.$$
Assume also that $f(0) = c$. Then
$$f(x) = \sum_{n=0}^{\infty} \frac{c}{n!}x^n \quad \text{for all } x \in \mathbb{R}.$$

Proof. Since $f' = f$ and f is differentiable, then f' is also differentiable. Induction shows that $f^{(n)}$ is differentiable and $f^{(n)}(x) = f(x)$ for all $x \in \mathbb{R}$. Also, $f^{(n)}(0) = c$, and f is bounded on $[x, 0]$ and $[0, x]$ (depending on whether $x > 0$ or $x < 0$). Then, for each $x \in \mathbb{R}$, for some ζ between x and 0,
$$\lim_{n\to\infty} |R_n(x)| = \lim_{n\to\infty} \left|\frac{f^{(n+1)}(\zeta)}{(n+1)!}x^{n+1}\right|$$
$$= \lim_{n\to\infty} \frac{|f(\zeta)|}{(n+1)!}|x|^{n+1}$$
$$= 0.$$
Lemma 8.5.4 completes the proof. $\square$

8.5.1. The Exponential Function. We have seen that the series $\sum_{n=0}^{\infty} \frac{x^n}{n!}$ converges absolutely, for all $x \in \mathbb{R}$, and uniformly on compact intervals in $\mathbb{R}$.

Definition 8.5.7. Define the **exponential function** $\exp x$, for $x \in \mathbb{R}$, by
$$\exp x = \sum_{n=0}^{\infty} \frac{x^n}{n!}. \qquad \Diamond$$

In particular this also gives Euler's constant e as defined in Exercise 2.2.16 by setting $x = 1$ so that
$$\exp 1 = \sum_{n=0}^{\infty} \frac{1}{n!} = e.$$

Theorem 8.5.8. *The exponential function has the following properties.*

(1) *$\exp x$ is infinitely differentiable on $\mathbb{R}$ and*
$$(\exp x)' = \exp x \quad \text{for all } x \in \mathbb{R}.$$
Furthermore, $\exp x$ is the only function such that $(\exp x)' = \exp x$ and $\exp 0 = 1$.

(2) *$\exp(x+y) = \exp(x)\exp(y)$ for all $x, y \in \mathbb{R}$.*

(3) *$\exp(-x) = \frac{1}{\exp x}$ for all $x \in \mathbb{R}$.*

(4) *$\exp x > 0$ for all $x \in \mathbb{R}$. Hence, $\exp$ is a strictly increasing function.*

(5) $\exp x$ *is the inverse of* $\ln x$:
$$\ln(\exp x) = x \text{ for } x \in \mathbb{R}, \text{ and } \exp(\ln x) = x \text{ for } x \in (0, \infty).$$

(6) $\lim_{x \to \infty} \exp x = \infty$, $\lim_{x \to -\infty} \exp x = 0$.

Proof. Since $\exp x$ is defined by a power series with an infinite radius of convergence, it is infinitely differentiable, and its derivative is given by

$$(\exp x)' = \sum_{n=0}^{\infty} \left(\frac{x^n}{n!}\right)' = \sum_{n=0}^{\infty} \frac{n x^{n-1}}{n!}$$
$$= \sum_{n=1}^{\infty} \frac{x^{n-1}}{(n-1)!} = \sum_{n=0}^{\infty} \frac{x^n}{n!}$$
$$= \exp x.$$

It is clear that $\exp 0 = 1$. The fact that it is the only function with these properties follows from Corollary 8.5.6.

Fix $y \in \mathbb{R}$, and let $f(x) = \exp(x + y)$. Note that $f(0) = \exp y$. The function f is differentiable on $\mathbb{R}$, and by the chain rule $f'(x) = f(x)$. By Corollary 8.5.6, f is infinitely differentiable and has the form

$$f(x) = \sum_{n=0}^{\infty} \frac{f(0)}{n!} x^n = f(0) \cdot \sum_{n=0}^{\infty} \frac{1}{n!} x^n = \exp y \exp x,$$

for all $x \in \mathbb{R}$.

Next we note that
$$1 = \exp 0 = \exp(x + (-x)) = \exp x \exp(-x),$$

so
$$\exp(-x) = \frac{1}{\exp x}.$$

From the definition it follows that $\exp x > 0$ when $x \geq 0$. When $x < 0$, we use the fact that $\exp(-x) = 1/\exp x$ to see that $\exp x > 0$ for all $x \in \mathbb{R}$.

Now define $g(x) = \ln(\exp x)$ for $x \in \mathbb{R}$. Since it is the composition of differentiable functions, it is differentiable. By the chain rule,

$$g'(x) = \frac{1}{\exp x} \cdot \exp x = 1.$$

As $g(0) = 0$, Question 5.2.9 implies that $g(x) = x$ for all $x \in \mathbb{R}$. Also, we know that $\ln : (0, \infty) \to \mathbb{R}$ is a bijection, so by Question 0.4.17 it follows that $\exp$ is its inverse.

Finally, we note that
$$\exp x = 1 + x + \sum_{n=2}^{\infty} \frac{x^n}{n!}.$$

Thus, $\exp x > 1 + x$ for $x > 0$. So $\lim_{x \to \infty} \exp x = \infty$. Also, as $\exp(-x) = 1/\exp x$, it follows that $\lim_{x \to -\infty} \exp x = 0$. □

8.5. Taylor Series

Remark 8.5.9. We can now generalize the definition of the power of a positive real number by an integer, or by a rational number, to the power of a positive real number by an arbitrary real number. First we recall that a^x was already defined in Exercise 1.2.12 for $a \in \mathbb{R}^+$ and $x \in \mathbb{R}$. We give another definition that connects it with the functions we have been studying. Recall that the constant e in Exercise 2.3.4 is $e = \sum_{i=0}^{\infty} \frac{1}{i!} = \exp(1)$. It follows by induction that

$$e^n = \exp(n) \text{ for } n \in \mathbb{N},$$

and one can use this to show (Exercise 8.5.2) that

$$e^x = \exp(x) \text{ for } x \in \mathbb{R},$$

where e^x is defined as in Exercise 1.2.12. Now we give a new, independent definition of the powers.

Let $a > 0$ and $x \in \mathbb{R}$. Define a^x by

$$a^x = \exp(x \ln a).$$

Then a^x is a differentiable and integrable function, and we have

$$(a^x)' = a^x \ln a.$$

This follows by the chain rule. As the composition of differentiable functions is differentiable, a^x is differentiable, and $(a^x)' = \exp(x \ln a) \cdot \ln a = a^x \ln a$.

Similarly, for $x > 0$ and $b \in \mathbb{R}$ one has

$$x^b = \exp(b \ln x).$$

This function is also differentiable and integrable, and by a similar argument we have the formula

$$(x^b)' = bx^{b-1}.$$

8.5.2. The Trigonometric Functions. We have seen that the exponential function is the unique differentiable function f that satisfies the equation $f'(x) = f(x)$ and $f(0) = 1$. One might consider other simple equations. This equation clearly implies $f''(x) = f(x)$, but the following equation is different:

$$f''(x) = -f(x).$$

We will see that this latter equation yields the sine and cosine trigonometric functions. We start with the following lemma.

Lemma 8.5.10. *Let $f : \mathbb{R} \to \mathbb{R}$ be a twice differentiable function such that*

$$f''(x) = -f(x) \text{ for all } x \in \mathbb{R}.$$

Then f is infinitely differentiable, and for all $x \in \mathbb{R}$

$$f(x) = \sum_{n=0}^{\infty} \frac{b_n}{n!} x^n,$$

where $b_n = f(0)$ when n is even and $b_n = f'(0)$ when n is odd.

Proof. Let $c_1 = f(0)$ and $c_2 = f'(0)$. We note that $f^{(3)}(x) = -f'(x)$ and $f^{(4)}(x) = f(x)$. So by induction, all derivatives $f^{(n)}(x)$ exist and the values appear in a repeating pattern,
$$f(x),\ f'(x), -f(x),\ -f'(x),$$
as n ranges over $0, 1, 2, 3, \ldots$. In particular, for $k \geq 0$,
$$f^{(4k)}(0) = c_1,\ f^{(4k+1)}(0) = c_2,\ f^{(4k+2)}(0) = -c_1,\ f^{(4k+3)}(0) = -c_2.$$

We calculate the remainder using Taylor's theorem. For each $x \in \mathbb{R}$, for some ζ between x and 0,
$$\lim_{n \to \infty} |R_n(x)| = \lim_{n \to \infty} \left| \frac{f^{(n+1)}(\zeta)}{(n+1)!} x^{n+1} \right| = 0,$$
since $f^{(n+1)}(\zeta)$ takes only four possible values. By Lemma 8.5.4, the Taylor series converges to the function. $\square$

By Lemma 8.5.10, the solution to $f'' = -f$ has a Taylor series of the form
$$f(x) = c_1 + c_2 x - \frac{c_1}{2!}x^2 - \frac{c_2}{3!}x^3 + \frac{c_1}{4!}x^4 + \frac{c_2}{5!}x^5 - \frac{c_1}{6!}x^6 - \frac{c_2}{7!}x^7 + \cdots.$$
We choose the two simplest solutions, one $c_1 = 0, c_2 = 1$, and the other with $c_1 = 1, c_2 = 0$, to define the following functions by their power series.

Definition 8.5.11. Define the **trigonometric sine and cosine** functions by
$$\sin x = x - \frac{1}{3!}x^3 + \frac{1}{5!}x^5 - \frac{1}{7!}x^7 + \cdots = \sum_{n=0}^{\infty} \frac{(-1)^n}{(2n+1)!} x^{2n+1}$$
and
$$\cos x = 1 - \frac{1}{2!}x^2 + \frac{1}{4!}x^4 - \frac{1}{6!}x^6 + \cdots = \sum_{n=0}^{\infty} \frac{(-1)^n}{(2n)!} x^{2n}$$
for $x \in \mathbb{R}$. $\diamond$

From these definitions we observe the following properties:
$$\sin 0 = 0, \cos 0 = 1, \sin(-x) = -\sin x, \cos(-x) = \cos x,$$
$$(\sin x)' = \cos x, (\cos x)' = -\sin x.$$
We state a refinement of Lemma 8.5.10, whose proof is left as an exercise.

Lemma 8.5.12. Let $f: \mathbb{R} \to \mathbb{R}$ be a twice differentiable function such that
$$f''(x) = -f(x) \text{ for all } x \in \mathbb{R}.$$
Then f is infinitely differentiable, and for all $x \in \mathbb{R}$
$$f(x) = c_1 \sin x + c_2 \cos x$$
for some constants c_1 and c_2.

We are now ready to prove the following identities.

Lemma 8.5.13. For all $x, y \in \mathbb{R}$, we have

(1) $\sin^2 x + \cos^2 x = 1$;

(2) $\sin(x + y) = \sin x \cos y + \cos x \sin y$;

(3) $\cos(x + y) = \cos x \cos y - \sin x \sin y$.

8.5. Taylor Series

Proof. Let $f(x) = \sin^2 x + \cos^2 x$. Then $f'(x) = 2\sin x \cos x + 2\cos x(-\sin x) = 0$. So $f(x)$ is a constant, which must be 1 as $f(0) = 1$.

Next fix $y \in \mathbb{R}$, and let $g(x) = \sin(x+y)$. Then $g'(x) = \cos(x+y)$ and $g''(x) = -\sin(x+y) = -g(x)$. Then by Lemma 8.5.12, $g(x) = c_1 \sin x + c_2 \cos x$ for some constants c_1, c_2. Also, $g'(x) = c_1 \cos x - c_2 \sin x$. One equation gives $g(0) = c_2$ and the other gives $g(0) = \sin y$. Similarly, $g'(0) = \cos y$ and $g'(0) = c_1$. So $c_2 = \sin y$ and $c_1 = \cos y$, showing part (2). The proof of part (3) is similar, and is left as an exercise. □

Now we try to understand some of the values of sine and cosine:

$$\sin 1 = (1 - \frac{1}{3!}) + (\frac{1}{5!} - \frac{1}{7!}) + \cdots > 1 - \frac{1}{6},$$

$$\cos 1 = (1 - \frac{1}{2!}) + (\frac{1}{4!} - \frac{1}{6!}) + \cdots > 0.$$

Using Lemma 8.5.13,

$$\cos 2 = \cos^2 1 - \sin^2 1 = 1 - 2\sin^2 1.$$

Since $\sin 1 > 5/6$, we have $2\sin^2 1 > 50/36 > 1$, so $\cos 2 < 0$. As cosine is continuous, it follows that there must be at least one zero of cosine between 1 and 2. So we can define

$$z = \inf\{x \in (1,2) : \cos x = 0\}.$$

This infimum exists since the set is bounded and nonempty. Also, since cosine is continuous, $\cos z = 0$. This number is important and is defined to be $\pi/2$. So

$$\cos\left(\frac{\pi}{2}\right) = 0 \quad \text{and} \quad 1 < \frac{\pi}{2} < 2.$$

From Lemma 8.5.13(1), we get $\sin\left(\frac{\pi}{2}\right) = 1$. Then using Lemma 8.5.13, we obtain

$$\sin \pi = \sin\left(\frac{\pi}{2}\right)\cos\left(\frac{\pi}{2}\right) + \cos\left(\frac{\pi}{2}\right)\sin\left(\frac{\pi}{2}\right) = 0.$$

Similarly, we obtain $\cos \pi = -1$, $\sin 2\pi = 0$, and $\cos 2\pi = 1$. A similar argument proves the following properties (see Exercise 8.5.7).

Lemma 8.5.14.

(1) *The fundamental period of* sin *and* cos *is* 2π: $\sin(x+2\pi) = \sin x$, $\cos(x+2\pi) = \cos x$.

(2) *For all* $x \in \mathbb{R}$, *we have* $\sin(x + \frac{\pi}{2}) = \cos x$ *and* $\cos(x + \frac{\pi}{2}) = -\sin x$.

8.5.3. The Gamma Function. We introduce the gamma function, a function defined by Euler that has many applications. The gamma function is a continuous function that interpolates the factorial function; i.e., it agrees with the factorial function on the integers. While one can imagine many continuous functions going through the points $(n, n!)$, it can be shown that the gamma function is the only function satisfying the properties of Theorem 8.5.16 and whose logarithm is a convex function; further details can be found in the Bibliographical Notes.

First write e^x for $\exp(x)$ for any $x \in \mathbb{R}$.

Definition 8.5.15. For $x > 0$ define the **gamma function** by
$$\Gamma(x) = \int_0^\infty e^{-t} t^{x-1} \, dt. \qquad \diamond$$

We start by showing that the integral gives a finite number for all $x > 0$. First note that for each $x > 0$, the function $g(t) = e^{-t} t^{x-1}$ is continuous on $[0, \infty)$, in fact differentiable, as it involves functions that have already been shown to be differentiable. When $x < 1$, we also note that we have an improper integral at $t = 0$. So we break it into two parts. First we consider t in the interval $[0, 1]$. We know that $e^t \geq 1$ for all $t > 0$, so $e^{-t} \leq 1$. Thus
$$\int_0^1 e^{-t} t^{x-1} \, dt \leq \lim_{s \to 0^+} \int_s^1 t^{x-1} \, dt = \lim_{s \to 0^+} \left(\frac{1}{x} - \frac{s^x}{x} \right) = \frac{1}{x} < \infty.$$

For the integral over the interval $[1, \infty)$, we first perform a calculation. We use L'Hôpital's rule to evaluate the limit
$$\lim_{t \to \infty} t^2 e^{-t} t^{x-1} = \lim_{t \to \infty} \frac{t^{1+x}}{e^t} = 0.$$
It follows that there exists $M \in \mathbb{N}$ so that for $t > M$,
$$|t^2 e^{-t} t^{x-1}| < 1 \quad \text{or} \quad e^{-t} t^{x-1} < \frac{1}{t^2}.$$
(Now it should be clear why we multiplied $e^{-t} t^{x-1}$ by t^2; in fact, t^α for any $\alpha > 1$ would have also worked.) Now we calculate,
$$\int_1^\infty e^{-t} t^{x-1} \, dt \leq \lim_{s \to \infty} \int_1^s \frac{1}{t^2} \, dt = \lim_{s \to \infty} \left(-\frac{1}{s} + 1 \right) = 1 < \infty.$$
Therefore, $\Gamma(x)$ exists and is finite for all $x > 0$.

Theorem 8.5.16. *The gamma function satisfies the following properties.*
(1) $\Gamma(x+1) = x\Gamma(x)$ *for* $x > 0$.
(2) *For* $n \in \mathbb{N}$, $\Gamma(n) = (n-1)!$.

Proof. We start with
$$\Gamma(x+1) = \int_0^\infty e^{-t} t^x \, dt.$$
We use integration by parts, choosing $f(t) = t^x$ and $g'(t) = e^{-t}$. Using that $f'(t) = xt^{x-1}$ and $g(t) = -e^{-t}$, we have
(8.32) $$\int_0^\infty e^{-t} t^x \, dt = \lim_{a \to 0^+} \lim_{b \to \infty} f(b)g(b) - f(a)g(a) - \int_0^\infty f'(t)g(t) \, dt.$$
After simplifying, we obtain
$$\int_0^\infty e^{-t} t^x \, dt = x \int_0^\infty e^{-t} t^{x-1} \, dt,$$
completing the proof of (1). Finally, it remains to compute
$$\Gamma(1) = \int_0^\infty e^{-t} t^0 \, dt = \int_0^\infty e^{-t} \, dt = \lim_{a \to 0^+} \lim_{b \to \infty} -e^{-b} + e^{-a} = 1.$$
Part (2) follows from part (1) and induction. $\square$

8.5.4. Products of Power Series. Let

$$f(x) = \sum_{n=0}^{\infty} a_n x^n \quad \text{and} \quad g(x) = \sum_{n=0}^{\infty} b_n x^n$$

be two power series with radius of convergence $R > 0$. We study a formula for the product

$$f(x)g(x) = \sum_{n=0}^{\infty} c_n x^n.$$

We need to prove that the product can in fact be represented by a power series and to find a formula for the coefficients c_n. This will also give a formula for the product of two infinite series, called the **Cauchy product**. As f and g are power series, by Remark 8.5.3 we can write them as

$$f(x) = f_n(x) + R_n(x) \quad \text{and} \quad g(x) = g_n(x) + S_n(x),$$

where f_n and g_n are Taylor polynomials of degree n, and R_n, S_n are the corresponding remainders and are such that

$$\lim_{n \to \infty} R_n(x) = 0, \ \lim_{n \to \infty} S_n(x) = 0, \text{ for all } |x| < R.$$

Now we find the formula for the c_n coefficients. We write

$$f_n(x) = a_0 + a_1 x + a_2 x^2 + a_3 x^3 + \cdots + a_n x^n,$$
$$g_n(x) = b_0 + b_1 x + b_2 x^2 + b_3 x^3 + \cdots + b_n x^n.$$

Then

$$f_n(x)g_n(x) = (a_0 + a_1 x + a_2 x^2 + \cdots + a_n x^n)(b_0 + b_1 x + b_2 x^2 + b_3 x^3 + \cdots + b_n x^n)$$
$$= a_0 b_0 + (a_0 b_1 + a_1 b_0) x + \cdots + (a_0 b_n + a_1 b_{n-1} + \cdots + a_n b_0) x^n$$
$$+ (a_1 b_{n-1} + a_2 b_{n-2} + \cdots + a_n b_1) x^{n+1} + \cdots + (a_n b_n) x^{2n}.$$

We see that the coefficients of the first $n+1$ terms are given by

$$c_k = \sum_{i=0}^{k} a_i b_{k-i},$$

and the remaining terms ($k = n+1, \ldots, 2n$) are given by

$$d_{k,n} = \sum_{i=k-n}^{n} a_i b_{k-i}.$$

Then we can write

$$f_n(x)g_n(x) = \sum_{k=0}^{n} c_k x^k + \sum_{k=n+1}^{2n} d_{k,n} x^k.$$

If we set $e_{k,n} = c_k$ for $k = 0, \ldots, n$ and $e_{k,n} = d_{k,n}$ for $k = n+1, \ldots, 2n$, then we have

$$f_n(x)g_n(x) = \sum_{k=0}^{2n} e_{k,n} x^k.$$

We are ready to prove the following theorem about products of power series.

Theorem 8.5.17. Let $f(x) = \sum_{n=0}^{\infty} a_n x^n$ and $g(x) = \sum_{n=0}^{\infty} b_n x^n$ be two power series with radius of convergence $R > 0$. Set $c_n = \sum_{k=0}^{n} a_k b_{n-k}$. Then the product $f(x)g(x)$ can be expressed as a power series with radius of convergence R and of the form

$$f(x)g(x) = \sum_{n=0}^{\infty} c_n x^n,$$

where

$$c_k = \sum_{i=0}^{k} a_i b_{k-i}.$$

Proof. As before, we write $f(x) = f_n(x) + R_n(x)$ and $g(x) = g_n(x) + S_n(x)$. Then

$$\begin{aligned}
f(x)g(x) &= f_n(x)g_n(x) + f_n(x)S_n(x) + g_n(x)R_n(x) + R_n(x)S_n(x) \\
&= f_n(x)g_n(x) + (f_n(x) + R_n(x))S_n(x) + g_n(x)R_n(x) \\
&= f_n(x)g_n(x) + f(x)S_n(x) + g_n(x)R_n(x) \\
&= f_n(x)g_n(x) + f(x)S_n(x) + (g_n(x) + S_n(x))R_n(x) - R_n(x)S_n(x) \\
&= f_n(x)g_n(x) + f(x)S_n(x) + g(x)R_n(x) - R_n(x)S_n(x) \\
&= \sum_{k=0}^{2n} e_{k,n} x^k + [f(x)S_n(x) + g(x)R_n(x) - R_n(x)S_n(x)].
\end{aligned}$$

Since $\lim_{n\to\infty} R_n(x) = 0$ and $\lim_{n\to\infty} S_n(x) = 0$, we have that

$$\begin{aligned}
f(x)g(x) &= \lim_{n\to\infty} \sum_{k=0}^{2n} e_{k,n} x^k \\
&= \lim_{n\to\infty} \sum_{k=0}^{n} e_{k,n} x^k \\
&= \lim_{n\to\infty} \sum_{k=0}^{n} c_k x^k \\
&= \sum_{k=0}^{\infty} c_k x^k. \qquad \square
\end{aligned}$$

The following corollary follows from Theorem 8.5.17 and Exercise 8.5.13, and it is left to the reader.

Corollary 8.5.18. Let $\sum_{n=0}^{\infty} a_n$ and $\sum_{n=0}^{\infty} b_n$ be two absolutely convergent series. Set

$$c_n = \sum_{k=0}^{n} a_k b_{n-k}.$$

Then the product $\sum_{n=0}^{\infty} a_n \cdot \sum_{n=0}^{\infty} b_n$ converges and equals

$$\sum_{n=0}^{\infty} c_n.$$

Exercises: Taylor Series

8.5.1 Give a direct calculation for the radius of convergence for the power series of $\sin(x)$ and $\cos(x)$.

8.5.2 Prove that $e^x = \exp(x)$ (see Remark 8.5.9 for the definition of e^x).

8.5.3 Prove that for all $x \in \mathbb{R}$,
$$\lim_{n \to \infty} \left(1 + \frac{x}{n}\right)^n = \exp x.$$

8.5.4 Let $f(x) = e^x$. Give another proof that $f'(x) = e^x$ by integrating term-by-term on the interval $[0, x]$ and then using the Fundamental Theorem of Calculus (first form).

8.5.5 Prove that for every $K > 0$ and $n \in \mathbb{N}$ there exists $x_0 \in \mathbb{R}$ such that $\exp(x) > Kx$ for all $x > x_0$.

8.5.6 Define the function
$$f(x) = \begin{cases} e^{-1/x^2} & \text{when } x \neq 0, \\ 0 & \text{when } x = 0. \end{cases}$$
Prove that $f(x)$ is infinitely differentiable but that it does not equal its Taylor series for all $x \neq 0$.

8.5.7 Prove Lemma 8.5.14.

8.5.8 Prove that $\cos x = 0$ if and only if $x = \frac{\pi}{2} + 2\pi n$ for some $n \in \mathbb{Z}$.

8.5.9 Define
$$\tan x = \frac{\sin x}{\cos x} \quad \text{and} \quad \sec x = \frac{1}{\cos x}$$
for $x \in \mathbb{R} \setminus \{\frac{\pi}{2} + 2n : n \in \mathbb{Z}\}$. Find formulas for $D_x \tan x$ and $D_x \sec$, and prove the identity
$$1 + \tan^x = \sec^2 x.$$

8.5.10 Define
$$\arctan x = \int_0^x \frac{1}{1+t^2}\, dt.$$
Prove that $\arctan x$ is infinitely differentiable on $(-1, 1)$ and has Taylor expansion
$$\arctan x = \sum_{n=0}^{\infty} \frac{(-1)^n}{(2n+1)} x^{2n+1}.$$

8.5.11 Evaluate $\int_0^{\infty} x e^{x^2}\, dx$.

8.5.12 Complete the details in the calculation of (8.32).

8.5.13 (Abel) Let $\sum_{n=0}^{\infty} a_n$ be a convergent series. Prove that $\sum_{n=0}^{\infty} a_n x^n$ converges absolutely for $|x| < 1$ and
$$\lim_{x \to 1^-} \sum_{n=0}^{\infty} a_n x^n = \sum_{n=0}^{\infty} a_n.$$

8.5.14 Prove Corollary 8.5.18.

8.5.15 Prove that the series $\sum_{n=0}^{\infty} (-1)^{n+1} \frac{1}{\sqrt{n+1}}$ converges conditionally and that when taking its Cauchy product with itself, the product does not converge.

8.6. Weierstrass Approximation Theorem

Polynomials are the simplest class of functions we have studied. We can think of power series as *infinite polynomials*. We have seen that if a function is represented by a power series, that power series must be its Taylor polynomial. At the same time, we have seen that not every function, even when infinitely differentiable, can be represented by its Taylor series. Surprisingly, every continuous function on a compact set can be approximated uniformly by a sequence of polynomials. So even a continuous but nowhere differentiable function can be uniformly approximated by polynomials. We have seen that if a sequence f_n converges uniformly to a function f on an interval I, the continuity and Riemann integration properties of f_n carry over to f. But differentiability properties do not carry over, as polynomials are differentiable and they can approximate a continuous but nowhere differentiable function.

Let
$$q(x) = (1 - x^2).$$

If we raise q to a high power n, the value of q^n for points around 0 will stay close to 1, but for points away from 0, the values of q^n will be small (as in Figure 8.7). Set
$$d_n = \int_{-1}^{1} q^n(x) \, dx$$
and
$$q_n(x) = \frac{q^n(x)}{d_n} = \frac{(1-x^2)^n}{\int_{-1}^{1}(1-x^2)^n \, dx}.$$

Then by definition the functions q_n integrate to 1: $\int_{-1}^{1} q_n = 1$.

Lemma 8.6.1. *Let $f : \mathbb{R} \to \mathbb{R}$ be a Riemann integrable function such that $f(t) = 0$ for all $t \notin [0, 1]$, and let $x \in \mathbb{R}$. The functions*
$$I_n(x) = \int_{-1}^{1} (1 - y^2)^n f(x - y) \, dy$$
are polynomials in x.

8.6. Weierstrass Approximation Theorem

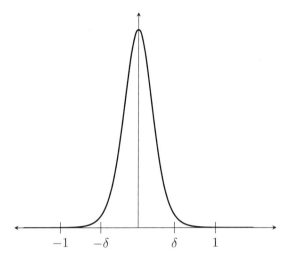

Figure 8.7. The functions q_n.

Proof. We know that $f(x-y) = 0$ when $x-y$ is outside the interval $[0,1]$. Now, if $0 \leq x - y \leq 1$, then $x - 1 \leq y \leq x$. So the integral becomes

$$I_n(x) = \int_{x-1}^{x} (1-y^2)^n f(x-y) \, dy.$$

To apply substitution, we write it as

$$I_n(x) = \int_{x-1}^{x} (1-(x-(x-y))^2)^n f(x-y) \, dy.$$

Now using substitution with $g(y) = x - y$, as $g'(y) = -1$, we obtain

$$I_n(x) = (-1) \int_{1}^{0} (1-(x-t)^2)^n f(t) \, dt = \int_{0}^{1} (1-(x-t)^2)^n f(t) \, dt.$$

The term $(1-(x-t)^2)^n$ expands into a polynomial in x of degree $2n$. After integrating with respect to t, we see that $I_n(x)$ is a polynomial in x. □

Lemma 8.6.2. *Let $\delta > 0$. Then*

$$\lim_{n \to \infty} \int_{-1}^{-\delta} q_n(x) \, dx = 0 \quad \text{and} \quad \lim_{n \to \infty} \int_{\delta}^{1} q_n(x) \, dx = 0.$$

Proof. First we note that if $\delta \leq x \leq 1$, then

$$(1-x^2)^n \leq (1-\delta^2)^n.$$

For our next estimate we first use the Bernoulli inequality to obtain

$$(1-x^2)^n \geq 1 - nx^2.$$

Then

$$d_n = \int_{-1}^{1} (1-x^2)^n \, dx$$
$$\geq \int_{0}^{1} (1-x^2)^n \, dx$$
$$\geq \int_{0}^{1/\sqrt{n}} (1 - nx^2) \, dx$$
$$= \frac{1}{\sqrt{n}} - \frac{n}{3}\frac{1}{n^{3/2}} = \frac{2}{3}\frac{1}{\sqrt{n}}.$$

Then, for $0 < \delta \leq x \leq 1$, as $(1-\delta^2) < 1$,

$$\lim_{n \to \infty} q_n(x) = \lim_{n \to \infty} \frac{(1-x^2)^n}{\int_{-1}^{1}(1-x^2)^n \, dx}$$
$$\leq \lim_{n \to \infty} (1-\delta^2)^n \frac{3}{2}\sqrt{n}$$
$$= 0.$$

Furthermore, this estimate shows that the convergence is uniform over $[\delta, 1]$. Therefore,

$$\lim_{n \to \infty} \int_{\delta}^{1} q_n(x) \, dx = \int_{\delta}^{1} \lim_{n \to \infty} q_n(x) \, dx = 0.$$

A similar argument shows that we also have

$$\lim_{n \to \infty} \int_{-1}^{-\delta} q_n(x) \, dx = 0. \qquad \square$$

Theorem 8.6.3. *Let $f : [0,1] \to \mathbb{R}$ be a continuous function such that $f(0) = 0$ and $f(1) = 0$. Let*

$$p_n(x) = \frac{I_n(x)}{d_n} = \int_{-1}^{1} \frac{(1-y^2)^n}{d_n} f(x-y) \, dy.$$

Then the sequence of polynomials (p_n) converges uniformly to f on $[0,1]$.

Proof. First extend f to all of $\mathbb{R}$ by setting $f(x) = 0$ for all $x \in \mathbb{R} \setminus [0,1]$. As f is 0 at the endpoints of $[0,1]$, it follows that this extension is continuous. By Lemma 8.6.1, each p_n is a polynomial in x.

8.6. Weierstrass Approximation Theorem

We start evaluating

$$|f(x) - p_n(x)| = \left| f(x) - \int_{-1}^{1} \frac{(1-y^2)^n}{d_n} f(x-y) \, dy \right|$$

$$= \left| \int_{-1}^{1} \frac{(1-y^2)^n}{d_n} f(x) \, dy - \int_{-1}^{1} \frac{(1-y^2)^n}{d_n} f(x-y) \, dy \right|$$

$$= \left| \int_{-1}^{1} \frac{(1-y^2)^n}{d_n} (f(x) - f(x-y)) \, dy \right|$$

$$\leq \int_{-1}^{1} \left| \frac{(1-y^2)^n}{d_n} (f(x) - f(x-y)) \right| dy$$

$$= \int_{-\delta}^{\delta} |q_n(y)(f(x) - f(x-y))| \, dy$$

$$+ \int_{-1}^{-\delta} |q_n(y)(f(x) - f(x-y))| \, dy$$

$$+ \int_{\delta}^{1} |q_n(y)(f(x) - f(x-y))| \, dy.$$

Since f is uniformly continuous on $[0,1]$, given $\varepsilon > 0$ there exists $\delta > 0$ such that when $|y| < \delta$, then

$$|f(x) - f(x-y)| < \frac{\varepsilon}{3}.$$

Then

$$\int_{-\delta}^{\delta} |q_n(y)(f(x) - f(x-y))| \, dy < \int_{-\delta}^{\delta} q_n(y) \frac{\varepsilon}{3} \, dy$$

$$= \frac{\varepsilon}{3} \int_{-\delta}^{\delta} q_n(y) \, dy$$

$$= \frac{\varepsilon}{3}.$$

Since f is continuous on $[0,1]$, it is bounded by a constant $M \in \mathbb{R}$. So, as f vanishes outside $[0,1]$, for all $x, y \in \mathbb{R}$,

$$|f(x) - f(x-y)| \leq 2M.$$

Now we use Lemma 8.6.2, applied to this δ, to obtain $N \in \mathbb{N}$ so that for all $n \geq N$ we have

$$\int_{-1}^{-\delta} q_n(x) \, dx < \frac{\varepsilon}{6M} \quad \text{and} \quad \int_{\delta}^{1} q_n(x) \, dx < \frac{\varepsilon}{6M}.$$

Then

$$\int_{-1}^{-\delta} |q_n(y)(f(x) - f(x-y))| \, dy \leq (2M) \int_{-1}^{-\delta} q_n(y) \, dy$$

$$< (2M) \cdot \frac{\varepsilon}{6M}$$

$$= \frac{\varepsilon}{3}.$$

Similarly, we obtain

$$\int_{\delta}^{1} |q_n(y)(f(x) - f(x - y))| \, dy < \frac{\varepsilon}{3}.$$

Then

$$|f(x) - p_n(x)| < \frac{\varepsilon}{3} + \frac{\varepsilon}{3} + \frac{\varepsilon}{3} = \varepsilon \text{ for all } x \in [0, 1].$$

Therefore, the sequence of polynomials (p_n) converges to f uniformly on $[0, 1]$. □

Theorem 8.6.4 (Weierstrass approximation theorem). *Let $f : [a, b] \to \mathbb{R}$ be a continuous function. Then there exists a sequence of polynomials (p_n) converging uniformly to f on $[a, b]$.*

Proof. By a change of coordinates and a translation, one can reduce the question to a function f defined on the interval $[0, 1]$ and such that $f(0) = f(1) = 0$. Then Theorem 8.6.3 completes the proof. □

Exercises: Weierstrass Approximation Theorem

8.6.1 Complete the second part in the proof of Lemma 8.6.2.

8.6.2 Let $f : [a, b] \to \mathbb{R}$ be a continuous function. Change the function f to a function $g : [0, 1] \to \mathbb{R}$ with $g(0) = g(1) = 0$ so that if g can be uniformly approximated by polynomials, then so can f.

8.7. The Complex Exponential

This is a short section where the reader is asked to provide several of the arguments. Sequences of complex numbers were covered in Exercise 2.1.19, but we will recall the definition here. It is a special case of convergence of sequences in metric spaces (for the reader who is familiar with the metric space sections). Suppose that for each $n \in \mathbb{N}$, z_n is a complex number. We say the sequence (z_n) converges to $z \in \mathbb{C}$ if for all $\varepsilon > 0$ there exists $N \in \mathbb{N}$ such that

$$n \geq N \text{ implies } |z - z_n| < \varepsilon.$$

The convergence of a series $\sum_{n=0}^{\infty} z_n$ of complex numbers is defined in a way analogous to the case of real numbers by considering the sequence of partial sums

$$S_n = \sum_{k=1}^{n} z_k.$$

Then we say that the series $\sum_{n=1}^{\infty} z_n$ converges when S_n converges. The reader will note that formally these definitions look like the definitions of convergence of sequences and series of real numbers, the difference being, instead of the absolute value for real numbers, we use the modulus of complex numbers. Because of this similarity, many of proofs of basic facts about series of complex numbers are similar

8.7. The Complex Exponential

to the real case and are left to the exercises. We can define absolute convergence of a series complex number in a analogous way to absolute convergence for real numbers, and also the Cauchy property for sequences and for series. For example, the Cauchy criterion for convergence (Theorem 7.2.12) holds with a similar proof. If a series is absolutely convergent, then it converges (Exercise 8.7.8).

For example, we verify that the geometric series

$$\sum_{n=1}^{\infty} z^{n-1} = \frac{1}{1-z}$$

converges for every complex number z that satisfies $|z| < 1$. In fact, by the same field properties of the real numbers that we used for the geometric series, we obtain

$$\sum_{n=1}^{k} z^{n-1} = \frac{1-z^k}{1-z}.$$

When $|z| < 1$, we have $\lim_{k\to\infty} |z|^k = 0$, which means that $\lim_{k\to\infty} z^k = 0$, so

$$\lim_{k\to\infty} \sum_{n=1}^{k} z^{n-1} = \lim_{k\to\infty} \frac{1-z^k}{1-z} = \frac{1}{1-z}.$$

The geometric series we have seen is an example of a power series. One can consider power series of the form $\sum_{n=0}^{\infty} a_n z^n$, where $z \in \mathbb{C}$ and $a_n \in \mathbb{C}$ for $n \in \mathbb{N}$. As in the real case, the power series has a radius of convergence, that is a number $R \in [0,\infty]$ such that it converges absolutely for all $|z| < R$ and diverges for $|z| > R$. The proofs are essentially the same as in the real case.

In the case of the geometric series we can in fact show absolute convergence since

$$\lim_{k\to\infty} \sum_{n=1}^{k} |z|^{n-1} = \lim_{k\to\infty} \frac{1-|z|^k}{1-|z|} = \frac{1}{1-|z|},$$

for $|z| < 1$. Since we know the limit does not exist for $|z| = 1$, it follows that the radius of convergence of this series is 1.

The series

$$\sum_{n=0}^{\infty} \frac{z^n}{n!}$$

has infinite radius of convergence, so it converges for every $z \in \mathbb{C}$ (Exercise 8.7.5).

Definition 8.7.1. We define the **complex exponential** by

$$\exp(z) = \sum_{n=0}^{\infty} \frac{z^n}{n!}.$$

$\Diamond$

We wish to evaluate e^z for $z = ix$. Note that

$$i^n = \begin{cases} 1 & \text{if } n = 4k, \text{ for some } k \in \mathbb{N}, \\ i & \text{if } n = 4k+1, \\ -1 & \text{if } n = 4k+2, \\ -i & \text{if } n = 4k+3. \end{cases}$$

Then

(8.33) $$\exp(ix) = \sum_{n=0}^{\infty} \frac{i^n x^n}{n!} = 1 + \frac{ix}{1} + \frac{(-1)x^2}{2!} + \frac{(-i)x^3}{3!} + \cdots$$
$$= \left(1 - \frac{x^2}{2!} + \cdots\right) + i\left(x - \frac{x^3}{3!} + \cdots\right)$$
$$= \sum_{n=0}^{\infty} \frac{(-1)^n}{(2n)!} x^{2n} + i \sum_{n=0}^{\infty} \frac{(-1)^n}{(2n+1)!} x^{2n+1}.$$

Thus when x is a real number, we obtain that for all $x \in \mathbb{R}$,
$$\exp(ix) = \cos x + i \sin x.$$
Then by letting $x = \pi$, we have the remarkable formula, known as **Euler's identity**,
$$\exp(i\pi) + 1 = 0.$$
We will use the following formula that we proved for the real exponential and that also holds for the complex exponential (Exercise 8.7.4),
$$\exp(z + w) = \exp(z) \exp(w).$$
Therefore,
$$\exp(2\pi i) = \exp(i\pi) \exp(i\pi) = 1.$$
It follows that for all complex numbers z,

(8.34) $$\exp(z + 2\pi i) = \exp(z).$$

Exercises: The Complex Exponential

8.7.1 Let $\sum_{n=0}^{\infty} a_n z^n$ be a power series of complex numbers. Prove that there exists $R \in [0, \infty]$ so that the series converges absolutely for all z such that $|z| < R$ and diverges for all $|z| > R$ (where in the case $R = \infty$ we interpret this as converging absolutely for all z).

8.7.2 Justify the calculations in equation (8.33).

* 8.7.3 Formulate and prove the analogue of Corollary 8.5.18 for the case of series of complex numbers.

8.7.4 Prove that for all $z, w \in \mathbb{C}$, $\exp(z + w) = \exp(z) \exp(w)$. (*Hint*: Use Exercise 8.7.3.)

8.7.5 Prove that the series $\sum_{n=0}^{\infty} \frac{z^n}{n!}$ has infinite radius of convergence.

8.7.6 Use equation (8.34) to give another proof that for all $x \in \mathbb{R}$, we have $\sin(x + 2\pi) = \sin x$ and $\cos(x + 2\pi) = \cos x$.

8.7.7 State and prove the Cauchy criterion for convergence similar to Theorem 7.2.12, and also for absolute convergence.

8.7.8 Prove that if a series of complex numbers is absolutely convergent, then it converges (i.e., if $\sum_{n=1}^{\infty} |a_n z^n|$ converges, then $\sum_{n=1}^{\infty} a_n z^n$ converges). (*Hint*: Use Exercise 8.7.7).

Appendix A

Solutions to Questions

Chapter 0

0.3.6: As this is an "if and only if" statement, we have to show two implications. We need to show that if $A \cup B = A$, then $B \subseteq A$, and that if $B \subseteq A$, then $A \cup B = A$. Assume first that $A \cup B = A$. To show that $B \subseteq A$, let $x \in B$. Then $x \in A \cup B$. Therefore, $x \in A$, showing that $B \subseteq A$. Next assume that $B \subseteq A$. It is clear that $A \subseteq A \cup B$. To show that $A \cup B = A$, it remains to show that $A \cup B \subseteq A$. Now let $x \in A \cup B$. Then $x \in A$ or $x \in B$. If $x \in B$, by assumption, $x \in A$. So in either case $x \in A$, meaning that $A \cup B \subseteq A$.

0.3.9: Let $x \in (A^c)^c$. Then $x \notin A^c$, which means that x is in A. Now let $x \in A$. Then it is not the case that $x \in A^c$, so $x \in (A^c)^c$.

0.3.11: Let $x \in A \cap B$. We need to show x is not in $(C \setminus A) \cup (C \setminus B)$. That means that it is not in $C \setminus A$ and is not in $C \setminus B$. But since x is in A it follows that it is not in $C \setminus A$, and similarly it is not in $C \setminus B$. Therefore, x is in $C \setminus [(C \setminus A) \cup (C \setminus B)]$. Now suppose that $x \in C \setminus [(C \setminus A) \cup (C \setminus B)]$. So x is not in $(C \setminus A) \cup (C \setminus B)$, which as we have seen means that $x \in A \cap B$.

0.3.12: We know that $\varnothing \subseteq A \cap \varnothing$. Now suppose $A \cap \varnothing$ is not a subset of $\varnothing$. Then there exists an element x of $A \cap \varnothing$. Then x is an element of A and an element of $\varnothing$. But this contradicts that $\varnothing$ has no elements. So $A \cap \varnothing$ is a subset of $\varnothing$.

0.3.16: Let $x \in A \setminus B$. Then x is in A and is not in B, so $x \in A \cap B^c$. Conversely, if $x \in A \cap B^c$, then x is in A and x is in B^c, so x is in A and x is not in B, which means x is in $A \setminus B$.

0.3.17: Let $x \in A \triangle B$. Then $x \in (A \setminus B) \cup (B \setminus A)$. If $x \in (A \setminus B)$, then $x \in A \cup B$. Also, as x is not in B, it is not in any subset of B, so it is not in $A \cap B$. Therefore, x is in $A \cup B \setminus (A \cap B)$. Now if x is in $B \setminus A$, then similarly x is in $A \cup B$ and is not in $A \cap B$. For the converse let $x \in A \cup B \setminus (A \cap B)$. Then x is in A or B but not in both. Suppose $x \in A$. Then x is not in B, so $x \in A \setminus B$. If $x \in B$, one similarly shows that $x \in B \setminus A$. Therefore, $x \in A \setminus B \cup B \setminus A$.

0.4.3: Let $a \in A$. As B is nonempty, we can choose $b \in B$. Then $(a, b) \in A \times B$. Thus (a, b) is in $B \times A$, so $a \in B$, which means $A \subseteq B$. Now let $b \in B$. Choose $a \in A$; then $(a, b) \in A \times B$. So $(a, b) \in B \times A$, which implies $b \in A$. Therefore, $B \subseteq A$, and thus $A = B$. The converse is clear from the properties of equality.

0.4.4: Here we must show that $A \times \emptyset$ is the empty set. Suppose $A \times \emptyset$ is not empty. Then it has an element, and this element must be of the form (a, b) where $a \in A$ and $b \in \emptyset$. But this contradicts that $\emptyset$ is empty. So $A \times \emptyset$ is the empty set.

0.4.17: Suppose that $g(f(x)) = x$ for all $x \in X$. To show that g is the inverse of f, we need to show that $g(y) = x$ if and only if $f(x) = y$. As f is surjective, there exists an element $a \in X$ such that $f(a) = y$. Then $g(f(a)) = x$, which by the assumption means that $a = x$; therefore, $f(x) = y$. Next suppose $f(x) = y$. Then $g(f(x)) = g(y)$, which again the by the assumption means that $x = g(y)$. Therefore, g is the inverse of f.

For the second part suppose that $f(g(y)) = y$ for all $y \in Y$. We start by assuming that $g(y) = x$. Then $f(g(y)) = f(x)$, so $y = f(x)$. Finally, assume $f(x) = y$. We know g is defined on $y \in Y$, and we wish to show that $g(y) = x$. Say $g(y) = z$. Then $f(g(y)) = f(z)$. So $y = f(z)$ and $f(x) = f(z)$. As f is injective, it follows that $x = z$, and this completes the proof.

Chapter 1

1.1.9: We note that $0 - (-1) = 1 > 0$, which means that $0 > -1$ or $-1 < 0$. Also, $-3 - (-2) = -1 < 0$, so $-3 < -2$. (Another short proof can be given based on Proposition 1.1.10.)

1.1.12: Suppose that $xy > 0$. If $x > 0$, then $1/x > 0$, so $(1/x)xy > 0$, which implies $y > 0$. If $x < 0$, then $1/x < 0$, so $(1/x)xy < 0$, which implies $y < 0$. Now suppose $xy < 0$. If $x > 0$, then $(1/x)xy < 0$, so $y < 0$. Similarly, if $x < 0$, then $(1/x)xy > 0$, so $y > 0$.

1.2.4: Suppose that b in F is an upper bound for the set S. Then $x \leq b$ for all $x \in S$. Thus $-x \geq -b$ for all $x \in S$, which means that $-b$ is a lower bound for $-S$. Similarly, if $-b \leq y$ for all y in $-S$, then y is of the form $y = -x$ for some $x \in S$. So $-b \leq -x$ for all x in S. Thus $b \geq x$ for all x in S, which means that b is an upper bound for S.

1.2.6: Let $a = \sup A$, $b = \sup B$, and $c = \max\{\sup A, \sup B\}$. Let $x \in A \cup B$. If $x \in A$, then $x \leq a$ and if $x \in B$, then $x \leq b$. Thus c is an upper bound for $A \cup B$. Now let d be an arbitrary upper bound for $A \cup B$. If $x \in A$, then $x \leq d$, and similarly if $x \in B$, then $x \leq d$. So $a \leq d$ and $b \leq d$. This means that $c \leq d$. It follows that c is the supremum of $A \cup B$. Now if $A \subseteq B$, then $A \cup B = B$, so $\sup B \geq \sup A$.

A. Solutions to Questions 285

1.2.8: Let β be an upper bound of $S \subseteq F$, and suppose $\beta \in S$. To show that $\beta = \sup S$, we need to show that if γ is an upper bound of S, then $\beta \leq \gamma$. But this is an immediate consequence of the fact that γ is an upper bound of S and $\beta \in S$.

1.3.5: Let B be a countable set, and let $A \subseteq B$. By Proposition 1.3.4 there is a function $g : A \to \mathbb{N}$ that is injective. Since B is a subset of A, there is a natural restriction of g to B: define $f : B \to \mathbb{N}$ by $f(b) = g(b)$ for all $b \in B$. Then f is also injective and Proposition 1.3.4 shows that B is countable. (As an exercise the reader should give a proof of this fact without appealing to Proposition 1.3.4.)

1.3.12: If $\bigcap_{n\in\mathbb{N}}\{i\in\mathbb{N} : i>n\}$ is not empty, then there exists $k\in\bigcap_{n\in\mathbb{N}}\{i\in\mathbb{N} : i>n\}$. Then for all $n \in \mathbb{N}$, $k > n$. Since $k \in \mathbb{N}$ we can consider $n = k+1 \in \mathbb{N}$. But then k is not greater than this n, a contradiction. Thus the set is empty.

1.3.15 : The idea of the proof is to reduce this to Theorem 1.3.14. Suppose first that Γ is countably infinite. Then there exists a bijection $f : \mathbb{N} \to \Gamma$. The sets $\bigcup_{\alpha\in\Gamma} A_\alpha$ and $\bigcup_{n\in\mathbb{N}} A_{f(n)}$ are the same. Let $B_n = A_{f(n)}$ for each $n \in \mathbb{N}$. Then each B_n is countable and by Theorem 1.3.14 the set $\bigcup_{n\in\mathbb{N}} B_n$ is countable, so $\bigcup_{\alpha\in\Gamma} A_\alpha$ is countable. Next, if Γ is not countably infinite, then there exists a bijection $g : J_m \to \Gamma$ for some $m \in \mathbb{N}$. For $n \in \{1,\dots,m\}$ define $B_n = A_{g(n)}$ and for $n > m$, let $B_n = \varnothing$. Then again the sets $\bigcup_{\alpha\in\Gamma} A_\alpha$ and $\bigcup_{n\in\mathbb{N}} B_n$ are the same, and Theorem 1.3.14 completes the proof.

1.4.9: Let $\alpha > [0]$. Then there exists $p \in \alpha$ such that $p > 0$, hence $\alpha \neq [0]$. Suppose now to the contrary that $-\alpha > [0]$. Then there exists q in $-\alpha$ such that $q > 0$. So $-q < 0$, which implies that for all rational $r > 0$, we have $-q - r < 0$. Thus $-q - r \in \alpha$, contradicting that q is in $-\alpha$. Therefore, $-\alpha < [0]$.

1.4.14: For part (a), let $r \in \alpha$. We need to show that $\alpha - [r] \geq [0]$. But $\alpha - [r] \geq \alpha + [-r]$. There exists $r' \in \alpha$ such that $r' > r$. Then $r' - r > 0$ and $r' - r \in \alpha - [r]$, hence $\alpha - [r] > [0]$. For the converse suppose $\alpha - [r] > [0]$. Then there exists $s \in \alpha$ and $t \in [-r]$ such that $s + t > 0$. Then $-t < s$. But $r < -t$, hence $r < s$, which implies $r \in \alpha$.

For part (b), suppose $\alpha > \beta$. Then $\alpha - \beta > [0]$. This means that there is an element of $\alpha - \beta$ that is positive. This element must be of the form $p + q$ where $p \in \alpha$ and $q \in -\beta$. Then $-q - r \notin \beta$ for some $r > 0$. Now let $x \in \beta$ and suppose that $x \notin \alpha$. Then $-q - r > x$, so $-r > x + q$ or $p - r > x + p + q > x$. Now we observe that $p - r \in \alpha$, this is so since $p \in \alpha$ and $r > 0$. But this gives a contradiction, so $x \in \alpha$. Therefore, $\beta \subseteq \alpha$.

Now suppose that $\beta \subseteq \alpha$ and $\alpha \neq \beta$. Let q be in α but not in β. Then $\beta \subseteq [q]$ and $[q] \subseteq \alpha$. By part (a), $\beta \leq [q]$ and $[q] \leq \alpha$. Since we know $\mathbb{R}^+$ is an ordered field, $\beta \leq \alpha$.

Chapter 2

2.1.3: Suppose we know that $\forall\,\varepsilon > 0\ \exists N \in \mathbb{N}$ such that $n \geq N \implies |a_n - L| \leq \varepsilon$. Let $\varepsilon_1 > 0$. Then choose $\varepsilon = \varepsilon_1/2$. Then for all $n \geq N$, we have that $|a_n - L| \leq \varepsilon = \varepsilon_1/2 < \varepsilon_1$. Thus $\lim_{n\to\infty} a_n = L$. The converse is immediate as $|a_n - L| < \varepsilon \leq \varepsilon$. The second part is similar.

2.1.5: The simplest proof is using the Proposition 2.1.4. As (a_n) converges to L, if (a_n) also converges to M, then by Proposition 2.1.4, $L = M$, a contradiction.

The second proof uses the definition. As $M \neq L$ we can set $\varepsilon = \frac{|M-L|}{2} > 0$. Since $\lim_{n\to\infty} a_n = L$, for this ε there exists $N \in \mathbb{N}$ so that $|a_n - L| < \varepsilon$ when $n \geq N$. The idea is that as a_n is close to L, it has to be at a certain distance from M. We make this formal by using an inequality. We have

$$|L - M| \leq |L - a_n| + |a_n - M| < \frac{|M-L|}{2} + |a_n - M|.$$

So

$$|a_n - M| > \frac{|M-L|}{2}$$

for all $n \geq N$, which means that (a_n) cannot converge to M.

2.1.12: Let $a_n = n$ and $b_n = \frac{1}{n} - n$. Then $a_n + b_n = \frac{1}{n}$ converges to 0, but (a_n) and (b_n) diverge. It is true that they diverge to $+\infty$ and $-\infty$, respectively, but one can also choose $a_n = (-1)^n n$, $b_n = \frac{1}{n} - (-1)^n n$.

2.1.13: There exists $M \in \mathbb{N}$ such that $|b_n| \leq M$ for all $n \in \mathbb{N}$. Let $\varepsilon > 0$. Choose $N \in \mathbb{N}$ so that $|a_n| < \frac{\varepsilon}{M}$ for all $n \geq N$. Then $|a_n b_n| \leq |a_n| M < \varepsilon$ for all $n \geq N$. Therefore, $\lim_{n\to\infty} a_n b_n = 0$. To see that (b_n) needs to be bounded choose $a_n = \frac{1}{n}$ and $b_n = n$. Then $\lim_{n\to\infty} a_n = 0$ but $\lim_{n\to\infty} a_n b_n = 1 \neq 0$. One could also have $(a_n b_n)$ diverging if $b_n = (-1)^n n$.

2.1.14: We write

$$\begin{aligned}(a_n c_n + b_n d_n) - L &= (a_n c_n + b_n d_n) - (c_n + d_n)L \\ &= (a_n c_n + b_n d_n) - c_n L - d_n L \\ &= (a_n - L)c_n + (b_n - L)d_n.\end{aligned}$$

We observe that, as (c_n) is bounded, $\lim_{n\to\infty}(a_n - L)c_n = 0$, and similarly

$$\lim_{n\to\infty}(b_n - L)d_n = 0.$$

Since $\lim_{n\to\infty} L$ exists, it follows that $\lim_{n\to\infty}(a_n c_n + b_n d_n)$ exists and equals L. (The only place where $c_n \geq 0$ is used is to show it is bounded.)

2.1.17: Suppose that for each $\alpha \in \mathbb{N}$, there is an integer $N \in \mathbb{N}$ such that $a_n > \alpha$ for all $n \geq N$. To show that $\lim_{n\to\infty} a_n = \infty$, let $\alpha \in \mathbb{R}$. By the Archimedean property we can choose $\alpha' \in \mathbb{N}$ such that $\alpha' > \alpha$. From the condition we know there is $N \in \mathbb{N}$ so that $a_n > \alpha'$ for all $n \geq N$. It follows that $a_n > \alpha$ for all $n \geq N$. So $\lim_{n\to\infty} a_n = \infty$. The converse is more direct. If $\lim_{n\to\infty} a_n = \infty$, then we can choose any $\alpha \in \mathbb{R}$, in particular we can choose it in $\mathbb{N}$.

2.1.18: Suppose (a_n) is a sequence such that $\lim_{n\to\infty} a_n = \infty$. Let $L \in \mathbb{R}$. Let $\varepsilon = 1$. Then there exists N such that for all $n \geq N$, we have that $a_n > L + 1$. So $|a_n - L| > 1$. Therefore, (a_n) does not converge.

Chapter 3

3.2.4: Let $A \subseteq \mathbb{R}$, and suppose $A \neq \varnothing$ so there exists $x \in A$. If A is open, there exists $\varepsilon > 0$ such that the open ball $B(x, \varepsilon)$ is a subset of A. But $B(x, \varepsilon)$ is uncountable, so A is uncountable.

3.2.17: It suffices to show that a closure point of $\bar{A}$ is a closure point of A. Let p be a closure point of $\bar{A}$. Then there exists a sequence $(x_n)_n$, with each $x_n \in \bar{A}$, that converges to p. For each n there exists a sequence $(y_{n,k})_k$, with each $y_{n,k} \in A$, converging to x_n. We need to find a sequence in A converging to p. For each $i \in \mathbb{N}$, choose $n_i \in \mathbb{N}$ so that $|p - x_{n_i}| < \frac{1}{2i}$. Now for each $n_i \in \mathbb{N}$, choose y_{n_i, k_i} so that $|x_{n_i} - y_{n_i, k_i}| < \frac{1}{2i}$. Let $a_i = y_{n_i, k_i} \in A$. We claim that $(a_i)_i$ converges to p. In fact let $\varepsilon > 0$. Choose $N \in \mathbb{N}$ such that $1/N < \varepsilon$. Then for $i > N$, $|p - a_i| = |p - x_{n_i} + x_{n_i} - y_{n_i, k_i}| \leq |p - x_{n_i}| + |x_{n_i} - y_{n_i, k_i}| < \frac{1}{2i} + \frac{1}{2i} < \varepsilon$.

3.5.2: Let $x \in K$. Then $x \in F_n$ for all $n \in \mathbb{N}$. Since $F_1 \subseteq F$, $x \in F$. Also, from the definition of F_n, $x \notin G_{n-1}$ for all $n \in \mathbb{N}$. So $x \notin \bigcup_{n \geq 0} G_n$. Thus $x \in F \setminus \bigcup_{n \geq 0} G_n$.

Now if $x \in F \setminus \bigcup_{n \geq 0} G_n$, then $x \in F$ and $x \notin G_n$ for all $n \geq 0$. We note that $x \in F_1$ since $F_1 = F \setminus G_0$. Thus $x \in F_2$ since $F_2 = F_1 \setminus G_1$. In this way, as $F_n = F_{n-1} \setminus G_{n-1}$, one shows by induction that $x \in F_n$ for all $n \geq 1$. Therefore, $x \in \bigcap_{n \geq 0} F_n = K$.

Chapter 4

4.1.3: Let $x \in A$ be an isolated point of A. Let $\varepsilon > 0$. Since x is an isolated point of A there exists $\delta > 0$ such that $B(x, \delta)$ is disjoint from A. Then for this δ, if $|x - y| < \delta$, then y cannot be in the domain of f. So the condition for continuity holds trivially.

4.2.2: Let $f : [1, 3]$ be defined by $f(x) = x$ when $1 \leq x \leq 2$ and $f(x) = x + 2$ when $2 < x \leq 3$. Then $f(1) < 3 < f(3)$, but there is no x such that $f(x) = 3$. Similarly, if $f : [1, 2] \cup [4, 5] \to \mathbb{R}$ is defined by $f(x) = x$, then f is continuous, but there is no x such that $f(x) = 3$.

4.5.9: Let (a_n) and (b_n) be Cauchy sequences. By the triangle inequality,

$$d(a_n, b_n) \leq d(a_n, a_m) + d(a_m, b_n)$$
$$\leq d(a_n, a_m) + d(a_m, b_m) + d(b_m, b_n).$$

Similarly,

$$d(a_m, b_m) \leq d(a_m, a_n) + d(a_n, b_n) + d(b_n, b_m).$$

Therefore,

$$d(a_n, b_n) - d(a_m, b_m) \leq d(a_n, a_m) + d(b_m, b_n),$$
$$d(a_m, b_m) - d(a_n, b_n) \leq d(a_m, a_n) + d(b_n, b_m),$$

which gives
$$|d(a_n, b_n) - d(a_m, b_m)| \leq d(a_n, a_m) + d(b_m, b_n).$$
Let $\varepsilon > 0$. There exists N_1 so that $d(a_n, a_m) < \varepsilon/2$ for all $m, n \geq N_1$, and there exists N_2 so that $d(b_n, b_m) < \varepsilon/2$ for all $m, n \geq N_2$. Hence, if $m, n \geq \max\{N_1, N_2\}$,
$$|d(a_n, b_n) - d(a_m, b_m)| < \frac{\varepsilon}{2} + \frac{\varepsilon}{2} = \varepsilon.$$
Thus $(d(a_n, b_n))$ is a Cauchy sequence.

4.5.11: Let $[(a'_n)] = [(a_n)]$ and $[(b'_n)] = [(b_n)]$. By the triangle inequality,
$$d(a'_n, b'_n) \leq d(a'_n, a_n) + d(a_n, b'_n)$$
$$\leq d(a'_n, a_n) + d(a_n, b_n) + d(b_n, b'_n).$$
Similarly,
$$d(a_n, b_n) \leq d(a_n, a'_n) + d(a'_n, b'_n) + d(b'_n, b_n).$$
Let $\lim_{n \to \infty} d(a_n, b_n) = \ell$ and $\lim_{n \to \infty} d(a'_n, b'_n) = \ell'$ (we know the limits exist as they are Cauchy sequences by Question 4.5.9). As $\lim_{n \to \infty} d(a'_n, a_n) = 0$ and $\lim_{n \to \infty} d(b'_n, b_n) = 0$, it follows that $\ell' \leq \ell$ and $\ell \leq \ell'$.

Chapter 5

5.1.16: We note that
$$\frac{f(q_n) - f(p_n)}{q_n - p_n} = \frac{f(q_n) - f(x)}{q_n - p_n} + \frac{f(x) - f(p_n)}{q_n - p_n}$$
$$= \frac{f(q_n) - f(x)}{q_n - x} \cdot \frac{q_n - x}{q_n - p_n} + \frac{f(x) - f(p_n)}{x - p_n} \cdot \frac{x - p_n}{q_n - p_n}.$$
Let
$$a_n = \frac{f(q_n) - f(x)}{q_n - x}, \quad b_n = \frac{f(x) - f(p_n)}{x - p_n}$$
and
$$c_n = \frac{q_n - x}{q_n - p_n}, \quad d_n = \frac{x - p_n}{q_n - p_n}.$$
Then $c_n \geq 0$, $d_n \geq 0$, $c_n + d_n = 1$, $\lim_{n \to \infty} a_n = f'(x)$, $\lim_{n \to \infty} b_n = f'(x)$. Also,
$$\frac{f(q_n) - f(p_n)}{q_n - p_n} = a_n c_n + b_n d_n.$$
By Question 2.1.14, $\lim_{n \to \infty} a_n c_n + b_n d_n = f'(x)$.

5.1.10: We first show f is continuous at a. We adapt the second proof of Proposition 5.1.9. Let $\varepsilon > 0$. We need to find $\delta > 0$ so that if $|x - z| < \delta$ and $x \in [a, b]$, then $|f(x) - f(a)| < \varepsilon$. Since the right-sided derivative limit exists, there exists $\delta_1 > 0$ so that
$$\left| \frac{f(z) - f(a)}{z - a} - f'^+(a) \right| < 1,$$
for all $z \in [a, b]$ such that $0 < |z - a| < \delta_1$. Then
$$\left| f(z) - f(x) - f'^+(a)(z - a) \right| < |z - a|.$$

A. Solutions to Questions

Therefore,
$$|f(z) - f(a)| \leq |f(z) - f(a) - f'^+(a)(z-a)| + |f'^+(a)(z-a)|$$
$$< |z-a| + |f'^+(a)(z-a)|$$
$$= |z-a|(1 + |f'^+(a)|).$$

Set
$$\delta = \min\left\{\delta_1, \frac{\varepsilon}{1+|f'^+(a)|}\right\}.$$

Then $|f(z)-f(a)| < \varepsilon$ whenever $z \in [a,b]$ and $|z-a| < \delta$. Therefore, f is continuous at a. We note that for the continuity of f at a we only need to consider values of z that are to the right of a, since by definition z must be in the domain of f. This is continuity at a from the left. Finally, the proof of continuity at b is similar.

5.2.8: Let $x \in (0,1)$. We calculate the derivative at x by $f'(x) = \lim_{h \to 0} \frac{f(x+h)-f(x)}{h}$. In the limit, h can be chosen sufficiently close to 0 so that $x+h$ is also in $(0,1)$. Therefore, $f(x+h) = 1$ and $f(x) = 1$, which implies that $f'(x) = 0$. Similarly, when $x \in (1,2)$, we obtain that $f'(x) = 0$. We have a function whose derivative is 0 on an open set G but is not constant on G. The set G is not an interval, hence Remark 5.2.7 does not apply.

5.2.9: Define $g(x) = f(x) - x$. Then g is differentiable on (a,b), and $g'(x) = f'(x) - 1 = 0$ for all $x \in (a,b)$. Then by Corollary 5.2.6 (see Remark 5.2.7) there exists $c \in \mathbb{R}$ such that $g(x) = c$ for all $x \in \mathbb{R}$. As $g(0) = 0$, then $c = 0$. So $f(x) = x + 0 = x$ for all $x \in G$.

Chapter 6

6.1.22: Let $\varepsilon > 0$. We may assume that $f(b) > f(a)$, as otherwise f would be a constant function on $[a,b]$, which we already know is Riemann integrable. Let $\delta = \frac{\varepsilon}{f(b)-f(a)}$. Let $\mathcal{P} = \{x_0, \ldots, x_n\}$ be a partition such that $\|\mathcal{P}\| < \delta$. Write $\Delta x_i = x_i - x_{i-1}$. Note that on each subinterval $[x_{i-1}, x_i]$ of the partition, f attains its maximum at x_i and its minimum at x_{i-1}. Thus $M_i = f(x_i)$ and $m_i = f(x_{i-1})$. We calculate

$$U(f,\mathcal{P}) - L(f,\mathcal{P}) = \sum_{i=1}^n f(x_i)\Delta x_i - \sum_{i=1}^n f(x_{i-1})\Delta x_i$$
$$= \sum_{i=1}^n (f(x_i) - f(x_{i-1}))\Delta x_i$$
$$\leq \sum_{i=1}^n (f(x_i) - f(x_{i-1}))\|\mathcal{P}\|$$
$$\leq \left(\sum_{i=1}^n f(x_i) - f(x_{i-1})\right)\|\mathcal{P}\|$$
$$= (f(b) - f(a))\|\mathcal{P}\|$$
$$< \varepsilon.$$

Consequently, by Theorem 6.1.20, f is Riemann integrable. A similar argument applies when f is monotone decreasing.

6.1.27: Since $f(x) = c$ is continuous, it is Riemann integrable on $[a,b]$. Therefore, we can apply Lemma 6.1.26. Let $\mathcal{P}_n$ be the uniform partition of $[a,b]$ with $n+1$ points, and let $\mathcal{P}_n^*$ be the sequence of corresponding left endpoints. Then the Riemann sums are

$$S(f, \mathcal{P}_n, \mathcal{P}_n^*) = \sum_{i=1}^n f(x_i^*)(x_i - x_{i-1})$$
$$= \sum_{i=1}^n c \frac{b-a}{n}$$
$$= c(b-a).$$

Chapter 7

7.1.14: We say that the infinite product $\prod_{n=1}^\infty a_n$ converges if the sequence of partial products

$$\prod_{i=1}^n a_i$$

converges. Let $P_n = \prod_{i=1}^n a_i$. So we are interested in the convergence of the sequence (P_n). Suppose (P_n) converges to L. Note that $a_n = P_n/P_{n-1}$ for $n \geq 2$. Then $\lim_{n \to \infty} a_n = 1$ provided (P_n) converges to a nonzero real number L. Here we note that it is important to assume that $L \neq 0$. Hence we revise our definition to say that an infinite product $\prod_{n=1}^\infty a_n$ converges if and only if its sequence of partial products (P_n) converges to a nonzero number. (This is a standard requirement for the convergence of an infinite product; in other words, it is standard convention to say that if the sequence of partial sums converges to 0, then the product diverges. In case there are finitely many terms a_n that are zero, then the product diverges for a trivial reason, and this case is usually considered as a special separate case.) Thus we can formulate a test for divergence for infinite products which says that if $\lim_{n \to \infty} a_n$ does not exist, or when it exists it does not equal 1, then the infinite product does not converge.

7.1.17: Let $S_n = \sum_{i=1}^n a_i$ and $T_k = \sum_{i=1}^k a_{2^i}$. Then $T_k \leq S_{2^k}$. Since the sequence (S_n) converges and the sequence (T_k) is monotone increasing, it follows that (T_k) converges; therefore, $\sum_{n=1}^\infty a_{2^n}$ converges.

The argument for the second part is similar, but here we have to be more careful with the upper bound and use that $a_i \geq a_{i+1}$. In fact, we will show that $\sum_{n=1}^\infty 2^{n-1} a_{2^n}$ converges, which is just the original series multiplied by a constant. Let $U_k = \sum_{i=1}^k 2^{i-1} a_{2^i}$. The idea for the estimate will come from the following case.

$$U_3 = a_2 + 2a_4 + 4a_8$$
$$= a_2 + a_4 + a_4 + a_8 + a_8 + a_8 + a_8$$
$$\leq a_2 + a_3 + a_4 + a_5 + a_6 + a_7 + a_8$$
$$\leq S_8 - a_1.$$

Similarly we can show that
$$U_k \leq S_{2^k} - 1.$$
Since the sequence $(S_{2^k})_k$ converges, it follows that (U_k) converges. Therefore, $\sum_{n=1}^{\infty} 2^{n-1} a_{2^n}$ converges.

Chapter 8

8.2.2 : Let $\varepsilon > 0$. There exists $N_1 \in \mathbb{N}$ such that $|f_n(x) - f(x)| < \varepsilon/2$ for all $x \in A$ and all $n \geq N_1$. Also, there exists $N_2 \in \mathbb{N}$ such that $|g_n(x) - g(x)| < \varepsilon/2$ for all $x \in A$ and all $n \geq N_2$. Let $N = \max\{N_1, N_2\}$. Then
$$|(f_n(x) + g_n(x)) - (f(x) + g(x))| \leq |f_n(x) - f(x)| + |g_n(x) - g(x)|$$
$$< \frac{\varepsilon}{2} + \frac{\varepsilon}{2} = \varepsilon,$$
for all $x \in A$ and all $n \geq N$. Hence $f_n + g_n$ converges uniformly to $f + g$ on A.

For the product we note that
$$|(f_n(x)g_n(x)) - (f(x)g(x))| = |(f_n(x)g_n(x)) - f_n(x)g(x) + f_n(x)g(x) - f(x)g(x)|$$
$$\leq |f_n(x)| \, |g_n(x) - g(x)| + |f_n(x) - f(x)| \, |g(x)|.$$
We know there exist constants $M_1 > 0$ and $M_2 > 0$ such that $|f(x)| < M_1$ and $|g(x)| < M_2$ for all $x \in A$. We may assume $M_2 > 1$. Let $\varepsilon > 0$. Choose $N_1 \in \mathbb{N}$ so that $|f_n(x) - f(x)| < \frac{\varepsilon}{2M_2}$ for all $x \in A$ and $n \geq N_1$. If $n \geq N_1$ and $x \in A$, we have
$$-\frac{\varepsilon}{2M_2} < f_n(x) - f(x) < \frac{\varepsilon}{2M_2},$$
$$-\frac{\varepsilon}{2} + f(x) < f_n(x) < f(x) + \frac{\varepsilon}{2},$$
$$-(M_1 + \frac{\varepsilon}{2}) < f_n(x) < (M_1 + \frac{\varepsilon}{2}),$$
$$|f_n(x)| < (M_1 + \frac{\varepsilon}{2}).$$

Choose N_2 so that
$$|g_n(x) - g(x)| < \frac{\varepsilon}{2(M_1 + \frac{\varepsilon}{2})} \text{ for all } x \in A \text{ and } n \geq N_2.$$
Then we have that for $n \geq \max\{N_1, N_2\}$, and for all $x \in A$,
$$|(f_n(x)g_n(x)) - (f(x)g(x))| < (M_1 + \frac{\varepsilon}{2})\frac{\varepsilon}{2(M_1 + \frac{\varepsilon}{2})} + \frac{\varepsilon}{2M_2} M_2$$
$$= \frac{\varepsilon}{2} + \frac{\varepsilon}{2}$$
$$= \varepsilon.$$

8.2.3: Let $\varepsilon = 1$. Then there exists $N \in \mathbb{N}$ such that $|f_n(x) - f(x)| < 1$ for all $x \in A$ and all $n \geq N$. In particular, $f_N(x) - 1 < f(x) < 1 + f_N(x)$. As f_N is bounded on A, it follows that f is bounded on A.

Bibliographical Notes

I learned analysis from the early edition of Rudin [40] and Dieudonné [12], and set theory from Halmos [18]. In the classes I have taught, I have used Bartle [2] and Krantz [23], and suggested as supplementary references Marsden and Hoffman [26], Thomson, Bruckner, and Bruckner [49], Morgan [30], and Rosenlicht [38]. Other books include Abbott [1], Fitzpatrick [13], Gaughan [15], Lay [25], Ross [39], and Schumacher [42]. Pugh [37], Sally [41], and Wade [50] cover additional topics, and can also be consulted after this book. Bressoud [6] offers a historical introduction to real analysis. Gelbaum and Olmsted [16] is a great source for counterexamples, and I have used some of them in the exercises.

For an introduction to set theory see Goldrei [17] or Halmos [18], where the student can also see a detailed construction of the natural numbers with proofs of the induction and recursion definition principles. Jech [20] and Herrlich [19] are great sources for the axiom of choice and its role in analysis and for additional study in set theory. For more on the Fundamental Theorem of Arithmetic see [27]. A surprising result, that was shown using the axiom of choice, is the Banach–Tarski paradox that roughly states that a sphere can be decomposed into a finite number of pieces and then put back together to make two copies of the original sphere; a discussion of this and other paradoxes can be found in Wagon [51].

For a history of open and closed sets see Moore [29]. Propp [36] discusses many properties equivalent to order completeness, and we have used it as inspiration for exercises related to this notion; see also Teisman [46]. For various proofs of the uncountability of the real line see [22].

For a history of the Cantor set and Cantor function see Fleron [14]. For a discussion on the set of continuity points of a function, see Thomson, Bruckner, and Bruckner [49]. I have followed Billingsley [5] in the proof that the Takagi function is nowhere differentiable. A more comprehensive survey of nowhere differentiable functions is in Thim [47]. If I were to include the proof that π is irrational, I would follow Niven [32]; the reader could also see Jones [21]. For proofs of transcendence of e and π see Burger and Tubbs [9]. The role of compactness in mathematics is discussed in Morgan [**31**]. The proof of Lemma 5.3.7 is from Bary-Soroker and Leher [**4**], and I thank Stewart Johnson for bringing this reference to my attention.

For the Henstock–Kurzweil integral, also called the generalized Riemann integral, see Bartle [**3**]. A short note on the harmonic series, with other references, is in [**24**].

For a history of the gamma function see Davis [**10**], which contains a comprehensive discussion of the gamma function; we have followed Rudin [**40**] and Buck [**7**] which contain additional details. Pinkus [**34**] discusses the Weierstrass approximation theorem and its history; a different proof using Bernstein polynomials can be found in Bartle [**2**]. For the Baire category theorem and Lebesgue integration, the reader is prepared after this course to consult Oxtoby [**33**]. There are many subjects the reader could learn after a first course in analysis. This book provides the background for [**43**], which contains an introduction to measure theory and ergodic theory, and to proofs of the Lebesgue theorems we have cited. A proof of the monotone convergence theorem for Riemann integrals without using Lebesgue theory is given in Thomson [**48**]. Many applications of analysis to dynamics can be found in Devaney [**11**]. The reader could also pursue applications to number theory in Miller and Takloo-Bighash [**28**]. A comprehensive introduction to further topics in analysis can be found in Simon [**45**]. Finally, we mention two books that can help the reader develop proof-writing techniques: Polya [**35**] is a classic that has many suggestions on how to approach a mathematical problem, and Burger and Starbird [**8**] also has directions on problem solutions.

Bibliography

[1] Stephen Abbott, *Understanding analysis*, 2nd ed., Undergraduate Texts in Mathematics, Springer, New York, 2015. MR3331079

[2] Robert G. Bartle, *The elements of real analysis*, 2nd ed., John Wiley & Sons, New York-London-Sydney, 1976. MR0393369

[3] Robert G. Bartle, *A modern theory of integration*, Graduate Studies in Mathematics, vol. 32, American Mathematical Society, Providence, RI, 2001. MR1817647

[4] Lior Bary-Soroker and Eli Leher, *On the remainder in the Taylor theorem*, The College Mathematics Journal **40** (2009), no. 5, 373–375.

[5] Patrick Billingsley, *Notes: Van Der Waerden's Continuous Nowhere Differentiable Function*, Amer. Math. Monthly **89** (1982), no. 9, 691, DOI 10.2307/2975655. MR1540053

[6] David M. Bressoud, *A radical approach to real analysis*, 2nd ed., Classroom Resource Materials Series, Mathematical Association of America, Washington, DC, 2007. MR2284828

[7] R. Creighton Buck, *Advanced calculus*, 3rd ed., McGraw-Hill Book Co., New York-Auckland-Bogotá, 1978. With the collaboration of Ellen F. Buck; International Series in Pure and Applied Mathematics. MR0476931

[8] Edward B. Burger and Michael Starbird, *The five elements of effective thinking*, Princeton University Press, 2012.

[9] Edward B. Burger and Robert Tubbs, *Making transcendence transparent: An intuitive approach to classical transcendental number theory*, Springer-Verlag, New York, 2004. MR2077395

[10] Philip J. Davis, *Leonhard Euler's integral: A historical profile of the gamma function.*, Amer. Math. Monthly **66** (1959), 849–869, DOI 10.2307/2309786. MR0106810

[11] Robert L. Devaney, *An introduction to chaotic dynamical systems*, 2nd ed., Addison-Wesley Studies in Nonlinearity, Addison-Wesley Publishing Company, Advanced Book Program, Redwood City, CA, 1989. MR1046376

[12] J. Dieudonné, *Foundations of modern analysis*, Pure and Applied Mathematics, Vol. X, Academic Press, New York-London, 1960. MR0120319

[13] Patrick M. Fitzpatrick, *Real analysis*, PWS Publishing Co., Boston, 1996.

[14] Julian F. Fleron, *A note on the history of the Cantor set and Cantor function*, Math. Mag. **67** (1994), no. 2, 136–140, DOI 10.2307/2690689. MR1272828

[15] Edward D. Gaughan, *Introduction to analysis*, 4th ed., Brooks/Cole Publishing Co., Pacific Grove, CA, 1993. MR1226448

[16] Bernard R. Gelbaum and John M. H. Olmsted, *Counterexamples in analysis*, Dover Publications, Inc., Mineola, NY, 2003. Corrected reprint of the second (1965) edition. MR1996162

[17] D.C. Goldrei, *Classic set theory: For guided independent study*, Chapman and Hall/CRC, Princeton, 1996.

[18] Paul R. Halmos, *Naive set theory*, The University Series in Undergraduate Mathematics, D. Van Nostrand Co., Princeton, N.J.-Toronto-London-New York, 1960. MR0114756

[19] Horst Herrlich, *Axiom of choice*, Lecture Notes in Mathematics, vol. 1876, Springer-Verlag, Berlin, 2006. MR2243715

[20] Thomas J. Jech, *The axiom of choice*, North-Holland Publishing Co., Amsterdam-London; American Elsevier Publishing Co., Inc., New York, 1973. Studies in Logic and the Foundations of Mathematics, Vol. 75. MR0396271

[21] Timothy W. Jones, *Discovering and proving that π is irrational*, Amer. Math. Monthly **117** (2010), no. 6, 553–557, DOI 10.4169/000298910X492853. MR2662709

[22] Christina Knapp and Cesar E. Silva, *The uncountability of the unit interval*, arXiv: 1209.5119 v2 (2014).

[23] Steven G. Krantz, *Real analysis and foundations*, Studies in Advanced Mathematics, CRC Press, Boca Raton, FL, 1991. MR1210958

[24] David E. Kullman, *What's harmonic about the harmonic series?*, The College Mathematics Journal **32** (2001), no. 3, 201–203.

[25] Steven R. Lay, *Analysis: An introduction to proof*, Prentice Hall, Inc., Englewood Cliffs, NJ, 1986. MR911681

[26] Jerrold E. Marsden, *Elementary classical analysis*, W. H. Freeman and Co., San Francisco, 1974. With the assistance of Michael Buchner, Amy Erickson, Adam Hausknecht, Dennis Heifetz, Janet Macrae and William Wilson, and with contributions by Paul Chernoff, István Fáry and Robert Gulliver. MR0357693

[27] Steven J. Miller and Cesar E. Silva, *If a prime divides a product...*, College Math. J. **48** (2017), no. 2, 123–128, DOI 10.4169/college.math.j.48.2.123. MR3626272

[28] Steven J. Miller and Ramin Takloo-Bighash, *An invitation to modern number theory*, Princeton University Press, Princeton, NJ, 2006. With a foreword by Peter Sarnak. MR2208019

[29] Gregory H. Moore, *The emergence of open sets, closed sets, and limit points in analysis and topology* (English, with English and German summaries), Historia Math. **35** (2008), no. 3, 220–241, DOI 10.1016/j.hm.2008.01.001. MR2455202

[30] Frank Morgan, *Real analysis*, American Mathematical Society, Providence, RI, 2005. MR2153931

[31] Frank Morgan, *Compactness*, Pro Mathematica **22** (2008), no. 43–44, 124–133.

[32] Ivan Niven, *A simple proof that π is irrational*, Bull. Amer. Math. Soc. **53** (1947), 509, DOI 10.1090/S0002-9904-1947-08821-2. MR0021013

[33] John C. Oxtoby, *Measure and category: A survey of the analogies between topological and measure spaces*, 2nd ed., Graduate Texts in Mathematics, vol. 2, Springer-Verlag, New York-Berlin, 1980. MR584443

[34] Allan Pinkus, *Weierstrass and approximation theory*, J. Approx. Theory **107** (2000), no. 1, 1–66, DOI 10.1006/jath.2000.3508. MR1799549

[35] G. Polya, *How to solve it: A new aspect of mathematical method*, Princeton Science Library, Princeton University Press, Princeton, NJ, 2014. With a foreword by John H. Conway; Reprint of the second (2004) edition [MR2183670]. MR3289212

[36] James Propp, *Real analysis in reverse*, Amer. Math. Monthly **120** (2013), no. 5, 392–408, DOI 10.4169/amer.math.monthly.120.05.392. MR3035440

[37] Charles C. Pugh, *Real mathematical analysis*, 2nd ed., Undergraduate Texts in Mathematics, Springer, Cham, 2015. MR3380933

[38] Maxwell Rosenlicht, *Introduction to analysis*, Dover Publications, Inc., New York, 1986. Reprint of the 1968 edition. MR851984

[39] Kenneth A. Ross, *Elementary analysis: The theory of calculus; In collaboration with Jorge M. López*, 2nd ed., Undergraduate Texts in Mathematics, Springer, New York, 2013. MR3076698

[40] Walter Rudin, *Principles of mathematical analysis*, 3rd ed., McGraw-Hill Book Co., New York-Auckland-Düsseldorf, 1976. International Series in Pure and Applied Mathematics. MR0385023

[41] Paul J. Sally Jr., *Fundamentals of mathematical analysis*, Pure and Applied Undergraduate Texts, vol. 20, American Mathematical Society, Providence, RI, 2013. MR3014419

[42] Carol Schumacher, *Closer and closer: Introducing real analysis*, Jones and Bartlett Publishers, Sudbury, Massachusetts, 2008.

[43] C. E. Silva, *Invitation to ergodic theory*, Student Mathematical Library, vol. 42, American Mathematical Society, Providence, RI, 2008. MR2371216

[44] Cesar E. Silva and Yuxin Wu, *No functions continuous only at points in a countable dense set*, arXiv:1809.06453 v2 (2018).

[45] Barry Simon, *Real analysis: A Comprehensive Course in Analysis, Part 1*, American Mathematical Society, Providence, RI, 2015. With a 68 page companion booklet. MR3408971

[46] Holger Teismann, *Toward a more complete list of completeness axioms*, Amer. Math. Monthly **120** (2013), no. 2, 99–114, DOI 10.4169/amer.math.monthly.120.02.099. MR3029936

Bibliography

[47] J. Thim. Continuous nowhere differentiable functions. *Department of Mathematics, Lulea University of Technology, Sweden, Master's Thesis*, 2003.

[48] Brian S. Thomson, *Monotone convergence theorem for the Riemann integral*, Amer. Math. Monthly **117** (2010), no. 6, 547–550, DOI 10.4169/000298910X492835. MR2662707

[49] Brian S. Thomson, Judith B. Bruckner, and Andrew M. Bruckner. *Elementary Real Analysis*. Prentice Hall, Upper Saddle River, New Jersey, 1997.

[50] Steven R. Lay, *Analysis: An introduction to proof*, Prentice Hall, Inc., Englewood Cliffs, NJ, 1986. MR911681

[51] Stan Wagon, *The Banach-Tarski paradox*, Encyclopedia of Mathematics and its Applications, vol. 24, Cambridge University Press, Cambridge, 1985. With a foreword by Jan Mycielski. MR803509

Index

absolute value, 60
accumulation point, 117, 123
additive inverse, 48
$\aleph_0$, 64
algebraic number, 69
alternating series, 232
antiderivative, 214
Archimedean property, 57
Axioms for Addition, 48
Axioms for Multiplication, 48

Banach Fixed Point Theorem, 169
belongs, 10
Bernoulli's Inequality, 53
biconditional, 7
bijection, 23
bijective, 3
binomial
   coefficient, 52
   formula, 52
Bolzano–Weierstrass theorem, 97
bounded
   above, 55
   below, 55
   in a metric space, 138
   set, 55
Bézout's identity, 38

Cantor function, 252
Cantor set, 130, 134
Cantor–Schröder–Bernstein theorem, 36
card(A), 33
cardinality, 33
cardinality $\aleph_0$, 64

Cartesian product, 3, 20
Cassini identity, 38
Cauchy complete, 141
Cauchy product, 273
Cauchy property, 104, 141
Cauchy sequence, 104, 141
Cauchy, uniform, 244
Cauchy–Schwartz inequality, 113
closed ball, 120
closed interval, 60
closed set, 116
closed, metric space, 122
closed, relative, 122
closure, 118, 123
closure point, 117, 123
codomain, 3, 22
common refinement partition, 200
compact, 126
compact metric space, 139
complement, 2, 15
complete, 141
completeness property, 57
complex addition, 75
complex conjugate, 75
complex exponential, 281
complex multiplication, 75
complex number, 74
composition, 3, 25
conclusion, 6
conditional, 6
conjunction, 5
connected, 136
connected component, 143

connected metric space, 141
contains, 10
continuous, 145
  at a point, 145
  on a set, 145
  uniformly, 164
continuously differentiable, 250
contraction, 168
contrapositive, 7
converge, 80
converge absolutely, 234
converge conditionally, 234
convergence, metric spaces, 122
converges pointwise, 241
converse, 7
convex function, 190
corresponding points, 198
cosine, 270
countable, 62
countable family, 66
countable open cover, 126
countably infinite, 62
cover, 126
cut corresponding to a rational number, 72
cut property, 103

Darboux's Intermediate Value Theorem, 189
De Morgan's laws, 17, 70
decreasing
  function, 152
  sequence, 89
Dedekind cut, 71
dense, 59
dense in a metric space, 142
derivative, 176
  higher order, 190
derived set, 117
differentiable at c, 176
Dini's theorem, 254
direct method of proof, 12
Dirichlet's funcion, modified, 152
Dirichlet's function, 146
disconnected, 136
discontinuous at a point, 145
discontinuous on a set, 145
disjoint, 15, 66
disjunction, 5
Distributive Axiom, 49
diverge to infinity, 88
divergent series, 223

diverges, 81
domain, 3, 22
Dominated Convergence Theorem, 252
dot product, 110
dyadic intervals, 95
dyadic rational, 95

element, 10
empty set, 2, 11
ε-close to L, 79
epsilon close, 145
equivalence class, 22
equivalence relation, 22
equivalent metric, 140
Euler's constant, 267
Euler's constant e, 106
Euler's identity, 282
existential quantifier, 8
exponential function, 267

Fibonacci sequence, 34
field, 49
finer partition, 200
finite set, 33
fixed point, 157
Fixed Point Theorem,
  Banach, 169
for all, $\forall$, 8
function, 22
  continuous, 145
  differentiable, 176
  Lipschitz, 152
  monotone, 152
  monotone decreasing, 152
  monotone increasing, 152
  period, 256
  Riemann integrable, 200
  strictly increasing, 152
fundamental period, 256
Fundamental Theorem of Arithmetic, 36
Fundamental Theorem of Calculus, 214

gamma function, 272
geometric series, 99
geometric sum, 53
global maximum, 155
global minimum, 155
greatest common divisor, 35

harmonic series, 231
Hausdorff metric, 170
Heine–Borel theorem, 127

higher order derivatives, 190
hypothesis, 6

image, 3
imaginary part, 75
implication, 6
implies, 6
includes, 10
increasing
   function, 152
   monotone increasing function, 152
   sequence, 89
induction, 29
inductive hypothesis, 29
infimum, 56
infinite set, 33
injective, 3, 23
integers, 2
interior, 119, 123
intersection, 2, 15, 66
interval, 60
inverse
   left, 27
   right, 27
inverse function, 24
inverse image, 3, 25
inverse image of a set, 25
invertible, 3, 24
irrational, 57
is in, 10
isolated point, 117
isometry, 168
iterate
   nth, 169
   second, 169

L'Hôpital's rule, 188
Lagrange remainder, 191
least element, 31
left inverse, 27
left limit, 163
left-sided derivative, 177
lim inf, 100
lim sup, 100
limit, 158
limit infimum, of sets, 70
limit point, 117, 123
limit supremum, of sets, 70
Lipschitz, 152
little "oh", 193
local extremum, 183
local maximum, 183

local minimum, 183
logarithmic function, 216
logical connectives, 4
lower bound, 55

mathematical induction, 29
maximum, 155
maximum element, 39
Mean Value Theorem, 184, 186
measure zero, 133
member, 2, 10
mesh, 199
metric, 110
metric completion, 172
metric space, 113
metric spaces, convergence, 122
metric, equivalent, 140
metric, topologically equivalent, 139
middle-thirds Cantor set, 130, 131
minimum, 155
modulus, 75
monotone
   function, 152
monotone decreasing
   function, 152
   sequence, 89
monotone increasing
   sequence, 89
monotone sequence property, 93
monotone sequence theorem, 92
monotonic
   sequence , 89
multiplicative inverse, 48

nth degree Taylor polynomial, 190
nth order derivative, 190
natural logarithmic function, 216
natural numbers, 2, 12
necessary, 6
negation, 4
nested intervals, 93
Nested Intervals Theorem, 93
nondecreasing
   sequence, 89
nonnegative integers, 12
norm, 110

one-to-one, 3, 23
one-to-one correspondence, 23
onto, 3, 23
open ball, 120
open ball in $\mathbb{R}$, 115

open cover, 126
open in a metric space, 120
open interval, 60
open relative to a subspace, 121
open set, 115
order, 52
Order Axioms, 50
order complete, 57
ordered pair, 3, 20

pairwise disjoint, 66
partition, 197
Peano axioms, 30
perfect set, 131
period, 256
    fundamental, 256
periodic function, 256
π, 271
positive set, 50
power series, 259
power set, 2
prime, 3
prime number, 35
proof by contradiction, 13
proof by contrapositive, 13
proper subset, 11

radius of convergence, 260
rational numbers, 2
real numbers, 47, 57, 71
real part, 75
real root of polynomial, 154
rearrangement of a series, 237
reflexive relation, 21
relation, 21
restriction, function, 24
Riemann sum, upper, 202
Riemann integrable, 200
Riemann integral, 200
    improper, 221
Riemann sum, 198
Riemann sum, lower, 202
right inverse, 27
right limit, 163
right-sided derivative, 177
Rolle's theorem, 184
root, 61

separable, 142
sequence, 64
    decreasing, 89
    increasing, 89

monotone decreasing, 89
monotone increasing, 89
monotonic, 89
    strictly decreasing, 89
    strictly increasing, 89
sequence of partial sums, 223
sequence of functions, 241
sequence of partial sums, 99
sequentially compact, 125
sequentially compact metric space, 138
series, 223
series, rearrangement of, 237
set difference, 14
set membership, 10
set minus, 14
sine, 270
statement, 4
strictly decreasing
    sequence, 89
strictly increasing
    function, 152
    sequence, 89
Strong Nested Intervals Theorem, 94
subsequence, 90
subsequential limit points, 100
subset, 2, 10
sufficient, 6
sum of a series, 223
superset, 10
supremum, 55
surjective, 3, 23
symmetric difference, 2
symmetric relation, 21

Taylor polynomial, 190
Taylor series, 265
Taylor series about c, 265
Taylor's theorem
    integral remainder, 219
    Lagrange remainder, 191
term of a sequence, 64
there exists, $\exists$, 8
Thomae's function, 152
topologically equivalent metric, 139
totally bounded, 142
totally disconnected, 143
totally disconnected subset of $\mathbb{R}$, 134
transitive relation, 22
triangle inequality, 110
trigonometric cosine, 270
trigonometric sine, 270

Index

ultrametric, 113
uncountable, 63
uniform convergence, 243
uniform partition, 198
uniformly Cauchy, 244
uniformly continuous, 164
union, 2, 14, 66
universal quantifier, 8
upper bound, 55

Weierstrass M-Test, 255
well-ordering principle, 31

Published Titles in This Series

36 **Cesar E. Silva,** Invitation to Real Analysis, 2019
33 **Brad G. Osgood,** Lectures on the Fourier Transform and Its Applications, 2019
32 **John M. Erdman,** A Problems Based Course in Advanced Calculus, 2018
31 **Benjamin Hutz,** An Experimental Introduction to Number Theory, 2018
30 **Steven J. Miller,** Mathematics of Optimization: How to do Things Faster, 2017
29 **Tom L. Lindstrøm,** Spaces, 2017
28 **Randall Pruim,** Foundations and Applications of Statistics: An Introduction Using R, Second Edition, 2018
27 **Shahriar Shahriari,** Algebra in Action, 2017
26 **Tamara J. Lakins,** The Tools of Mathematical Reasoning, 2016
25 **Hossein Hosseini Giv,** Mathematical Analysis and Its Inherent Nature, 2016
24 **Helene Shapiro,** Linear Algebra and Matrices, 2015
23 **Sergei Ovchinnikov,** Number Systems, 2015
22 **Hugh L. Montgomery,** Early Fourier Analysis, 2014
21 **John M. Lee,** Axiomatic Geometry, 2013
20 **Paul J. Sally, Jr.,** Fundamentals of Mathematical Analysis, 2013
19 **R. Clark Robinson,** An Introduction to Dynamical Systems: Continuous and Discrete, Second Edition, 2012
18 **Joseph L. Taylor,** Foundations of Analysis, 2012
17 **Peter Duren,** Invitation to Classical Analysis, 2012
16 **Joseph L. Taylor,** Complex Variables, 2011
15 **Mark A. Pinsky,** Partial Differential Equations and Boundary-Value Problems with Applications, Third Edition, 1998
14 **Michael E. Taylor,** Introduction to Differential Equations, 2011
13 **Randall Pruim,** Foundations and Applications of Statistics, 2011
12 **John P. D'Angelo,** An Introduction to Complex Analysis and Geometry, 2010
11 **Mark R. Sepanski,** Algebra, 2010
10 **Sue E. Goodman,** Beginning Topology, 2005
9 **Ronald Solomon,** Abstract Algebra, 2003
8 **I. Martin Isaacs,** Geometry for College Students, 2001
7 **Victor Goodman and Joseph Stampfli,** The Mathematics of Finance, 2001
6 **Michael A. Bean,** Probability: The Science of Uncertainty, 2001
5 **Patrick M. Fitzpatrick,** Advanced Calculus, Second Edition, 2006
4 **Gerald B. Folland,** Fourier Analysis and Its Applications, 1992
3 **Bettina Richmond and Thomas Richmond,** A Discrete Transition to Advanced Mathematics, 2004
2 **David Kincaid and Ward Cheney,** Numerical Analysis: Mathematics of Scientific Computing, Third Edition, 2002
1 **Edward D. Gaughan,** Introduction to Analysis, Fifth Edition, 1998